Many years ago C. G. Jung warned us that without an "imagination for evil" we modern human beings were in danger of becoming instruments of the very evil we abhor. Glen Slater's brilliant and passionate analysis of online culture and its insidious seductions of hyperreality, virtual companions, and cyber presences—all run by artificial intelligence—opens up that imagination in ways that are both terrifying and illuminating. To become conscious of these dehumanizing forces in our midst and how to combat their dissociative effects on the inner life of the soul should be a major focus of all depth psychological training in the 21st century. I cannot emphasize strongly enough the importance of this book.

Donald Kalsched, Ph.D., author of *The Inner World of Trauma* (Routledge, 1996) and *Trauma and the Soul* (Routledge, 2013).

Slater brilliantly argues that the future posthumanists are promoting will sever our ties with the deeply human basis of being. To head in this direction is to become psychically "numbed," a state of minimal being that is already pervasive to the point of being normative. Numbing makes us incapable of being satisfied with the actuality of events. We take residence in left-hemisphere cortical processes, becoming half-brained and cyborg-like. To resist these cyborg prospects, Slater extracts from the psychology of C. G. Jung the most useful bits. An exploration of the natural richness of the psyche and its fabulous imaginative power comprise the natural antidote to a bleak future defined by posthumanism.

Ginette Paris, Ph.D., author of *Pagan Meditations* (Spring Publications, 1986), *Wisdom of the Psyche* (Routledge, 2007), and *Heartbreak: New Approaches to Healing* (Mill City Press, 2011).

JUNG VS BORG

FINDING THE DEEPLY HUMAN IN A POSTHUMAN AGE

GLEN SLATER

WINTER PRESS
A Spring Publications Imprint

Jung vs Borg: Finding the Deeply Human in a Posthuman Age

Copyright © 2024 by Glen Slater and Winter Press
All rights reserved

Published in the United States by Winter Press
Arroyo Grande, CA
www.winterpresspublishers.com

Winter Press is an imprint of Spring Publications
www.springpublications.com

First edition 2024

Library of Congress Control Number: 2023920753
ISBN: 978-1-7362057-1-6

Cover design by Jonas Perez
Interior book design by Andrew Tennant

*For Safron,
whose love and patience
have inspired and guided this endeavor
to draw together drive, feeling, and word.*

CONTENTS

Introduction   1

PART 1: ORIENTATION

1   Polarities   17

Four Shadows of Technology . . . The Future and the Past . . . Gods and Demons . . . Nature and Civilization . . . Utopias and Dystopias . . . Vision and Memory

2   Parts   54

The Religious Function . . . And the Hive-Mind . . . Becoming Cyborgs . . . Posthumanism . . . The Borg . . . The Deeply Human . . . Inflection Point

3   Gaps   95

Significant Others . . . Cyberspace Cadets . . . First the Earth, Now the Psyche . . . Silent Spring Again . . . The Technologist . . . And the Historian . . . Story and Meaning . . . Deeper Awareness

4   Tinkering   127

    *Beneath the Surface . . . Missing the Sacred . . .
    Mad Scientist Syndrome . . . Posthuman
    Check-In . . . Heidegger . . . The Hunter and the
    Shaman . . . Two Paths*

PART 2: DOWN THE RABBIT HOLE

5   Numbed   165

    *Two Levels of Dissociation . . . The Disconnect of
    Mind and Body . . . Fragmentation and Self
    Invention . . . MOMA . . . Hermetic Intoxication . . .
    "The Constitution of Knowledge" . . . Medicated
    Numbing . . . Division and Integration*

6   Borged   208

    *Reduced . . . Automata to Androids . . . The Quest
    for Immortality and Perfection . . . Robot Love . . .
    The Spiritualization of Cyberspace . . . Scientific
    Presentism . . . Revenge of the Nerds . . .
    Algorithmic Halos . . . Flesh, History, World . . .
    Stepping Off Point*

7   Half-Brained   250

    *Hemispheric Relations . . . Neurotic
    Implications . . . Turn-ons and Turn-offs . . . Half-
    Brained Ideas . . . Reading the Brain . . . Why
    Intelligence is Not Enough*

8   Paternity   282

    *Postmodern Base Notes . . . Conditions and
    Philosophies . . . Philosophical Posthumanism . . .
    Behaviorism . . . Cognitive Psychology and
    Computation . . . Weaving Strands*

PART 3: BELOW GROUND

9 Kurzweil's Dreams 313

Depth Psychology and Collectivity . . . The Backstory . . . Does Kurzweil Dream of Butler's Sheep? . . . Looking in All Directions . . . Western Civilization's Two Streams

10 Tin Men 345

Everyday Psychopathy . . . Calculating Psychopaths . . . Losing Heart . . . And Looking For a Heart . . . Inner Voices and Human Will

11 Re-sinking the *Titanic* 368

Contrasting Descents . . . The Return of the Titans . . . Titans Among Us . . . Lopez-Pedraza Goes to Battle

PART 4: RESTORATION

12 Conscious Computing 393

Not So Fast . . . What is Consciousness? . . . Specific Drivers of Human Consciousness . . . One Giant Turing Test . . . House versus Home . . . Turning to Ash . . . The Conscious Cosmos

13 Co-Creation 429

Co-creative Responsibility . . . The Third Act of the Modern Story . . . Learning Curve . . . Imagination . . . Algorithm or Archetype? . . . Anthropocene Citizenry . . . Starting Close to Home . . . Looking Further Away

14 Reanimated 468

Gaia, Spirit, and Matter . . . Crafting, Sensing, and Returning . . . A New Counterculture

Epilogue: Between the Lines   493
   *Psyche, Soul, Eros . . . Soul-Making . . . The Harari
   Gap . . . The Symbolic-Imaginal Life . . . In
   Conclusion*

Bibliography   513
Acknowledgments   525

INTRODUCTION

In 1962 the biologist Rachel Carson exposed the indiscriminate use of pesticides and the harm this was inflicting on ecosystems and other species, including humans. Her book *Silent Spring*[1] was met with significant resistance, not only because she confronted corporations and governments with matters they preferred to ignore, but because she revealed a syndrome in scientific research and technological development: By compartmentalizing the problems they aim to solve, scientists and technologists frequently fail to see the bigger picture or serve the greater good. Carson's book eventually shifted collective awareness about environmental damage caused by practices and attitudes that had gone largely unquestioned.

A half-century later, with a better understanding of pollutants, toxins, and greenhouse gases, we have become warier about solving environmental problems with compartmentalized approaches, more savvy about the interconnected character of the natural world, and more open to integrative ways of thinking. Corporations may be more powerful than ever but their environmental impact is more closely monitored. Whether or not we have turned the ecological corner, there is more consciousness about the way we relate to the world around us and the actions required to avert a climate catastrophe.

However, the world *within* us, the inner life of thought and emotion, is another matter. Here integrative understanding has been resisted. The human psyche, the ecosystem of the mind, with its own structures and dynamics, our relation to which is surely as significant as our relation to the outer world, now faces its own significant disruption, one that essentially parallels the syndrome Carson described.

As the ill-effects of twentieth-century industrialism still play out, the ill-effects of twenty-first-century post-industrialism are just coming into focus. At the same time whole economies are turning to data management, and artificial intelligence (AI) has begun taking over human decision-making, disinformation is spreading, social fabrics are eroding, geopolitical structures are collapsing, and the depths of human nature are becoming harder to recognize and protect. Abundant information has furthered scientific knowledge but not human understanding. It has instead left us dazed, confused, and disoriented. Attention, motivation, identity, self-image, and the capacity to reason clearly and imagine deeply are all impacted. Beyond the fragile bonds of globalization and the early promise of the World Wide Web, individuals, communities, and even nations are fragmenting. While the advantages of digital technologies are impressed upon us at every turn, the rapid entry of these technologies into every aspect of life is evidently impacting our ecology of mind.[2] In my view, a vast psycho-social failure is becoming the most pronounced feature of the digital age, poised to surpass climate disruption as the most pressing crisis of the twenty-first century.

Recapitulating last century's approach to the outer environment, today's tech corporations have focused on another natural resource: human interiority. Personal information has become the latest material to be mined and algorithms have been deployed to manipulate behavior and generate revenue. These developments have turned our habits of mind into the earth's most valuable commodity. As in the past, these efforts are driven by motivations of power and profit. Unlike the past, such efforts are also altering the very means by which we perceive and understand what is happening to us. As we do not yet realize the degree to which the attempt to exploit the outer world is being repeated in the inner world of perception, thought, and emotion, we have entered a particularly critical moment in history. The technosphere now overlays the ecosphere and we cannot help but inhale its post-industrial gases. Virtuality has begun to displace reality, making the ground of human existence hard to discern.

Under the guise of keeping us connected while maximizing advertising dollars, social media is making us more isolated, less knowledgeable, collectively tribal, and progressively egotistical, resulting in an erosion of social skills and a diminished desire for human contact. Screen addiction is tied to an epidemic of anxiety and depression among digital natives. Elsewhere, medical science has convinced us emotional problems are caused by faulty neural mechanisms and can be treated solely with chemicals, and short-term psychotherapies approach the mind as if it is a computer. The human psyche and the nature of the whole person are not only suffering in this technospheric environment, the suffering has itself been given over to technological solutions, resulting in a vicious cycle.

Most recently, AI has also begun to replace workers across a wide range of occupations, and astute technologists and social commentators have begun to question its impact on culture. Modeled on a selective understanding of human cognition, this narrowly conceived form of intelligence also serves as a catalyst for those who think computers will eventually surpass us. With generative AI instantly distilling online information, writing college-level essays, and replacing the work of journalists and storytellers, a displacement of human intelligence is indicated. Moreover, when children start to treat virtual assistants like playmates and the rest of us allow Siri and Alexa to mediate much of what we do, relating to machines as persons will alter perceptions of where real thought resides. By skewing educational philosophies away from disciplines concerned with understanding the human condition and emphasizing STEM (science, technology, engineering, mathematics) disciplines, many educators also seem convinced the future lies in learning how to service these machines and put aside the study of human nature. As these trends play out, our appreciation for the sort of intelligence and consciousness that stands behind authentic wellbeing and cultural enrichment appears to be diminishing. Some are even suggesting civilization may not need us—at least not the version of us that corresponds to our naturally evolved and acculturated human form.[3]

As I will set out in the chapters that follow, the penchant for digital ways of relating, expanding faith in AI, and the one-sided education designed to service these things are combining to generate reductive conceptions of psychological life. We are, in particular, discounting the *deeply human*.[4] I employ this term to designate the essential qualities of human experience, which extend from the instinctual patterns that shape basic behavior to the timeless values that mold the cultural imagination. The deeply human anchors the vertical axis of inner understanding; it grounds the ecology of mind. It is also what connects us to the more-than-human. In our era, however, an almost exclusive dedication to a horizontal axis of data gathering threatens this verticality and grounding. This is leading to a world drowning in information and thirsting for understanding.

In essence, just as the Industrial Revolution changed the outer world and threatened the devastation of nature-at-large, a post-industrial revolution is changing the inner world and risks devastating human nature. And just as yesterday's futurists dreamt up schemes to depart a wasted earth, today's futurists are scheming to depart from the human form altogether—departure via the radical alteration of minds and bodies.

Couched in terms such as *enhancement* and *augmentation* and lauded in the technoscience community this trend also signals an emerging crisis, one yet to be approached with the systemic awareness the environmental crisis eventually necessitated. At the core of this new crisis is a compartmentalizing of human nature that invites radical mental and physical change without consideration of the bigger picture or the greater good. Those extrapolating this trend foresee a *posthuman*[5] world in which the attempt to keep up with advancing AI will turn us into human–machine hybrids or *cyborgs* (cybernetic organisms). As this cyborg existence gets underway, there are indications of accelerating social upheaval and even a growing normative madness. Last century's loss of soul is turning into this century's soul absence—an absence posthuman antidotes to human shortcomings overlook. In addition to the planet, with such a massive disruption of the human psyche, we too appear to be flirting with a form of systemic collapse.

This book presents my understanding of these themes, the dynamics behind them, and the implications they portend. It describes the present and impending effects of the ideas and innovations surrounding posthumanism and their prospective outcomes. The book goes on to describe a countercultural movement that has been running in the background of our technological exploits, a movement that may renew our contact with the deeply human and provide an alternate vision for directing our godlike power to transform this life.

———

In the chapters ahead I frequently return to one psychological phenomenon and its crucial role in the presently burgeoning technocracy. As we stand on the slippery slope of digital mind control and posthuman possibility, we are also adopting a style of unconsciousness that enables these prospects: the dissociative style. Whether occasioned by stress or trauma, overwhelming amounts of information and stimulation, or the simple inability to be with ourselves, dissociation fragments and numbs thoughts and feelings, eroding contact with elemental aspects of our nature and diminishing the capacity to resist the compartmentalizing proclivities of science and technology. In the digital age, this dissociative style has begun to permeate all aspects of life, fueling a number of vicious cycles, which turn on feelings of alienation, addictive tendencies, the pursuit of perfection, and the denial of death. As each vicious cycle turns, on one side we lose connection with what makes us most deeply human, and on the other we obsessively search for further technological escape routes. Far from the perpetual promise of becoming more advanced, such a dynamic suggests a heady mix of progress and pathology.

Altering the one-sided path we appear to be on will require overcoming this dissociative style. Something akin to the integrative attention to the earth process we have seen in response to climate disruption will be needed. However, dissociative dynamics and deliberate tinkering with our psyches is also going to impact our ability to grasp what is disruptive. Already today the numbing of pain and suffering is so pervasive that the ability to feel an absence of wellbeing and integrity is diminished. Our very perceptions of reality and habits of mind are thus at stake in this new inner world crisis, making the consequences ultimately greater. Both emotional detachment and radical alteration of the mind will end up compromising the ability to recognize what ails us.

Whereas this threat to the collective psyche is ominous, preparations to meet such a threat have also been underway. As we have become more immersed in technology and the psyche has produced symptoms of fragmentation, a compensatory stream of thought has been activated. Concerned with integration rather than compartmentalization, this stream has been evident in many disciplines focused on the rich and complex breadth and depth of the human condition. Fittingly enough, this stream began to flow alongside early industrialization and may be understood, in part, as a response to the psychological impact of this historical shift. First most evident in the Romantic Philosophy of the nineteenth century, the thinking and sensibilities this stream has fostered derive from one notion in particular: existence cannot in the end ignore or escape the patterns and rhythms of nature. At the heart of this stream of thought has been the psychology of the unconscious. And the most comprehensive and compelling expression of this psychology has been that put forward by the Swiss psychiatrist C. G. Jung.

While attending to other currents of thought resonant with the psychology of the unconscious and suggestive of a nexus of inner and outer ecologies, I will argue that Jung's comprehension of the depths of human nature constitutes an incisive counterpoint to the assumptions of posthumanism and to the dissociative bubble that presently fosters these assumptions. The title of this book, *Jung vs. Borg*, recognizes this counterpoint.

Jung sheds light on the self-regulating nature of the psyche and the archetypal forms behind this—forms we may choose to overlook but cannot ultimately dismiss. These forms pertain to brain structure, anatomy, and evolved patterns of perception and behavior. But they also reflect the larger rhythms of the cosmos and are seemingly woven into the fabric of life itself. Jung pointedly demonstrates that even as we have embraced reason and science, the archetypal world of non-rational impulses and religious ideas have continued to unconsciously influence our thoughts and actions. Similarly, he reveals how fundamental dichotomies of life, such as the future and the past, creation and destruction, as well as the great polarities of Western consciousness—spirit and matter, mind and body, and inner and outer—all stand in need of bridging if existence is to progress with any kind of psychological integrity. In bringing forth such understandings, Jung also exposes the way one-sided social attitudes act as critical catalysts of psychological disturbance. He insists that the parts of the psyche we diminish or reject only make unconscious claims on us by reappearing in symptomatic and often destructive ways.

Such insights into the psyche inevitably generate grave concerns about our godlike sense of supremacy over nature. Overconfidence in our own ingenuity and a corresponding lack of reverence for the timeless forces acting on the psyche, which the ancients described as gods, leads to scientific and technological overreach. All the while, the deeper psyche remains beholden to these gods. Since Jung's death fifty years ago, our godlike position has become more strident as well as more challenged, pointing to the showdown this book aims to hasten. Moving from reshaping the earth to redesigning human nature, the meta-lessons of our Promethean hubris remain largely unlearned and leave us ill-prepared for the tests ahead.

By focusing on Jung's understanding of the human psyche and by juxtaposing this with our cyborg prospects, I aim to draw attention to the unconsciousness, dissociation, and one-sidedness involved in radical human redesign and the path we seem to be clearing for this. I will argue the rapid growth of certain psychological symptoms, such as the epidemic of depression and anxiety we find among digital natives and the distortions of reality the online world has produced, are harbingers of a possible future, and that the eclipse of the actual world by the virtual world has coincided with an increase in more serious forms of psychopathology, including anti-social behavior, narcissism, and psychopathy. I will also argue these problems invite us to expand our understanding of what makes us most deeply human.

Godlike power necessitates godlike responsibility. This begins with self-awareness, which is rooted in a sober consideration of the human psyche, the most critical part of which is shedding light on the shadow side of our willful pursuits. Comprehending how futurism blinds us to the presence of the past, how creativity and idealism cloak destructive tendencies, and how the attempt to transcend the instinctive basis of mind only invites a regressive resurgence of base motivations becomes imperative. In the approach to such matters, it is Jung who provides the right tools. Plans for ultimate change require ideas about ultimate things. On the psychological front at least, the most effective ideas are offered by Jung.

In the final section of the book, I consider what a godlike responsibility would entail and argue we adopt a *co-creative* stance going forward—a conscious shaping of existence based on a greater familiarity with our deeper nature and a careful consideration of the feedback loop our innovations and interventions set in motion. This stance urges us to listen to the way the psyche as a whole responds to the prospect of radical change, taking special note of those expressions of the cultural imagination that lodge themselves in our minds. Weaving in and out of my treatment is thus an attempt to grasp how these expressions—artistic, literary, and cinematic—may be approached as collective dreams and brought to bear on the choices before us. I demonstrate the way this and other kinds of deep listening can aggregate into reliable and convincing sources of guidance and generate an awareness of being partnered in creation by something beyond an egoic vision. Co-creation concedes that will and reason alone cannot form a seat of wise agency—a far more expansive consciousness is required. This is the venture before us.

The book is divided into four parts. Part 1, *Orientation*, sets out key considerations and navigational aids. It begins with "Polarities," the primary one being between the future and the past. Our obsession with the future and neglect of the past sets up a number of further polarities—between the divine and the demonic, civilization and nature, utopia and dystopia. In their most distilled form, these polarities return us to vision and memory, two dimensions of the human psyche that must remain conversant lest we become divided against ourselves. Consideration of key "Parts" of our subject matter follows, including the role of the unconscious, which is the presence of the past within us, potentiated by our failure to remember. The allure of human–machine hybridization and posthuman thinking is then set out before the chapter describes what is at stake if we sever ties with the deeply human basis of being. The third chapter, "Gaps," considers the present and pending effects of digital technology on social and psychological life. Here we see a widening divide between fabrication and authenticity, a loss of more self-aware and soulful modes of living, and an increase in anxiety and depression.

This first part of the book concludes by examining the "Tinkering" impulse, a human attribute that is likely instinctive. Nonetheless, given the power we possess to transform the earth and ourselves, how this propensity is situated among other impulses and associated behaviors presents a critical concern, especially as our tinkering is aiming to augment some aspects of our nature and displace others.

Part 2, *Down the Rabbit Hole*, presents my psychological critique of posthumanism. It opens by describing processes of dissociation and being psychically "Numbed," which has become pervasive to the point of being normative. Dissociation is a psychological defense mechanism that compartmentalizes experience; numbing avoids the sensation of inner conflict, which gives way to a lost sense of inner substance and complexity. Such numbing promotes the influence of the hive mind and encourages the turn to the hyperreal. In short, we start living beside ourselves and fail to be satisfied with the actuality of events. I present the case that posthumanism finds traction because the world is succumbing to this phenomenon. That is, the inclination to radically alter mind and body derives in large part from the prevalence of psychological anesthetization.

Considered in the context of dissociation and numbing, the cyborg transition represents something other than a mere embrace of technical possibility, placing the collision with the deeply human in a very different light. Mechanized bodies and computerized minds, the search for perfection and the grasp for immortality may be traced to the cultural-historical dominance of certain ideas. But our compulsive turn to these ideas is driven by becoming alienated from the deeper reaches of the human psyche. Rather than resolving the existential alienation of the twentieth century, this alienation has instead become normalized, and the fantasy of radically altering ourselves is evidently built upon this. In all these ways we are being "Borged"—conditioned for the human–machine merger by a technocratic juggernaut few know how to control. Instead of facing the complexity of the psyche and enduring states of fragmentation in more conscious ways, we're electing to take the ultimate escape route by literally running away from ourselves.

The neuromania of brain science is one influential outgrowth of this escapist neurosis, dominating the way we think about minds and playing a central role in the posthuman outlook. When psychological life is reduced to the neural activity of a proverbial "brain in a vat," imagined capable of generating thoughts and perceptions apart from the nexus of sensation and feeling that belongs to the entire nervous system, the body, the phenomenal world, relational life, and culture, a view takes over that is effectively mindless. In perhaps the ultimate irony, this reductive approach is also apt to ignore total brain function, basing models of cognitive function in humans and intelligence in computers on left-hemisphere cortical processes. Our approach to human psychology has in these ways become "Half-Brained." Whereas a number of neuroscientists have adopted a holistic grasp of the body–mind connection, reaffirming many depth psychological concepts in the process, most posthumanists are convinced disembodied minds will one day exist in cyberspace and recreated at will—flesh, instinct, and emotional life becoming unnecessary.

The final chapter of Part 2 examines the academic seeding of the radical human makeover. It traces the "Paternity" of posthumanism to the unlikely convergence of *postmodernism* and mainstream cognitive-behavioral psychology, both of which have compromised our contact with the deeply human. Postmodernism has scrambled and undermined the very notion of human nature, fetishizing alienation in the process; mainstream psychology, forever compensating for a scientific inferiority complex, has eagerly adopted a computational and programmable model of mind that sidelines major preoccupations of the psyche such as beauty, creativity, and spirituality. For the generations marked by the turn of the millennium, the search for meaning and purpose has been delegitimized by many postmodern thinkers, who view the human condition as an epiphenomenon of social contexts and a pastiche of subjectivity. This search has also been outsourced by psychologists ensconced in the technocracy, who offer mechanistic mind hacks and promote compartmentalized, robotic views of human nature.

Part 3, *Below Ground*, explores the twin birth of modern technology and the psychology of the unconscious, demonstrating how efforts to understand the psyche constitute a cultural-historical counterpoint to an industrialized world. Each chapter contemplates a form of psychological disturbance associated with the rise of technoscience.

In "Kurzweil's Dreams" I review the frequently overlooked late nineteenth-century coincidence of the Industrial Revolution and the rise of conditions such as railway neurosis, hysteria, and neurasthenia. This period also witnessed nascent fears of machines overrunning us. A century or so on, the reach of technology and the psychological symptoms may look different, and the fantasies of machine takeovers have become more vivid, but the essential pattern is highly similar. Ray Kurzweil may be the most prominent and prolific advocate of the key tenets of technological posthumanism.[6] His work is thus at the forefront of the drive to perfect ourselves through redesign until the flaws of body and mind can be excluded. But the unconscious reaction to this drive already appears to be every bit as problematic as the advent of neurosis in the industrial age.

Parallels between depictions of wayward artificial intelligences in the future and human psychopaths in the present are the focus of the next chapter. "Tin Men" lack hearts. Worse than insanity, which at least draws attention and treatment, psychopathy hides behind the "mask of sanity,"[7] with its central feature being an absence of empathy. Exhibiting a malignant form of dissociation, the calculative and soulless psychopath turns compartmentalization into an efficient and effective mode of being. If matters of the heart and the ability to mourn are doors to the deeply human, the psychopath effectively closes these doors. Increasingly prominent in corporate life and social leadership, psychopathy also concerns the AI community in their study of non-human forms of intelligence: Whether in fact or in fiction, the psychopathic potential of sentient computers, humanoid robots and augmented humans stalks the posthuman project.

In "Re-sinking the *Titanic*" I contemplate a more systemic subversion of psychological life and the cultural battle growing around this. Over a century ago, the most infamous exercise in modern technological hubris occurred right as Freud and Jung's early work was breaking into collective awareness and exposing the underbelly of the rational will. "Why do we have psychology?" Jung once asked, before providing his own answer: "Because we are already strangled by our rational devices."[8] Just as the story of the *Titanic* never rests, the titanism it epitomizes never departs, resurfacing whenever technological development and abstract systems ignore existential limits and life patterns. It is a state in which enduring values and everyday perceptions are eclipsed by spectacle, speed, excess, and emptiness. titanism may be the ultimate enemy of our reach for the deeply human and thus the ultimate obstruction to imagining alternative ways forward. If we are to recover an ecology of mind, discerning and deflecting titanic tendencies may be our biggest challenge.

If Parts 2 and 3 awaken us to the psyche's response to a wired way of life, Part 4, *Restoration*, describes the promise of such awakening. I begin by asking what "Conscious Computing" would look like. The nature of consciousness may preoccupy those hoping to create sentient machines, but this interest also invites more study of the faculties that generate self-awareness and form the basis of intersubjective relations, which may in the end serve to remind us of the distinct and irreplaceable elements of humanness.

An expansion of consciousness is imperative if we are to make responsible use of the transformative power in our hands. Our impact on the earth process has been formalized in the naming of the Anthropocene epoch, a development that accentuates the interdependence of inner and outer realities and necessitates bringing greater awareness to our actions. Yet what shall come to guide these actions? As we are forced to account for values and goals beyond those supporting the technocracy and we combat compartmentalization with holistic thinking, we are still left to ponder what might counter reckless desire and blind will. With divine guidance largely beyond our secular vision, we are left to look within for something deeper than our controlling inclinations and to look without to perceive the guidance of nature's intelligence. Guidance must ultimately come from dialogue with these marginalized sources of knowledge and from the cultural imagination, which shapes this dialogue. The result will be a "Co-creation"—a partnership between innovation, self-knowledge, and a cosmology befitting this age. Such a co-creative process will mitigate and shape our technologies as well as generate opportunities for spiritual renewal.

Moving beyond objectification of the earth and of the bodies and minds that inhabit it would be one of the fruits of this co-creative process. Many fields that question the reductionism accompanying the rise of science and technology are also revisioning the bonds of inner and outer life. A "Reanimated" view of existence is the result. Among other attributes, such a perspective counters the commodification of all things that is currently consuming us. Not a metaphysics or mysticism as such, this reanimated view validates the phenomenal field between us and the surrounding world, offering means to recognize the presences, agents, and subjects we regularly encounter beyond ourselves. Mystery has its place, as do reverence and awe—even a feeling for the sacred. A stance of contemplation and care would be the outcome. James Lovelock's Gaia hypothesis is one expression of such a reanimated view, one that incorporates rather than departs from science. Ecopsychology, biological complexity, and systems theories, post-Cartesian philosophies as well as critical strands of depth psychological thinking also suggest a necessary reanimation of the world. Not a regressive return to earlier forms of enchantment, reanimation conveys a confluence of spirit and matter—the very means by which a sense of soul is generated.

Counteracting many assumptions and aims of technocracy and posthumanism, such a reanimation may also constitute a new countercultural correction and provide a philosophically rigorous vision of sensibilities that surfaced in the 1960s and 1970s in more intuitive and unwieldy ways. Loosening the grip of the mega-machine and reclaiming the soul of life were also aims of this original counterculture a half-century ago. What burned brightly yet faded rapidly left behind a poetics of psycho-spiritual reckoning, which continues to linger and inspire. Something akin to a more methodical and epistemologically sound countercultural movement may well nudge our current course toward a necessary correction.

A final chapter in the form of an epilogue draws out what lies "Between the Lines" of the book. Here I review the underpinnings of my approach, contextualizing Jung's psychology of the unconscious in the history of modern psychology and considering how knowledge of the unconscious has become an existential necessity. This knowledge reveals that fantasies, dreams, and nightmares in the recesses of our minds unconsciously determine future events. By vivid contrast, the prevailing proclivity to attend only conscious aims and develop select forms of intelligence will almost certainly create a civilization in which nature, culture, and imagination will eventually evaporate, making the rule of the computer all but certain. Whether such a civilization will be civilized or even viable is the question that currently hangs over us—the question the pages ahead are dedicated to engaging.

Notes

1 Rachel Carson, *Silent Spring* (New York: Houghton Mifflin, 1962).
2 I borrow this phrase from Gregory Bateson, whose critical contribution to a number of disciplines is presented in his most prominent book, *Steps to an Ecology of Mind* (New York: Jason Aronson, 1972). Bateson was, ironically, one of a small handful of scientists who started the field of cybernetics, which, in turn, has developed into today's artificial intelligence (AI).
3 See Hans Moravec, *Mind Children: The Future of Robot and Human Intelligence* (Cambridge, MA: Harvard University Press, 1988).
4 The deeply human pertains directly to the archetypal patterns of the psyche. These forms are evident in biology and the evolutionary process, on the one hand, yet reflected in the images and ideas of enduring cultural value, on the other. Timeless and universal concerns with love, creativity, spirituality, beauty, community, maturation, and so on, belong to and are studied via this instinctive-imaginal-ideational spectrum. The deeply human connotes both these primary concerns and our contemplation of them.
5 The term *posthuman* has complex and even contradictory origins. Here I am employing it primarily to describe the prospect of redesigning ourselves by fusing human biology and digital technology. *Posthumanism* is the intellectual and cultural movement that has developed around this prospect, and leans toward its support and promotion. While posthumanism has also developed into a philosophical

critique of anthropocentrism and classical humanism—connoting a movement beyond humanism—I do not employ the term in this sense, mainly because such critical concerns may be engaged without simultaneously advocating for human–machine hybridization. This noted, in Part 2 I argue that undermining of humanistic ideas is partly responsible for technological posthumanism. The terms *transhuman* and *transhumanism*, which are employed more by technologists than philosophers and cultural critics, are broadly synonymous with posthuman and posthumanism, though the preferred terminology is debated in the inner circles of these movements.
6 Kurzweil associates his thinking more with transhumanism than posthumanism, arguing that technology will extend the essence of human existence, even as embodied, phenomenal life gives way to AI. However, he is widely considered a primary thought-leader in technological posthumanism. See Ray Kurzweil, *The Singularity is Near: When Humans Transcend Biology* (New York: Viking, 2005), 374.
7 This phrase originated with Hervey M. Cleckley, *The Mask of Sanity* (Augusta, GA: E. S. Cleckley, 1988). Originally published in 1941, this work also initiated the wider use of the term "psychopath."
8 C. G. Jung, *The Earth Has a Soul: The Nature Writings of C. G. Jung*, ed. Meredith Sabini (Berkeley, CA: North Atlantic Books, 2002), 149.

Part 1
ORIENTATION

Chapter 1
POLARITIES

Only a new, imaginative, religious, moral and social context for science and technology will make it possible to weather the storms that seem to be closing in on us.

Robert Bellah[1]

There is only one psyche, in relation to which all conflict is endopsychic...

Norman O. Brown[2]

The digital age has placed before us one spectacular and unsettling vision of the future, a vision of radically remaking ourselves by fusing our minds with artificial intelligence, fitting our bodies with robotic parts, and stepping into virtual realities poised to displace everyday life. On the horizon this vision perceives the remnants of human existence giving way to new kind of sentience, a displacement we ourselves will instigate, letting technology take us to a place we may not be needed. Once free of the flesh and in reach of immortality, our mechanized and digitized offspring will spread a wholly different way of being throughout the universe. The species formerly known as human will then be everywhere—and nowhere.

Among all the visions that have aimed to change the course of civilization, with a gravity so extreme and implications so sweeping, this one has no peer. By assuming a godlike role in the only known experiment with conscious life in the universe, the prospect of deliberately altering evolution and upending existence itself has been placed before us. What aspects of human nature might endure and at what point will we become something altogether different? Having passed from sci-fi novels and films into the overt concerns of think tanks, corporations, and government agencies, this question now looms large. One historian claims we are already well on the road to becoming *Homo deus*.[3]

Posthumanism, the movement that anticipates this human–machine merger—a future as cyborgs and what may follow—has become the unofficial religion of many technologists who have also demonstrated their capacity for social engineering. As we adapt to the innovations of these technologists, we find ourselves approaching the same altar and surrendering to this mother of all makeovers. In this respect posthumanism already permeates our lives and has begun to direct the way of the world. We are, however, ill-prepared for the implications of this radically altered existence.

Human–machine hybridization has been imagined for some time, but present-day innovations and socio-economic patterns are preparing the world for its realization. Futurists are extrapolating such trends and describing ways of being that are so fully wired the boundaries of the skin and the by-ways of the mind will no longer define us. The slippery slope involved has ethicists deployed to tech corporations, philosophers contemplating the prospect of computer consciousness, and sociologists considering how the enhanced and unenhanced among us will create even larger chasms between the haves and have-nots.

Some describe this change as an inevitable, even natural, extension of the evolutionary process. As Mark C. Taylor puts it, in this new phase, "technogenesis is anthropogenesis, and anthropogenesis is technogenesis."[4] Devices are being seen as extensions of biological function and computers as extensions of mental function; terms such as "organic" and "self-organizing" are being seamlessly woven into conceptions of where innovation is taking us; human and artificial intelligence are being imagined as co-equal; the language of software and hardware is infiltrating how we think of living things; data and information are being touted as the new building blocks of being. Others vigorously challenge this outlook, suggesting a cyborg future would be more devolutionary than evolutionary and declaring that nothing about our tinkering is predestined.[5]

Yet beyond calls to preserve our biological inheritance and humanistic ways, it is difficult to discern a competing vision or coherent set of values capable of redirecting technology's march toward this radical change. The World Wide Web has provided nothing less than an alternate web of life. The information economy that maintains this web drives innovation according to ideologies and dynamics that effectively define modern civilization—capitalism, enterprise, corporatization, and military-industrial complexes. In a world replete with incentives to embrace every technological advance, we find ourselves ensconced in a global technocracy that has no interest in questioning new things. In the industrialized world at least, we have subsequently begun handing more of our lives over to AI and have begun thinking that becoming cyborgs is the inevitable next step. We already live inside the machine; why not let the machine live inside us? While the future may not be predestined, other paths forward appear to be rapidly shrinking. Resistance seems futile.

As I will argue throughout this text, this dramatic shift in the relationship with our own nature is poised to compromise the human psyche in a way that promises to make climate disruption look like a warm-up act. In the twentieth century we faced a social, environmental, and political flux that altered everything *around* us. In the twenty-first century, AI, robotics, and other data-driven technologies are gearing up to alter everything *within* us.

Nonetheless, a number of critical inflection points still lie ahead, necessitating moments of collective soul-searching that can also impact collective vision. So, although it may seem like a rising tide, this pending upheaval of our inner lives is not predestined. Many of us think of science as an inevitable accumulation of knowledge, and technology as the methodical application of this knowledge, with the outcomes of both becoming impossible to avoid. But this is not the case. Both science and technology are shaped by critical values. How do we weigh the value of experimentation versus continuity when it comes to the fundaments of being? Is increasing communication more important than sustaining community? How have we come to regard the earth as a resource to control rather than a realm to tend? Are humans just complex machines or something much more? Are logic and measurement more essential than art and poetry? The knowledge we embrace and how we use that knowledge have always depended on how different cultures and eras have answered these questions. When we see a world of objects rather than animated presences and conceive of an earth without agency and intelligence, we generate technologies that are apt to manipulate and exploit. By contrast, when the world is seen with awe and reverence and the earth is recognized as the regulator of all living things, technologies are developed that tend to preserve, protect, and attune us to patterns of being both fundamental and mysterious.

Along these lines, a posthuman future appears to be less a logical outcome of predictable processes of discovery and innovation and more a projection of mechanized and computational conceptions of ourselves and the world. This makes the cyborg less a literal depiction of a future state of being and more a culminating image of current ways of thinking—an image of all the ways we are departing from ourselves. What we see on the horizon is thus nothing less than a questioning of our essential being, reflecting the grip of certain attitudes and conceptions of life. There is a war brewing within us—an endopsychic face-off between human and posthuman images of being and becoming.

To reimagine the future and find our footing on the slippery slope we have started down, we must realize it is not only our scientific and technological ways of being that have brought us to this point, and the desire to fabricate bodies and enhance minds is not only a capitulation to innovative and economic trends. Traction on this slippery slope is hard to find because we are also losing grip on the deeper reaches of our humanity, to the point of being unsure about what it means to be human. These sources of inner uncertainty and disconnection from the psychic and somatic givens of life also fuel fantasies of radical change. Competing views of identity, society, and human nature mean there is less holding us back from wanting to redesign ourselves. Whether prompted by the spirit of the age, generational desires, or simply conceptions of a better us, we have arrived at a place where everyone and everything seems primed for an ultimate makeover. Posthumanism cannot be separated from these matters either.

Disconnection from the deeply human is also permitting digital technology to dominate our lives in ways that are generating some very dark clouds. Widespread enthrallment with cyberspace together with innovations that exploit psycho-social disorientation have produced a number of disruptive and damaging side effects: expanded access to unlimited information has not resulted in being more informed; the spread of disinformation has demonstrated that algorithms and machine learning do not always align with the greater public interest; a more connected globe has not resulted in the triumph of liberal democracy; unlimited internet access has not increased wisdom. While the efficient sharing of data and rapid dissemination of knowledge have proven a great benefit in many fields, the informational free-for-all has offset this benefit by generating distortions of reality powerful enough to shift geopolitical landscapes. Infodemics hosted by the technosphere are proving to be as problematic as epidemics hosted by the biosphere, and nation-states are using the internet to undermine their adversaries. Once a Promised Land, cyberspace has become a battlefield, filling the collective psyche with polarization and intolerance.

Extrapolating some of these dark clouds, the cultural imagination has also generated a steady stream of expressions that make our impending merger with the machine look anything but smooth. Images of computer systems going rogue and humans turning into automatons fill our windows on the future. Beyond being fearful reactions to the new, these images express an unrest in the collective unconscious and their relation to actual trends and the propensities driving these trends needs exploration. A challenge before us is thus the cogent relating of these expressions of the cultural imagination to our contemporary state of mind.

These dark and destructive visions are not without modern historical precedent. They also return us to a century-old question: Why are logic and science alone not enough? Or, as the sociologist Anthony Giddens puts it, "Why has the generalizing of 'sweet reason' not produced a world subject to our prediction and control?"[6] Giddens' partial answer is that "one of the most characteristic features of modernity is the discovery that the development of empirical knowledge does not in and of itself allow us to decide between different value positions."[7] Although scientific knowledge constitutes a pinnacle of human achievement and offers disciplined pathways to consensus understanding, in the absence of other ways of knowing it can be fragmenting and misleading. As psycho-social forces governing our thoughts and actions make clear, mathematical intelligence—whether our own or the artificial variety—cannot alone construct a viable civilization. Untethered from faculties such as emotional intelligence, aesthetics, ethics, artistic creativity, and self-reflection, all of which do generate values and moral orientation, calculative knowledge will not alone foster better societies or richer cultures. It may even prove dangerous: the way a one-sided rationalism can generate compensatory and destructive bouts of irrationalism has been evident throughout the modern age. If, as Kevin Kelly, the tech guru and founder of *WIRED* magazine, posits, "The technium wants what evolution began,"[8] we should be prepared for deep and enduring conflict generated by everything the technologist is apt to overlook. As we embrace artificial intelligence, we will still have to contend with the human condition in its totality; our psyches will make sure of this, at least for as long as they can.

These shortcomings are consistently lost on those betting our existence on technological innovation. Often unable to think beyond the bubble of scientism, posthumanists in particular consider our pending merger with machines to be a matter of lining up the innovative breakthroughs and bracing ourselves for the existential equivalent of a jump to light speed. This jump has already been named "The Singularity."[9] A turning point like no other, it is imagined to be the moment AI surpasses human intelligence and takes over all important decision-making and future innovation. It has also been referred to as "The Great Transition."[10] Both the inevitability and the blessings of this approaching moment have been extolled by a number of posthuman utopians. But there are going to be a lot of speed bumps between where we are now and any Singularity. Until human nature has been neutralized, it is going to complicate matters. Such strife may turn out to be useful.

Reflections on the complications ahead may well begin with the following realization: computers can make their way into every aspect of existence, but where computers go, debased motivations, distorted values, and manipulation can also follow. With the exploitation of personal information buried in the fine print of corporate user agreements and governments all over the world preparing for cyber warfare, are we really ready to smooth the way for our minds and bodies to become fully wired? Are we going to let a company such as Meta turn everyday life into an algorithmically structured "Metaverse"? Are we willing to let atrophy faculties and sensibilities that cannot be digitally enhanced? And if the line between the virtual and the real continues to blur and the embodied becomes the fabricated, instigating a loss of the deeply human, what faculties will remain to expose the techno-traps? Who or what will grapple with questions of meaning and purpose?

This conversation must move beyond earlier commentaries on digital life and posthuman possibility, wherein the enchantments of cyberspace and redesigned bodies were treated as cultural curiosities and postmodern thought-experiments. We are also well past entertaining the prevalent notion that online connectivity is a recipe for community building, with the internet positioned to unite the world. Too many tricksters, gremlins, and ghosts have been unmasked; the flux of possibility has released the flow of primitive impulses; globalization has given way to culture wars.[11] Preliminary reports on attachment to our devices and the associated rewiring of our brains are already in: two decades of soaring rates of depression, anxiety, and suicide among digital natives; widespread cyber addictions; increased distractibility; rampant disinformation; surreptitious use of personal data; the erosion of social and relational life; geopolitical upheaval. The vision of joining mind and computer can no longer be considered without facing digital technology's already sizable dark-lands.

Four Shadows of Technology

Modernity is comprised of at least four distinct yet overlapping technological shadows. The first originated in the ruptured lifestyle and exploitation of workers that occurred with the Industrial Revolution; the second was the development of killing machines for the more efficient extermination of our fellow humans during two world wars, followed by the annihilating potential of nuclear weapons; the third has been the devastation of the natural world, resulting in climate mayhem and animal extinction. Now a fourth is before us, involving the way cyberspace is reshaping all aspects of life, with the familiar promise of increased control and freedom delivering an equally familiar cocktail of power, greed, and dehumanization.

Minimizing or overlooking these shadows simply allows their dark potential to build. Post-industrial living has not ended the plight of factory workers in developing countries; both superpowers and rogue nations continue to stoke the war-machine; and the impact of a warming planet affects all. In many parts of the world the dehumanizing effect of these things is only just taking hold, especially with globalization, enabled by information technologies, ensuring that each of these techno-shadows are simultaneously realized.

The notion that technology only provides tools and that how we use them is the real problem is a common rationalization of the pickle we are in. It is a view oblivious to the profound shift that has taken place in our relationship to technology. Contemplating the Industrial Revolution, Henry David Thoreau famously observed, "Men have become the tools of their tools."[12] Nearly a century later, in the lead-up to the Information Era, playing on Thoreau's words, Marshall McLuhan's view of technology was crystalized with the reflection "We become what we behold. We shape our tools and thereafter our tools shape us."[13] A half-century on, with our constant reach for the smart phone, social media feedback loops, and an online world guided by machine learning, our tools are doing some remarkable shaping. Commenting on digital devices, MIT researcher Sherry Turkle has suggested that *we* have now become *their* app.[14] With all this tooling around, the logic of the posthuman outlook is easily apparent: the machines have already taken charge. Even when tools could be picked up and put down again, vast numbers ended up using them badly; now these tools mediate most everything we do.

Technologized life exhibits a consistent pattern: seeking more control, we end up being controlled. The rise of the machines, an idea going back to the nineteenth century,[15] is effectively well underway, with rogue computers and wayward robots of the sci-fi imagination merely being overt depictions of covert reality. As think tanks are currently working on how to keep AI "friendly,"[16] stories of machine takeovers can no longer be dismissed as technophobic, because they serve to remind us of a critical universal dynamic, which Heraclitus referred to as the rule of *enantiodromia*—the sudden turning of something into its opposite. Fantasies of machine takeovers reflect an already existent yet largely unconscious slavery to the machine and a loss of agency as our minds meld with online worlds. Such fantasies present the flip side of technology's mesmerizing effect and our passive drift into a posthuman trance state. They are our collective dreams—no, nightmares—challenging the path we are already on. The potential for destruction and misery wrought by our own inventions may not be new, but the mind–computer marriage, the growth of AI, and our immersion in cyberspace has raised the stakes.

To orient the treatment of our theme, the remainder of this chapter will be devoted to describing several critical polarities operating in technology and posthumanism—polarities that may become pivot points showing us a way forward, but which currently exist as psychosocial divisions seemingly speeding our demise. These polarities are the future and the past, gods and demons, nature and civilization, and utopia and dystopia. Each exposes a one-sidedness that has generated the shadow problem of technology, and each clearly shows what we are neglecting at our peril. Together they point to one overarching polarity: *vision and memory*. This division essentializes the shadow problem of our current innovative posture while also directing us toward a primary pivot around which today's psyche turns, in relation to which the battle lines within us will be drawn.

As we delve into this vital connection between vision and memory, the psychology of C. G. Jung becomes our essential guide. More than any other thinker of the modern age, Jung showed how past, present, and future seek binding in the human psyche, to be torn apart only at the risk of losing our minds. By comprehending the conscious–unconscious, mind–body, spirit–matter divisions the modern mind indulges, and by grasping how these divisions may be overcome, Jung's work provides the counterpoint to the cumulative one-sidedness behind our plans for human redesign. His understanding of the psyche essentializes a wider stream of thought that has been flowing in a different direction to the one today's technology seems to be taking us. This contrasting stream of thought preserves inner and outer ecologies, furthers our understanding of the fully and deeply human, and insists that the being and soul are placed alongside growth and innovation.

The Future and the Past

By becoming the tools of our tools, we succumb to the secular religion of futurism. Entranced by the horizon, we look for meaning in all that is new and seek purpose in constant forward movement. In this futurist mode, innovation and progress justify themselves. Risks may abound, but the abiding belief is when problems appear, technical solutions will follow, suggesting little need to understand the source of these problems. Turning to the old and to origins come to be regarded as quaint, academic, or even regressive exercises. In the futurist faith, looking back is tantamount to sin and looking within is suspect. This is a faith so pervasive, we do not even know we have adopted it.

Futures markets determine economic prosperity, education is geared to the cutting edge, and work requires constant reinvention. Even entertainment is replete with tales of tomorrow. In contrast to our forebears, who spent much of their time telling ancestral stories and preserving traditions through ritual remembrance, our rituals have devolved into filling out planners, ensuring software is updated and checking retirement accounts. In seemingly every spare moment, we are eyeing the future. Rather than looking to the wisdom of the past for guidance, we believe science and technology will show us the way. Even living in the present in an ethical and responsible way entails thinking about our impact on the future, which again turns our mind to techno-fixes. As one cultural commentator puts it, "modern civilization is a shrine to the future,"[17] while another says "we live predominantly in the future and for the sake of the future."[18]

For those of us most subject to this futurist faith, irrespective of where we might still cling to the vestiges of tradition, our sense of where agency lies has undergone universal change. For in this new religiosity and its fierce focus on the frontier the future has fallen squarely into our hands. Prayer to the overtly divine has been replaced with meditation on our own power. It is a faith in our own godlike capacity to remake the world and, just over the horizon, to remake ourselves. Trust is no longer in God as such, but in our own ingenuity, as well as in the "thinking" our computers have begun to do for us.

A fear of this new divinity or divine newness has also followed. We could alter the course of creation, destroy ourselves, or do both at once. While human creativity has become the source of all blessing, and civilization has geared itself accordingly, our godlikeness has also become the most likely source of our future destruction. With the expectation of "a happier and more secure social order" on one side, and a "fraught and dangerous" world on the other, we have arrived at the sharpest extremity of what Giddens has called "the double-edged character of modernity."[19] Tending to split off rather than connect each side of the picture, with half our time spent cultivating an ability to control and transform things and almost as much time dealing with the side effects of doing so, we have not come to terms with the way each proclivity gives rise to the other: control turns into overreach; overreach courts catastrophe; the prospect of catastrophe demands more control.

This pattern only deepens the religiosity of our futurist stance. It recapitulates the polarity of the old-time religion, with the godly realm defined by an endless battle between light and dark. Once again we confront a god capable of both love and vengeance; once again we are left unsure whether the better angels can match the fallen angel. This polarity in the cultural imagination and in our psyches leaves us poorly equipped to contemplate the ways our creative and destructive potentials play off one another and must ultimately be understood as deeply entwined. The question that thus haunts our futurism is, what kinds of gods will we be?

In the American context, the beginnings of such a futurist faith were conspicuous in the Puritan origins of enterprise and pragmatism, undergirded by New World expectations and the blessings of material progress imagined as manifest destiny. This soon became a way of life involving a one-sided embrace of all things ingenious, innovative, and solution-oriented. But, even as the civilizing impulse was regularly compromised by its uncivilized twin, faith in progress endured and allowed for little awareness of the long shadow of cultural-historical catastrophes and periods of deep regression. As James Hillman puts it, "We fast-forward to avoid replay, rewind, reflect, regret. We flee to the future so as to deny the carnage of the past, the record of our history."[20] Elsewhere Hillman asks, "Is the Future an American Addiction?" He writes, "Other members of Future's gang are called Development, Progress, Acceleration, New, Technology, Prophecy—and two very rosy and seductive forms: one called Hope; and the other, Escape."[21] This addictive, idealist, and often escapist character of American futurism present from the start is amplified in this time of virtuality, permeated as it is with an ethos of not only determining one's future, but also one's reality.

Today, each new gadget emanates from a line of cities and towns on a thin strip of land along the western edge of North America. Both geographically and imagistically, this is also the edge of Western civilization—the post-industrial cutting-edge. It is also where we find an unconscious perpetuation of the frontier myth, which the humanistic psychologist Rollo May calls "the crucial myth . . . distinctively American," featuring the "lone pioneer . . . the expression of the destiny of America" and, of course, the "lone cowboy."[22] It is as far as the New World and manifest destiny have come. From Seattle to Silicon Valley to Sunset Boulevard, new visions and visions of the new reflect back through the rest of the world, orchestrating a cultural and economic colonization the likes of which the world has never seen. Landing a computer on every desk and putting a smart phone in every hand, this slice of the New World has converted the screen from a tool for communication to the means by which we daily affirm our existence. Oddly enough, this existence needs a lot of affirming. Yet as much as our screens mirror ourselves back to us, what appears there never quite squares with where we are or even who we are.

Cyberspace has in all these ways become the new Wild West, a realm accessible from any place, where opportunities abound and the horizon goes on forever. Cyberspace is thus a spiritual realm too, making the devices that allow our entry objects of entrancement. The coin of this realm is novelty and anticipation, shifting the cult of the new from things as such to states of mind saturated with constant expectation. In what Herbert A. Simon conceived of as the attention economy,[23] which Michael Goldhaber subsequently called "the natural economy of cyberspace,"[24] what we are mostly paying for is a feeling of possibility, a constantly renewing promise of what may come, a feeling made all the more palpable by the instability of what is. *Becoming rather than being* has captured the soul.

Gods and Demons

The inclination to imagine the future is no doubt part of the human condition. This is not the problem. Looking ahead without looking behind or within is the problem, because it splits an existential constant. This lack of existential integrity is also what sustains our dark potential, for the devil tends to get his hands on what we neglect. Such a proclivity not only condemns us to repeat the past, as Santayana suggested,[25] it exercises an even more difficult dynamic: when history remains unconscious, its repetitions are exaggerated, to the point that what we leave behind can possesses us—which the return of generational wounds and tribal resentments frequently demonstrates. The constant search for a brighter future can in this way blind us to our underlying motivations. Failing to become conscious of what is trying to break into awareness, the situation rapidly spins out of control.

We have been primed for this dynamic. Neglecting the darkness within us belongs to Western monotheism and to Christian idealism in particular and is betrayed by the omnipresent feeling the devil still has plenty of room to move. As we are likely to be possessed by what is not made conscious, this bedevilment is a religious phrasing of a psychological truth. Even further back, in the polytheistic conceptions of the ancient Greeks, we find that the hubris of displacing the gods invited divine punishment, with the gods personifying the principles of existence we ignore at our peril. The meta-questions of today are hardly separable from these themes: What will fall into the devil's hands? How far can we push our technological hubris? What form might the divine punishment take? If the gods are the forces we neglect because of an ignorance of the past at work within us, they will surely find more dramatic ways to curb our enthusiasms.

Such an injection of uncertainty into our futurist faith could initiate new levels of awareness and curtail our more excessive tendencies. However, when there's nothing to rely on save our own devices, we tend to suppress any uncertainty and double-down on our futuristic ideals and their technocratic vehicles. This now default pattern recapitulates the psychology of the religious fundamentalist, who represses doubt and clings to a formulaic faith. As the faith becomes more rigid, the doubt is disowned and projected onto the "faithless," who then become subject to paranoid enmity. All pious and perfection-seeking religious movements are haunted by this specter of an autonomous evil, and once we understand that gods and demons are enduring metaphors for psychological realities, we'll see our futurist religious faith is no exception; displacing the divine onto our own sense of power and plans to remake the world just cloaks these enduring religious dynamics in a different garb. Yet the psychological reality is difficult to escape: in designing a New Jerusalem we remain stalked by fear and doubt; with no more metaphysical cards left to play, we are left holding a hand in which all coming evil can only come from us.

Modern history already provides an appalling record of mass destruction instigated by our own genius. Nietzsche's declaration of the death of God was quickly followed by attempts to fill the vacuum of ultimate meaning with secular movements and philosophies offering ultimate solutions for a desacralized world. Overcompensating for our loss, such movements became governing ideologies. Capitalism, socialism, communism have offered answers to questions of human purpose that have at times become more totalizing than their dogmatic religious forerunners. The chief instigators of these systems are the new prophets, guiding us to Promised Lands of our own making. But as many astute observers have noted, the answers to questions of ultimate meaning and purpose these systems have provided have proven at least inadequate. Communism compromised individuality; socialism compromised initiative; capitalism compromised the greater good. The materialism within them all compromised the spirit of most everything. By refusing to recognize how and why these systems fail, the true believers just double their efforts, pointing to the inherent problem of all "isms": they promise total solutions that are inevitably unfulfilling. This in turn leads to denial, rationalization, and ideological purity—a dynamic that gives rise to fanaticism on the one hand and the aggressive suppression of critics on the other.

Something all these modern ideologies share is being built upon the seemingly neutral arenas of science and technology—*seemingly*, because once they are drawn into the service of these ideologies and those orchestrating them, science and technology quickly lose their neutrality. Among others, Theodore Roszak has described this dynamic at length, already observing in the 1970s that "All technocracies we see abuilding in the world today are in the hands of what Veblen called 'saboteurs'—those who undertake a 'conscientious withdrawal of efficiency' from the industrial process for their own selfish advantage." Yet, he quickly notes, "such sabotage can, with the aid of enough co-opted expertise, become remarkably 'efficient' in its own right—certainly at mystifying the public and defending vested interests."[26] He goes on to describe the way the "suave technocracy" of America,[27] the "vulgar technocracies" of "collectivist societies,"[28] and the "teratoid technocracies" accompanying the "racist and nationalist megalomania"[29] of catastrophes like Nazi Germany illustrate the degrees to which science and technology can be coopted and misdirected in such ways.

With so many examples of the extreme misuse of technology throughout the industrial age, it is time that those involved in post-industrial technoscience developed a sense of responsibility for what happens when their work leaves the lab. After noting that "Marxists should ask themselves what it was about the teachings of Marx that paved the way to the gulag," Yuval Noah Harari contends, "scientists should consider how the scientific project lent itself so easily to destabilizing the global ecosystem, and geneticists in particular should take warning from the way the Nazis hijacked Darwinian theories."[30] The more totalistic modern ideologies have become, the more they have also involved the scientific rationalization of the oppression and eradication of those deemed either inferior or enemies of the system. As Raymond Tallis observes in this context, "To see people as machines—genetically determined or programmable—is no light matter."[31] Their objectification easily follows. With the greatest evils of the past century having been perpetrated at this nexus of political ideology, scientific reductionism and technological destruction, this volatile mix is a form of evil we need to keep our eye on.

As the world has moved from the industrial to the post-industrial phase, some of the hard divisions between capitalism, socialism, and communism might have softened, resulting in the blended socio-political systems we find in Scandinavia, China and, haltingly, in Russia. Scientific and technological interests, guided by multinational corporations, are freer to take on ideological casts of their own. Scientism and technologism are thus coming to exert their own totalistic claims on us—claims that have begun to eclipse socio-political differences. One measure of this change is the determinative power of the new global technocracy, which currently dwarfs the power of any particular nation-state. We need look no further than the so-called New Atheism to perceive the powerful notion that science should be placed at the center of any viable philosophy of life. One of the most significant challenges before us is thus how to remain scientific without becoming scientistic and construing all problems as needing technical solutions.

As technocracy itself displaces political ideologies, it forces these ideologies to battle for control of the online world. And whereas the spread of communication technologies first appeared to correspond to the spread of liberal democracy, a darker array of movements has appeared, especially since various extremist and terrorist groups began using the internet to recruit and spread propaganda. More recently we have seen the rise of authoritarian leaders, willing to manipulate facts and spread falsehoods to assume power and attack political enemies. Aside from providing a platform for such phenomena, the technocracy that controls the internet also controls social connectivity, professional networking, and most forms of media and entertainment, all of which guide the way people think and interact. Across the globe, in developed, developing, and undeveloped countries alike, there is a migration to an online territory that knows no borders and is subject to the influence of tech corporations as much as nation-states. Migration to this *new* New World has only just begun, as have the attempts to dominate it.

Our futurist religion is thus a dualistic one—a turn to the light followed by a growing darkness. What may be darkest of all is this unconscious religiosity, one in which the faithful masses have little idea of what they have got into. This unconsciousness makes all the difference, for as we take hold of our newfound powers, we are gripped by powers we do not understand. Molded into the psyche but far from mind, this unconscious religiosity belies the belief we are harnessing data and information in a logical and reasonable manner. It means we are blind to the ritualistic, sacrificial, creedal, and dogmatic modes that shape our enthrallment with the technosphere.

Amidst this futuristic fervor a vague sense of the ancients is also present and we may well suspect our hubris. For others hubris is a way of life, thoughts of divine punishment grow distant, conscious religious attitudes grow fainter, and everything is drained of any conception of forces beyond the human will. When we can no longer grasp where we are missing the mark or offending the gods, our Promethean exploits are, indeed, unbound. The suffering that results is rationalized away or placed in reductionistic boxes, and we lack the sense to perceive or appreciate the greater powers to which we are still subject. An intoxicated deification comes to obscure an intractable demonization.

Nature and Civilization

Human beings live at the intersection of nature and civilization. Throughout history, the instinctive basis of being and the unrelenting call to transcend this have sometimes torn us apart. But the heat of their collision has also forged moments of profound symphony. We endure the collision and thrive on the symphony, realizing the heart of existence lies in the ability to sustain this heat until these constants can converse with each other. By joining natural impulses and rarefied ideals, we find friendship and love, make homes and communities, and engage in satisfying work and pastimes. Even if we often regress, the quest for stronger and more just societies endures. This dynamic drives the arc of history.

What is innate, inherent, and vital, the organic essence of things, as well as "the power, force, principle etc. that seems to regulate this,"[32] is the crux of nature. That which has been developed and organized with deliberation, advancing a "high order,"[33] is the crux of civilization. Nature attempts to return us to what we inherently are; civilization drives us towards what we aspire to become. Nature reminds us we are bodies, comprised of organs, subject to physical limitations; civilization keeps pushing us beyond such limitations. Nature is concerned with what is given and beholden to the great rounds of existence—life and death, light and dark, order and chaos, creation and destruction—binding us to the earth, and the earth to the cosmos. Civilization is concerned with principles that elevate us beyond what is given and its realization offsets our mortality, effectively binding us to movements and institutions that will outlive us. To be fully human is to become part of this great oscillation, remembering our natural inheritance while simultaneously reshaping and overcoming this inheritance. At different times we may find ourselves favoring one side of this oscillation, and entire lives can be lived embracing or rejecting historical shifts in either direction. Ultimately, however, without the capacity to endure the collision between nature and civilization and understand the predicament this generates, we also lose the capacity to know what makes us essentially human.

I set out this elemental pattern of humanness because there is much talk today of departing from it by transcending the bounds of nature. But it is hard to see how we get to such a point of departure without sundering humanness in two: as technology dominates civilization, nature becomes more unfamiliar; we then perceive natural things as unpredictable and in more need of regulation and taming; this, in turn, promotes technologies that grant more control. A vicious cycle results.

Although the challenge presented by this vision of transcendence is peculiar to this historical moment, it turns out we have a deep well of navigational aid to draw on. We have been working on the confluence of nature and civilization for some time, and *culture* is the fruit of these efforts. Culture hosts and transforms the collision of nature and civilization. It provides a "cognitive network" manifesting as "language, myths, core metaphors, know-how, hierarchy of values, and worldview."[34] At the very center of culture we find the symbolic life and associated narratives of meaning, which is evident in the anthropologist Clifford Geertz's definition: "an historically transmitted pattern of meanings embodied in symbols, a system of inherited conceptions expressed in symbolic form by means of which (humans) communicate, perpetuate, and develop their knowledge about and attitudes towards life."[35] Robert J. Lifton's pithy formulation is that "Culture is inseparable from symbolization."[36]

To express events and experiences symbolically is what distinguishes human beings from other animals, even other primates. Language itself displays this capacity, spanning everything from the most rudimentary patterns of speech to the most sophisticated and compelling literary and poetic expressions. An early human paints a bull on the wall of a cave and contemplates invisible worlds; a child says "no" and turns raw emotion into social currency; a young musician writes a rebellious song that starts a protest movement; a poet describes a wasteland, helping generations to reflect on the excesses of progress. We make culture through metaphor and analogy, modes of thought crucial to language and visual expression, from the earliest ritual performances and mythic figures, to the most sublime novels and films. Words, symbols, metaphors, and analogies, arranged in artistic, spiritual and literary expressions, draw upon and then re-present deep responses to life, forming bridges between nature and civilization.

Although culture and civilization are often seen overlapping and are, at best, mutually supportive, they are distinct. Culture preserves and protects the most profound images as well as the need to imagine. It gives rise to collective values, reminds us of what has been significant over time, and generates stories that act as binding agents. Culture is the means by which we express and record *what it means to be human and the presence of the past within us*. Civilization builds things. It organizes people, consolidates power and makes economic progress. It is concerned with supplying water and removing sewage, providing goods and services, and, in today's world, with maintaining the internet: it is about *making things work and keeping things moving*. Yet culture and civilization are also interdependent. Without the structures of civilization, culture-making lacks containers; without cultural values, civilization lacks guiding principles. Whereas the artifacts of culture are threatened when civilizations collapse, when we stop making culture, civilizations succumb to base motivations such as power and growth for their own sake.

Starting thirty to forty thousand years ago, and clearly evident ten thousand years ago, the fundaments of life began to be channeled into some discipline or design. Over time the need for community, friendship, and intimacy was met by the building of social structures. Images were generated and objects were made that both metabolized and enhanced deeper patterns of existence. Religious practices and philosophies were born, which attempted to negotiate the competing claims of animality, practicality, and angelicity. Through millennia of trial and error, we created pathways between basic drives and lofty aspirations. It is this process of negotiation and creation that has become the basis of culture, our enduring response to the predicament of life.

Cultural artifacts thus express the inner life of collectives, constituting a record of human reflexivity and providing the basis of continued thought and reflection. Such artifacts are the primary means by which *we think about our thinking*—that most distinctive feature of our humanness. The imaginative products of the cultural process also serve an evaluative function, drawing together body and mind in a way that allows us to feel into our thoughts and actions. Artists, musicians, poets, and mystics bind the poles of nature and civilization accordingly. Surely this is what Shakespeare was doing with his incomparable descriptions of contemplated actions, captured in his declaration via Hamlet that the purpose of art is "to hold, as 'twere, the mirror up to nature." We can see why Bloom attributes to Shakespeare "the invention of the human."[37] Yet the great artists and teachers of all cultures have been occupied with the same, contemplating the existential givens and critical values that guide us. The cultural record thus anchors our conception of who and what we are as a species, providing our greatest source of guidance when navigating the way forward. The relationship to this record and the body of expression it contains truly define us.

Because the significance of our cultural inheritance is determined by sensual and symbolic faculties, which make us embodied presences in a phenomenal world, a database can never supplant culture. As culture is the accumulated record of the transformation of nature, it is these natural faculties that allow us to understand and build on this inheritance—faculties that cannot be fabricated. Sensual life is peculiar to our anatomy, which has evolved as such, just as our symbol-making is peculiar to the mind–earth relationship, which has developed over millennia. Neither can be reduced to data or be artificially replicated.

None of this suggests the cultural process is not choppy sometimes. Insight comes slowly. The most enriching expressions take time to surface and sometimes decades to receive fitting recognition. Formulas fail because culture is a living process. Without also sensing the spirit of the times and where it is pointing, for example, turning to the past can be regressive, forgetting that how we remember is speaking to where we stand. A diversity of peoples and places also necessitates variations in cultural responses. Indigenous peoples who have managed to preserve their lands have kept the civilizing impulse close to the rhythms and limits of nature; urbanized peoples are more apt to live at some distance from these rhythms and limits. Both individually and collectively, our various dispositions, locations, educations, and exposure to other cultures position us differently on the spectrum of nature and civilization, with different beliefs about how to best hold the tension and what cultural forms we need to embrace. These differences are important, and they are sustainable as long as there is dialogue between them.

Human history is, however, full of schemes and philosophies that pull us to the extremes of nature or civilization. And once our eggs are all in one basket we become prone to other sorts of other polarizations—spirit or matter, mind or body, science or art. In modern history the civilizing impulse has tended to dominate, so denying the deeper needs of nature and overriding the less predictable path of culture-making has been conspicuous. Utilitarianism has taken precedent. In this regard authoritarian and totalistic systems have been worst of all, suppressing cultural processes and cutting people off from their own natures.

Such repressive movements must ultimately come to face the rejected part of the picture, which grows in the cracks or on the fringes and forces its way back in. Yet whereas the collision between these great poles of existence cannot ultimately be avoided, a great deal of damage can occur as the repressed returns. When things are pushed into the unconscious, terrible internal divisions and external scapegoating occur.

That is, when the culture-making process is rejected, we are apt to make some other group carry our failures for us. In early modern history indigenous peoples suffered because they carried the European's and American's rejected relationship with nature. With the rise of totalitarian regimes any group that did not sign on to the formula on offer could be targeted. Those who displayed a rich, coherent culture were often deemed threats and were sometimes eliminated. Today's so-called "culture wars," especially in the form of book-banning and other attempts to erode social discourse, undermine what Rauch has called "the constitution of knowledge,"[38] threatening the cultural process. We don't yet know where these wars are headed.

The destructive outcome of such attempts to repress cultures present a giant lesson in trying to escape these essential parameters of the human condition. Curiously, however, that lesson is not only slow to be learned; it continues to be met with remarkable resistance. Posthumanism seeks to avoid the nature–civilization collision by transcending it altogether. If the above dynamics hold, it is also likely to severely disrupt or altogether obliterate the cultural process.

To the extent we remain biological beings, nature is, in some manner, always with us. As the majority of people move to urban environments, however, nature at large only reaches us in sublimated forms. Wild nature, in particular, exists only at a distance, and is experienced indirectly—via gardens, in designs inspired by nature, by watching documentaries or going to the zoo. For many, the larger rhythms of nature are no longer directly experienced; they are encountered only by means of culture-making, which provides connective links to the patterns of nature. For the urban dweller, exposure to the imagery of vast pristine wilderness can contribute to the feeling of psychic balance, even if these places are rarely if ever accessed. In such a way, the presence of nature becomes very much dependent on the quality of culture. As one commentator of posthumanism indicates, "nobody has ever really contested their interdependence."[39]

How well culture functions in its mediating role thus becomes critical. When "making things work and moving forward" (civilization) stops conversing with "what it means to be human" and "the presence of the past within us" (culture), our lives become reduced to component parts in the machinery of bureaucracy and industry. Civilization then no longer serves anything beyond its own expansion and no longer functions as a vessel for culture-making. Processes of reflective criticism, artistic response, and spiritual contemplation, which generate cultural values, become window dressing rather than existential mirrors held up to the politicians, corporations, and technologists doing all the moving and shaking. As funding for the arts and humanities and for the preservation of plants and animals is cut, educational programs designed for competition in a growing global technocracy take precedence. Expedience, profit, and utility take over, draped in the abstract and vacuous garb of progress. Cities lose their soul, peoples lose their character, and the strength of nations is measured by GDP and internet speed. The earth-process and the cultural process lose their integrity. Becoming cyborgs then starts to look like an inevitable next step.

Utopias and Dystopias

We read Rumi, Blake, and Melville because of the way culture filters the products of the imagination, leaving us with poems, stories, and symbols that convey meaning and guidance. We contemplate the work of Van Gogh, Kahlo, and Picasso and refer to the characters of *Star Wars* because of the same. These images orient our thinking. Even when this filtering system becomes clogged and the cultural process appears to break down, some dogged expressions can still alert us to what we might otherwise avoid seeing. The imagination becomes insistent, working around our conscious intentions.

One expression of this imagistic insistence concerns the extrapolation of present trends producing images of extreme potential. Such images have a prophylactic function, forcing reflection and providing opportunities for cultural change. Herein we may locate another polarity, which is the way visions of the future produce both utopian dreams and dystopian nightmares, and the way these products of the imagination can shed light on deep flaws in our thinking. Here too we come upon some of the immediate implications of splitting the future and the past.

Utopian expectation expands as contact with history and introspection contract. In influential sectors of today's society, especially in the tech world, planning frequently comes wrapped in the search for perfection, with technology expected to reengineer our shortcomings and fast-track human evolution. More broadly, the promise of unending progress governs our fantasies and inflates our expectations, which also distances us from enduring difficulties. Dams of darker possibility then grow larger. When these dams break, dystopian expectation kicks in. Not only does this compensatory expectation signal that these utopian–dystopian propensities are two faces of the same configuration, it reveals the function of dystopian stories, which is to keep our idealistic plans in check.

The veracity of this pattern is not just evident in the products of the imagination: it has been clearly demonstrated by historical events, showing how images shape actions. Dictatorships and totalitarian schemes have been selling utopias and generating dystopias for some time. These dystopian outcomes of utopian plans are as impressive as they are disturbing. Sometimes the dam does not break and the darkness seeps into the system from the start, often in full view, yet compartmentalized in the mind. So-called "cultural revolutions" have thus started with the killing of millions and proceeded with thought-control and oppression, even while promising glorious futures. The heady mix of philosophical idealism, group identity, expansion of power, and the elimination of those who would question the program is enough to prevent any flaw or failure from fully registering: dissociation promotes the duplicity. Cults display the same dynamics, with the idealization of the cult leader and their teaching generally at complete odds with the oppressed, compartmentalized lives of the cult members. Indeed, cult leaders and manipulative politicians often use the same play book. Anyone who fully awakens to this contradiction has to be banished or eliminated, so remaining asleep, even with eyes open, becomes the only option. The geopolitical landscape remains subject to such patterns.

While sold as ultimate solutions to social fragmentation and/or spiritual disorientation, utopian visions always involve existential escape, achieved through a psychological failure to recognize the needs of the fully human. Modern utopias in particular are premised on the notion that humans can conform to mechanistic programs without causing psycho-social damage. But none endure without such damage occurring.

As we consider the landscape of today's technological idealism, especially the concern with perfecting minds and bodies to find paths to immortality, we may seem worlds away from these overtly problematic political experiments and cult movements. However, the cultural imagination keeps insisting on the utopian and dystopian potential. With the course of civilization in the grip of technocracy and increasingly directed by posthumanists, the genre of posthuman dystopia keeps expanding. How we read this phenomenon is critical, as tales of technological overreach may also be invitations to heal the deeper divisions of today's collective psyche.

History has shown we cannot rule out the enactment of these utopian–dystopian inclinations. Yet this enactment occurs primarily because the desires and fears behind these inclinations are not reflected upon and traced to their psychological roots. We literalize fantasies of perfecting ourselves to avoid the conundrums of existence. Plans to transcend human nature are concretized by scientists and technologists without recognizing how much of the impetus for these plans is concerned with avoiding aspects of our nature that do not fit mechanical and computational paradigms. Becoming enslaved or destroyed by machines may turn out to be a real threat, but this escalating fantasy also reflects the way technocracy and attachment to our devices are already a threat to our psychological integrity. To the extent we wallow in their literal prospect, dystopian narratives can also be escapist. If they only serve to affirm our nihilism or offer some kind of End Times spiritual by-pass, we will miss their commentary on the cultural-historical challenge right before us. Literalizing coming darkness can also be a failure of reflection and deeper insight.

Psychological splitting is thus not confined to the actual implementation of utopian plans and denial of their dystopian outcomes; it also belongs to the failure to confront the inner divisions that generate these polarized prospects in the first place—the failure to grasp the two faces of this one configuration. The task, then, is to become psychologically responsive to the images of utopian and dystopian futures in order to prevent their blind actualization—to embrace the prophylactic function of the imagination and cultural discourse. Such an embrace would, for example, help us see that the imagination builds opposing and medicinal elements into the utopias and dystopias it envisions. Utopian stories have their black holes, just as dystopian stories display pools of light, forming something like a Yin-Yang of possible futures.

In H. G. Wells's *A Modern Utopia* we find a society that could achieve harmony with technological advance, yet it does so only through a legalistic conformism and narrow concept of freedom. This society is also maintained by banishing its rebels to islands. Tapping this motif, Aldous Huxley makes it a key feature of his dystopian novel *Brave New World*. In a later utopian story, *Men Like Gods*, Wells creates a society in a parallel universe several thousand years into the future wherein liberty plays a more central role. The protagonist and others access this perfected world through a time warp. Comparison with early twentieth-century humanity is unavoidable and animosity between the two results. In the end, even as this utopian world plants seeds of change in the present, the intolerance of human foibles is pervasive. Before the protagonist returns, one of the utopian elders asks, "What could Utopians do with the men of Earth?" He then offers, "You are too like us [for us] to be patient with your failures . . . We might end by exterminating you." Perfection necessitates intolerance.

Klaatu, the extraterrestrial visitor in the 1950s classic sci-fi film *The Day The Earth Stood Still*, offers something of the same possibility. The other planets he represents have agreed to be policed by a race of robots, which prevent all aggression and violence, but only through their own threat of aggression and violence. Earth, he suggests, must follow suit or be destroyed. All utopias appear to come at a similar high cost.

Even drawn on the blank canvas of the creative mind, open to whatever the imagination conjures, utopias seem possible only through an intolerance of flaws and a rejection of misfits. The repressing of emotion is also a pervasive theme in utopian fiction, and it is a notable one in *Brave New World*. It is thus instructive to observe the key role the life of the emotions plays in dystopian stories. Here inner flaws and outer misfits frequently become critical to pathways of insight and redemption. Emotional outrage at mechanisms of destruction, empathy in the face of raw survival, and the intensifying bonds of community are also prominent features. In cinematic stories in particular, renegades show up to either expose dystopias in the making or to preserve a sense of humanity once these dystopias take hold. Many of these renegades are women—a thinly veiled commentary on what is typically devalued in the run-up to the dystopian realization: Sarah Conner in *The Terminator*, Leia Organa in *Star Wars*, Trinity in *The Matrix*, Katniss Everdeen in *The Hunger Games*, Imperator Furiosa in *Mad Max Fury Road*. Refusing to succumb to fallen worlds and dehumanization, these characters unite, outsmart, and outfeel their oppressors, carry the conscience of oppressed communities, and bear witness to the polarizations and hypocrisy that set the destructive forces in motion. While their battles are imagined beyond the here and now, they portray confrontations with what threatens to our deeper humanity today. To understand them as such leads us to question those forces that might well carry us to the end of these utopian and dystopian streets.

One utopian theme is the attempt to control nature, making nature's revenge a driver of dystopian outcomes. How we relate to nature turns out to be critical. The degree to which we perceive and validate the feedback loop of natural responses to our actions, coming from within and without, corresponds to the degree to which we may disrupt utopian expectations and defuse dystopian potentials. The utterances of nature anchor us to reality and make us less inclined to transcend or erode that reality. We have already learned that building cities, redirecting waterways, and turning landscapes into resources generates ecological pushback, which has, however ploddingly, shifted our attention. Such pushback has necessitated a more conscious dialogue, reversing a tendency to ignore or repress such signs of nature's intelligence. However, we continue to live largely oblivious to the natural patterns permeating all things, including ourselves: the deeper patterns of the psyche in particular still elude us.

The history of Western civilization has instilled a quest to go beyond nature in pursuit of abstract principles and rarefied modes of existence. Yet this may not be the full story. There is much to suggest another plotline is playing out before us, one that may, among other things, be more inclusive of non-Western peoples and their better ear for the voice of nature. With so many pathways of demise before us, some kind of shift in the narrative seems necessary. We are coming to see, for example, that the undeniable improvements to the everyday lives of many, and the advances in science and medicine that protect us from the unpredictable ravages of the wild, do not come without problematic side effects. When the rhythms of nature are too far removed, vitality is lost and disorientation sets in, with inner life becoming chaotic and unpredictable. Two millennia of Christian emphasis on the otherworldly, the sins of the flesh, and the eradication of nature-religions has created a psycho-spiritual remove, yet the supposed unruliness of nature we have tried to transcend has only turned into an insatiable monster within. As problematic as it is not to recognize this monster in ourselves, the outer enmity caused by its projection onto a series of convenient scapegoats has come with untold costs. The values of the Enlightenment laid on top of two millennia of Christian morality has not been enough to prevent the modern torture and death of those who have carried this projection. Somehow between the Enlightenment and the present day, the pursuit of reason has become disconnected from the effort to comprehend human nature. Though it was a combination of scientific reason and humanistic thought that gave birth to the modern world, the latter has come to be seen as something optional. We are still in the grip of this mistake.

A split from outer nature that is, simultaneously, a split from inner nature, has become omnipresent. The dystopias we associate with outer destruction usually turn around this kind of inner division. The familiarity with what makes us fully human recedes, and inhuman impulses step in to fill the void.

Contemporary dystopian tales convey this dynamic in the most impressive ways. The inhuman factor, whether in the form of literal machines, mechanistic bureaucracies, abstract systems, or sheer brutality, takes over when the fully human is repressed or forgotten. States of unconsciousness pervade in which people become distant from themselves, eventually unable to feel their essential needs. In what has become a modern collective dream, or nightmare, these situations only come to an end when some conscious awareness or act of awakening disrupts the system, even if this occurs amidst great catastrophe. In keeping with the dual meaning of "apocalypse," such an end can also be a revelation.

Meditating on apocalyptic outcomes is not only revelatory; it may be necessary to our self-knowledge and avoidance of literal destruction. As Guggenbühl-Craig notes, "a progress myth that is not complemented by a myth of Armageddon is clearly dangerous and harmful. Collectively, it causes one to overestimate oneself and then fall prey to *hubris*." He adds: "As a consequence, in the course of our enthusiastic progress we overlook and neglect present suffering."[40] An over-identification with progress also prevents us from seeing what ails us.

Much turns on our capacity to grasp these psycho-social patterns. We need to see that images of possible worlds and failed experiments are calls to negotiate present-day hopes and fears in more conscious ways. Remaining conscious of what is imperfect—the ordinary, everyday suffering that is an inevitable part of mortal life—is a key element of this. Ideals work well if ordinary flaws and weaknesses are not suppressed and if grace and good fortune are valued alongside sheer willfulness. Without such humilities, we become divided against ourselves, unable to see there is no perfection or paradise that accords with being human or mortal. Moreover, when the pursuit of progress involves a disregard for human integrity and already evident suffering, our self-destructive tendencies only escalate and threaten widespread destruction. How many times must we see the utopian dream become the dystopian nightmare? How blind will we remain to the tales of technological advancement resulting in political and social regression? In each instance, it is a failure to grapple with our totality of being and inner nature that drives the outcomes. As Eisler points out, "Be it in Frank Herbert's *Dune* or George Lucas's *Star Wars*, what we frequently find is actually the social organization of feudal emperors and medieval overlords transposed to a world of intergalactic high-tech wars."[41] Such images suggest that advanced technologies will in no way ensure advanced societies; they may even be prone to a compensatory regression.

If utopias begin with aspirations for a higher plane of existence, the dystopias they instigate are defined by becoming possessed by base motivations. That is, dystopias manifest the lowest rung of impulses, especially all-consuming drives for power, which eventually result in inhuman regression: the perfect social system turns into a hellish bureaucracy without room for humanness; those who cannot embrace the supposed ideals and principles are subjugated, banished, or annihilated; brainwashing becomes necessary to keep the masses oriented to the cause. When we imagine futures in which machines enslave us, the machines are caricatures of our own tendency to colonize and control. In all cases, it is the return of the deeper, uncivilized parts of our nature that sets the whole sequence in motion.

In religious cults, the utopian ideal serves as a brittle outer shell that houses and hides rigid hierarchical structures, denials of emotion, and abuses of power. The contradiction between the espoused ideas and the underlying reality becomes so glaring that only processes of dissociation and other psychological defense mechanisms can maintain it. Nation-states with monolithic political structures and authoritarian leadership require the same psychodynamics. This form of utopianism presents the most danger, especially as our capacity for dissociation grows, which is the case today in the United States, where some groups are failing to acknowledge events taking place right before them.

Assembling the themes of this chapter thus far, I would describe the resulting picture of the world this way: a futurist religiosity, built upon scientism and technologism, ignoring the influence of the past and elevating becoming over being, is fueling a drive to redesign ourselves and change the course of human evolution. As this religiosity takes hold, the corrosive effects and destructive potentials of our digital lives have become more evident, though not yet fully felt, widely considered, or well understood. As the former becomes more utopian and the latter more dystopian, we confront a polarization that betrays a profound unconsciousness of what is behind our radical plans and what our initial migration into cyberspace and nascent merger with computers are actually doing to us.

The utopian–dystopian theme thus becomes a punctuation point for actual developments and a portal into where the imagination is directing our attention. In the techno-utopian scenario we merge with an artificial superintelligence made possible by the exponential growth in computational power and swap a biological basis of being for an engineered one, perfecting and extending existence. In the techno-dystopian scenario, the smart machines we design start behaving like monsters from the id and either they or their own offspring enslave or extinguish us. The first scenario describes the start of a slippery slope we are already on; the second scenario describes what may come. But it may just be a metaphorical expression of where we already stand.

Vision and Memory

The above themes provide lines of orientation for my critique of posthumanism: futurism and the neglect of the past, godlike certainties and their demonic counterpoints, civilization and nature, utopian plans and dystopian outcomes are all divided because the cultural bridges between them have been weakened. The widespread tendency to embrace one side of these pairings and neglect the other suggests a sizable split in the psyche—a division in the ecology of mind that is a dangerous fault-line in our approach to reality. This danger has three main components: first, it distorts our thinking about our thinking; second, it opposes holistic awareness and psychological integrity; third, it sets up the destructive return of each neglected aspect, because in the psyche nothing of real significance ever leaves us—the repressed always returns.

Underlying all of these matters is *a failure to tether vision and memory*. Without such a tether, our pursuits are prone to be driven by what we forget. It is in just this way that posthumanism blindly recapitulates preoccupations of the human past, its vision shaped by what it has forgotten. This means we are approaching the most radical prospective changes to existence ever conceived with a wholly inadequate comprehension of what we are doing or why we are doing it.

May tells us that "memory can liberate us from attachment, from desire or attachment to the wrong things."[42] This point is worth pondering at length. In our growing attachment to all things technological, *vision* is everything; technologists, we are told, are visionaries. Yet what animates these visions from behind is overlooked. The more this is the case—the narrower and more obsessive the vision—the more rigid and compulsive its enactment becomes. If we are to mitigate, if not prevent, our "desire or attachment to the wrong things," we are thus tasked with a corrective remembering—of cultural wounds, of forgotten ideas, of misplaced human needs, and of old myths that shape even the newest thinking.

May has more: "Memory is our internal studio, where we let our imaginations roam, where we get our new and sometime splendid ideas, where we see a glorious future that makes us tremble."[43] Remembering not only creates understanding; it helps us see what is necessary as we move forward. Importantly, memory is not a database or a video recording of the past: it recollects the past in ways that are germane to what is before us. This is not to say that memories are always misshapen, but that their shape is determined by the way immediate concerns activate processes of recollection. That is, memory is not time travel; it is looking back from where we stand. Perhaps it is, as May seems to suggest, the needs of the future that memory serves—memory as a necessary partner to vision.

His final point in this context is that "memory and myth are inseparable . . . Memory can, according to Dante, form the past into any myth, any story, any hope. Dante believed that memory can lead to God via myth."[44] As we remember, we are also discerning the mythos at work in our logos, the universal patterns animating our most compelling ideas.

This mythical background of our thinking is affiliated with the deep human need to story life; life stories remind us of deep human needs. Whether we are conscious of it or not, myth and story play a major role in how we think about our thinking, to the point some have suggested our species would be more fittingly named *Homo narrans*—the storytelling human.[45]

Myth, as Cassirer set out in some detail, is also at the very core of culture.[46] Culture returns to myth in the same way physical science returns to mathematics, because myth is the original language of the imagination, giving consistent shape to perception and thought, circling the most enduring values and plot points of the human story. As we remember the lives of those who have gone before us we also mythologize them, whether describing heroic or spiritual quests, the trials of love, the pursuit of a dream, or the protection of the tribe. We remember stories and ideas that tell us how to live in relation to the recurring conundrums of life. The widespread use of the term "myth" as a synonym for falsity reflects the lost appreciation for myth as metaphorical description of the deeper reaches of being, revealing cross-cultural patterns in thinking and behavior. Perhaps this is why Guggenbühl-Craig bluntly claims, "without mythology the humanities are dead."[47]

To find the presence of the past in our visions of the future via this memory-myth continuum would loosen the grip of futurism and prevent attachment to the wrong things. To embrace remembering as a psychocultural check on envisioning would uncover the way posthuman dreams reenact mythic dramas, revealing motivations we only think we have left behind. To seek perfection, defy the gods, and effectively become spiritual beings is nothing new. But to understand the mythic background of such notions is to reveal such goals are less rational than we assume, making us humbler and less driven in the face of such pursuits. We would slow down these efforts and put a rearview mirror on the juggernaut of technocracy.

This turn to memory involves one further element. Alongside the temporal dimension of remembering there is a spatial one too, recalling the dimensionality of experience and the breadth of being. The narrow focus contemporary life demands makes us prone to forget aspects of our own nature as well as how we are imbedded in nature at large. A meta-theme of mythology is to remind us of the propensity to be absorbed by some god or principle as we neglect others. Nietzsche considered modern life to be a battle between Apollo and Dionysus, with the latter ending up second best. With mercurial speed, invisibility, and slippery communication, it is Hermes that is omnipresent in the digital era, but recognition of his handiwork in our frenzied connectivity is mostly absent. That we no longer see gods in our enthrallments makes us blind to what shapes our views. Indeed, we are as prone to be just as unconscious of what we identify with as we are of what we suppress.

To take up May's point that memory and myth go together is also to perceive the way myth brings together the temporal and spatial aspects of remembering. We are reminded that primordial, archetypal forms, which myth personifies as gods, are always present, and that even the newest visions cannot escape their influence. Hillman contends that "There is no place without Gods and no activity that does not enact them. Every fantasy, every experience has its archetypal reason. There is nothing that does not belong to one God or another."[48]

Depth psychology has done much to restore our awareness of the gods in the form of psychic realities. Comparative religion and literary studies, if not the humanities in general, also provide us with constant reminders of this, which depth psychologists regularly draw upon.

Critics have argued the emphasis on mythic-archetypal forms can neglect cultural diversity and overlook the constantly evolving character of socio-cultural conditions, with the so-called "post-structuralist" approach becoming quite influential. However, one habitual misunderstanding in this argument involves the failure to appreciate the way archetypal forms are expressed with widely varying, culturally-specific content. As the artistic, literary, and cinematic worlds continually demonstrate, the return to enduring patterns of the human situation does not preclude creative, contemporary, and cross-cultural renderings. Once more we are invited to behold the way the most imaginative and stirring expressions of the cultural imagination function as a bridge between what is innate and what evolves. The most significant of these expressions combine the timeless and universal with the timely and unique.

The call to remember as we envision is thus not a call to revert to the way things once were. It is not traditionalism or some other attempt to revive the ways of the past. Traditionalism is not remembering but pseudo-remembering—an attempt to bind the present to the past without awareness of the movement in the psyche prompting this. Traditionalism is thus a neglect of the spatial memory, a denial of what impels it, which is often the fear of expansive knowledge and breaking tribal bonds. Religious fundamentalism illustrates this dynamic, attempting to place narrowly selected, literal interpretations of traditional texts at the center of contemporary life, while concealing the wounds, confusion, and anxiety caused by secularization and liberalism in the process. Aspects of being such as sexuality, emotional intelligence, and aesthetic pleasure are then readily split off. The pursuit of perfection and hypocrisy inevitably follow. Everything driving this defensive pseudo-remembering becomes even more unconscious, making adherents cling to the tradition even more tightly.

Whether we are conscious of it or not, our envisioning is always a remembering and our remembering is always an envisioning. The lack of consciousness is not innocuous, however; for when memory narrows, so does vision, and the powers we pretend to understand tend to have their way with us.

It is no exaggeration to claim the course of history now depends on attending to this vision–memory relationship. To avoid the escape to the future we are called to remember the past. Failure to do so will harden the futurist position, which will then produce regressive socio-political reactions based on rigid and reductive accounts of the past. By maintaining the vision–memory relationship we may thus avoid the escapist futurism and the regressive traditionalism that are both presently nipping at our heels.

If the syndrome I have laid out in this chapter is as crucial as it seems, it prompts the question of where we will find streams of thought and cultural expressions that will meet this challenge and turn vision and memory to face one another. Surely such a crucial relationship has already been considered at length; surely the series of polarities I've set out have been subjected to other kinds of scrutiny; surely some cultural process has already been at work on this. It turns out that this is indeed the case. The one-sidedness of our futurism and technological overreach has generated a number of corrective perspectives, some of which have been stalking the modern project from the start. And at the core of this corrective movement we may locate the depth psychology of C. G. Jung.

Vision and memory are direct outgrowths of the conscious and unconscious dimensions of the human psyche and are subject to the dynamic relations between these dimensions. To the extent our way forward is hobbled by our failure to remember, it is Jung's understanding of the unconscious and the task of making this dimension of the psyche more conscious that offers a most pertinent means of overcoming this failure—a recovery of the presence of the past within us that provides an antidote to the futurism that grips us. As Jung puts it, "there exist certain unconscious contents which make demands that cannot be denied, or send forth influences with which the conscious mind must come to terms, whether it will or no."[49] Restated for the contemporary moment, whereas our consciousness has been captured by an expanding technocracy and its hive-minded symptomatology, the unconscious demands we come to terms with the enduring patterns of nature and timeless cultural values that would curtail these trends. Such demands are rooted in the unfathomable depths of our being.

Jung developed his psychology by observing the individual and collective psychodynamics of late modernity. His opus describes the psychological need to recover the fully and deeply human, calling for the overcoming of polarization and the embrace of holistic thinking well before other disciplines did so. He understands psychological wellbeing in terms of how well this call is heeded, and he sees psychological disturbance and suffering in terms of its rejection. The danger, as he sees it, occurs "when free reign is given to an intellect that has grown estranged from human nature," and his recommendation is "to return, not to Nature in the manner of Rousseau, but to [one's] own nature . . . to find the natural man again."[50] This approach to the psyche counters the mechanization and automation of life and opposes attempts to reduce and systematize the human condition, such as we find in the radical behaviorism of John Watson and B. F. Skinner, which, as we will see, has come to play an outsized role in the theoretical and economic background of posthumanism.

For Jung, the psychology of the unconscious involved a corrective remembering of the deeper layers of existence in a world gripped by the knowledge and agency that science and technology provided. He observed the newfound power of industrialization generating a psychological hangover, and it was this specific hangover his psychology aimed to attend. This compensatory relationship between the rise of industrialization and the concepts of depth psychology has been noted by the field's historians,[51] but the significance of this relationship has not been widely acknowledged, and we have barely begun to consider the role an understanding of the unconscious may play in the post-industrial world. However, if one thing is clear, such a psychology of the unconscious preserves a vertical axis of understanding amidst the endless expanse of horizontal information. It is thus to Jung's work and its relation to key elements of the futurist-posthuman outlook we now turn.

Abbreviation

CW   *The Collected Works of C. G. Jung*, R. F. C. Hull trans., H. Read, M. Fordham, G. Adler and W. McGuire eds. vols. 1-20 (London: Routledge & Kegan Paul, 1953-1979). Cited by volume number and paragraph.

Notes

1 Robert N. Bellah, *The Broken Covernant: American Civil Religion in Time of Trial* (New York: Seabury Press, 1975), xiv.
2 Norman O. Brown, *Love's Body* (New York: Vintage Books, 1966), 163.
3 Yuval Noah Harari, *Homo Deus: A Brief History of Tomorrow* (New York: HarperCollins, 2017).
4 Mark C. Taylor, *Intervolution: Smart Bodies Smart Things* (New York: Columbia University Press, 2021), 151.
5 See Jaron Lanier, *You Are Not a Gadget* (New York: Alfred A. Knopf, 2010); Bill McKibben, *Enough: Staying Human in an Engineered Age* (New York: Times Books, 2003); Charles T. Rubin, *Eclipse of Man: Human Extinction and the Meaning of Progress* (New York: New Atlantis Books, 2014).
6 Anthony Giddens, *The Consequences of Modernity* (Stanford, CA: Stanford University Press, 1990), 151.
7 Ibid., 154.
8 Kevin Kelly, *What Technology Wants* (New York: Viking, 2010), 270.
9 Ray Kurzweil, *The Singularity is Near: When Humans Transcend Biology* (New York: Viking, 2005).
10 Russell Blackford, "The Great Transition." In Max More and Natasha Vita-More eds., *The Transhuman Reader* (Oxford: Wiley-Blackwell, 2013), 421–429.
11 See David Brooks, "Globalization is Over. The Global Culture Wars Have Begun." *The New York Times*, April 8, 2022.
12 Henry David Thoreau, *Walden* (Springfield, OH: Crowell, 1899), 61.
13 Often mistakenly attributed to McLuhan himself, this quote comes from an article about McLuhan's views on technology by Father John Culkin, who was a Professor of Communication and friend of McLuhan's. See J. M. Culkin, "A Schoolman's Guide to Marshall McLuhan." *Saturday Review*, March 1967, 70.

14 Sherry Turkle, "TEDcUIUC – Sherry Turkle – Alone Together," YouTube Video, 16:24, March 25, 2011. https://www.youtube.com/watch?v=MtLVCpZIiNs
15 See, for example, Samuel Butler, *Erewhon* (New York: Dover Publications, 2002). Originally published in 1872.
16 See James Barrat, *Our Final Invention: Artificial Intelligence and the End of the Human Era* (New York: St. Martin's Press, 2013).
17 Derek Thompson, "How Civilization Broke Our Brains." *The Atlantic*, January/February, 2021.
18 Bruno Maçães, *History Has Begun: The Birth of a New America* (New York: Oxford University Press, 2020), xi.
19 Giddens, *Consequences*, 10.
20 James Hillman, *Philosophical Intimations* (Thompson, CT: Spring Publications, 2016), 382.
21 Ibid., 392.
22 Rollo May, *The Cry for Myth* (New York: W. W. Norton, 1991), 93–94.
23 Herbert A. Simon, "Designing Organizations For An Information-Rich World." In *Computers, Communications, and the Public Interest*, edited by M. Greenberger, (Baltimore, MD: The Johns Hopkins Press, 1971).
24 Michael H. Goldhaber, "Attention Shoppers." *WIRED*, December, 1997. See also Charlie Warzel, "The Internet Rewired Our Brains. This Man Predicted it Would." *The New York Times*, February 7, 2021.
25 George Santayana, *The Life of Reason: Reason in Common Sense* (New York: Collier Books, 1962), 284.
26 Theodore Roszak, *Where the Wasteland Ends* (New York: Anchor Books, 1973), 38–39.
27 Ibid., 41.
28 Ibid., 42.
29 Ibid., 43.
30 Yuval Noah Harari, *21 Lessons For the 21st Century* (New York: Spiegel & Grau, 2018), 217.
31 Raymond Tallis, *Why the Mind is Not a Computer* (Charlottesville, VA: Imprint Academia, 2004), 26.
32 *Webster's New World Dictionary*, College Edition, 1960.
33 Ibid.
34 Jeremy Lent, *The Patterning Instinct* (New York: Prometheus Books, 2017), 25.
35 In Richard A. Shweder and Robert A. LeVine, *Culture Theory: Essays on Mind, Self, and Emotion* (Cambridge UK: Cambridge University Press, 1984), 1.
36 Robert J. Lifton, *The Protean Self: Human Resilience in an Age of Fragmentation* (Chicago: The University of Chicago Press), 13.
37 Harold Bloom, *Shakespeare: The Invention of the Human* (New York: Riverhead Books, 1999).
38 Jonathan Rauch, *The Constitution of Knowledge* (Washington, DC: The Brookings Institution Press, 2021).
39 Stefan Herbrechter, *Posthumanism: A Critical Analysis* (New York: Bloomsbury, 2013), 8.
40 Adolf Guggenbühl-Craig, *The Old Fool and the Corruption of Myth* (Dallas, TX: Spring Publications, 1991).
41 Raine Eisler, *The Chalice and the Blade* (New York: Harper & Row, 1987), 185.

42 May, *Cry*, 70.
43 Ibid.
44 Ibid.
45 This term appears to have been independently proposed by Kurt Ranke, a German ethnographer and Walter R. Fisher, a communications theorist.
46 See Ernst Cassirer, *An Essay on Man: An Introduction to a Philosophy of Human Culture* (New Haven, CT: Yale University Press, 1944).
47 Guggenbühl-Craig, *Old Fool*, 38.
48 James Hillman, *Re-Visioning Psychology* (New York: HarperCollins, 1975), 168-169.
49 CW 8, par. 713.
50 CW 11, par. 868.
51 See Henri F. Ellenberger, *The Discovery of the Unconscious* (New York: Basic Books, 1970); George F. Drinka, *The Birth of Neurosis* (New York: Touchstone Books, 1984).

Chapter 2
PARTS

Idea and thing come together, however, in the human psyche, which holds the balance between them.

C. G. Jung[1]

A century ago, in an essay entitled "The Spiritual Problem of Modern Man," C. G. Jung considered the impact of futurism on the psyche, declaring modern people had become "unhistorical" and stood "before the Nothing out of which All may grow." He described this outlook as a "Promethean sin."[2]

As the myth goes, the Titan Prometheus, whose name means "forethought," furthers human agency through his transmission of technical knowledge to humans. He then ends up stealing fire from the Olympian gods and is condemned to torturous punishment. Playing off the link poets and other cultural observers had already made between Prometheus and modernity, the sin to which Jung referred is this mythic figure's hubris, reenacted in modern technological overreach and futurism. To be Promethean is to forget the past, along with its ancestral voices and sense of divine order and to forge ahead with willful abandon and inventive cunning, an approach that risks the wrath of the gods. Prometheus thus represents both the triumph and tragedy of human endeavor, epitomized by power over nature, the temptation to use this power to excess, and the lingering specter of divine punishment. The mythologist Carl Kerényi goes so far as to call Prometheus "an archetypal image of human existence."[3]

Writing this essay between the First and Second World Wars, Jung had seen significant devastation, much of it attributable to military technology. But he was yet to witness the catastrophe Hitler would create. Alongside the disconnection from the past, much of the spiritual problem he describes in this essay pertains to the pivots I set out in the previous chapter. He contends, for example, that neither "Christian Idealism"[4] nor "the ideals of material security, general welfare and humanitarianism"[5] appear to alter our destructive impulses, and that "every step forward in material 'progress' steadily increases the threat of a still more stupendous catastrophe."[6] Here Jung also puts his finger on an essential dynamic of the utopian–dystopian turn, referring to what "Heraclitus called the rule of enantiodromia (a running towards the opposite)," which "steals upon modern man through the by-ways of his mind."[7] This is all to say that despite its sophistication, knowledge, and Christian values, the modern psyche cannot seem to contain its barbarous shadow. This would subsequently come into full view with Nazism and the Holocaust.

Against this dark weave Jung did, however, locate a contrasting thread, which he saw as countering the problem of Western superiority and its "technical proficiency."[8] This was the simultaneous emergence of psychological interest and search for spiritual renewal. For Jung, the exploration of the unconscious, together with the burgeoning fascination with Eastern philosophy, altered states, mind–body connections, and indigenous wisdom were counter-weights provoked by the Promethean trends. He thus concludes his essay with a quote from Hölderlin: "Where danger is, arises salvation also."[9]

Prometheus, it seemed, could also be an awakener.[10] Modern people thus contain within them both a drive for excess and a corresponding call to become more conscious. Yet whether there can be enough consciousness to mitigate the excess is the question still playing out before us, one that rests on our capacity to understand the unconscious background of our thinking and behavior. For in practice, becoming more conscious means shining a light on where we are most unconscious and resistant to self-understanding.

After Freud's well-known exposure of what Victorian morality had pushed into the personal unconscious, Jung turned our attention to the collective unconscious, a neglected underworld comprised of the "patterns of instinctual behavior"[11] or *archetypes* and their associated symbolic forms—often represented by the gods and other recurring motifs of the mythic and religious imagination. As Hillman puts it, "the unconscious, so newly found, was in fact a palace left from antiquity and the Renaissance, still inhabited by the surviving pagan Gods and once called the realm of *memoria*."[12] Jung saw this hidden presence of the past within us as the primary shaper of our conscious experience. The collective unconscious is not, as commonly understood, a storehouse of particular images and ideas, but a dimension of the psyche containing imaginal, perceptive, behavioral, and ideational patterns, coinciding with the organization of human anatomy, brain structure, and adaptation to the earthly habitat. These psycho-dynamic patterns create the continuity of human experience. Jung described the archetypes of the collective unconscious as "mode(s) of apprehension,"[13] often evident in dreams, fantasies, and stories where recurring themes align with supreme values. Stages of maturation, interests of love and community, calls to adventure, battles with darkness, spiritual quests, and similar themes have always shaped human life and its cultural reflections. In the present context it is well to note that such archetypal patterns structure the mythos we find blended into logos of contemporary thinking—albeit most often in an unconscious manner. To describe the proclivities of modern society as Promethean is a case in point: the desire to join or go beyond the gods is itself an enduring archetypal motif, appearing in most mythologies. However, archetypal impulses work to mitigate one another, making identification with one configuration problematic: to forget the collective unconscious, or what Jung later termed the "objective psyche," is to forget our lives must contend with an array of impulses as well as restraints.

In his treatise on the social meaning of Jung's psychology, Ira Progoff contends that Jung "makes it his principle that all analysis must start from the primary fact of the social nature of man," a position that has "two main roots."[14] The first is his "conception of the psyche," which "he derives ... not from individual experience, but from the great communal experiences." And the second is his "intellectual resources," which are, in this context, traceable to "the spirit of Durkheimian sociology."[15] Progoff states, "To Jung the social is essentially the unconscious, and more particularly, the deeper layers of the Collective Unconscious." Here "he [Jung] is following his more fundamental idea that 'consciousness comes from the unconscious.'"[16] Progoff is drawing out the way contemporary society and its aims derive from a reservoir of past psycho-social imperatives whose depth of influence remains largely unknown to us.

Becoming more conscious of the archetypal patterns of the collective unconscious thus becomes central to Jung's psychology. As he was able to discern, the symbols and images produced by these patterns formed the basis of psychological transformation. That is, symbolic experiences, both individually and culturally, can transform instinctual impulses: symbols can redirect behaviors. Jung's first major book, written to distinguish his views from Freud's, came to be translated as *Symbols of Transformation*.[17] It is here that Jung wrote of religious iconography and the gods in particular as "libido analogies"[18] and about the function of symbols in terms of the "canalization of libido."[19] Archetypal symbols and the shifts in attitude and action they instigated were in these ways foundational.

Even an AI researcher such as George Zarkadakis argues for our need to be familiar with these "cultural universals" and associated "primal river of stories . . . in which the visible and invisible formed an uninterrupted continuum, where everything had a soul, a mind, and intelligence."[20] For Jung, it is this background of the psyche that our identification with Promethean forethought has neglected the most, resulting in the conceit of the mind as a blank slate, to be inscribed as we determine—a view that goes back to the influential work of British empiricists such as Locke and Hume.[21] Few other images offer those looking to redesign the mind with such a blanket invitation to do so. This neglect of the psyche's underlying structures is effectively a repression of the presence of the past within us—an unconsciousness of our deeper nature. Like Zarkadakis, I believe recognition of these structures of the mind to be vital as we embrace artificial forms of intelligence and consider remaking ourselves. Such recognition would be the primary way to offset our Promethean sin and some of the dangers this invites. Perhaps most important of all, through Jung we come to see that it is via this archetypal dimension of experience that the poles of nature and spirit are conjoined and prevent our becoming overly identified with either one, which is the frequent source of danger today. It is in this way that the archetypal basis of mind anchors the vertical mode of understanding and defines our contact with the deeply human. And it is this archetypal basis of mind and associated modes of understanding that juxtapose the horizontal collection of information that defines the digital age and the posthuman vision that extends from it.

The Religious Function

Estranged from the archetypal world in our conscious outlook, two primary psycho-social problems appeared in the first decades of the twentieth century, which are still with us. The first of these is the displacement of the religious impulse. When humanistic concern was liberated from the creeds of the past, and the overt grip of Western religion loosened, an erosion of symbolic and ritualistic modes of being took place. Unbeknownst to the architects of the Enlightenment, however, this erosion of overt religiosity did not translate into an erosion of the religious impulse itself: the need to feel connected to larger mystery and purpose and to sense one's participation in a cosmic order never left; the energy once associated with religious life just went in search of new molds.

Science, technology, economics, and several large-scale socio-political experiments eagerly swooped in. Secular values of utility, pragmatism, and progress assumed quasi-religious status, proposing new pathways to meaning and purpose. The rational mind was well served by this turn, but the non-rational and instinctual modes of being, which ultimately require mythic thinking, were not. As Robert Bellah puts it:

> The substitution, in an effort to de-mythologize the political system, of a technical-rational model of politics for a religious-moral one does not seem to me to be an advantage. Indeed it only exacerbates tendencies that I think are at the heart of our problems.[22]

Reflecting in his autobiography on those impacted by this phenomenon, Jung wrote:

> If they had lived in a period and in a milieu in which man was still linked by myth with the world of the ancestors, and thus with nature truly experienced and not merely seen from outside, they would have been spared this division within themselves. I am speaking of those who cannot tolerate the loss of myth and who can neither find a way

to a merely exterior world, to the world as seen by science, nor rest satisfied with an intellectual juggling with words, which has nothing whatsoever to do with wisdom.[23]

This cultural-historical change delivered Jung a long line of distressed patients in search of deeper meaning. He himself also suffered the loss of myth, leading him to shoulder the full implications of what Nietzsche had earlier identified as the death of God. Jung thereby both witnessed and succumbed to the central psychological malady of the modern era, caught between the diminishing returns of traditional religion and the innate demand for symbol and myth. Amidst this syndrome and the research it prompted, Jung identified a psychological path out of this existential conundrum and spiritual crisis; as he tersely noted, "morbidity drops away the moment the gulf between the ego and the unconscious is closed."[24] He had discovered that the archaic roots of the mythic world and the possibility of a viable spiritual orientation resided in the collective unconscious. The path to this unconscious world would, however, require *psychological* thinking, not the theological or metaphysical thinking of past ages. Having been dismissed by reason, myth and religion could be approached as phenomena derived from the archetypal patterns of the psyche, in need of symbolic understanding and metaphorical interpretation. Closing "the gulf between the ego and unconscious" thus had a great deal to do with developing such an approach.

The death of God reflected a dramatic cultural shift; it was itself a mythopoetic expression of the death of an older style of religiosity and a literal approach to spiritual concerns. But it was not the death of religious and spiritual concerns as such, and these changes represented an existential disruption that was not without a deeper purpose.

In the West at least, the medieval world, which provided a springboard to the Renaissance and Enlightenment, had been a hot mess of human misery. The life of the intellect might have gained some momentum, but the life of the body was toil and torture. The Gothic vision of Christianity not only attempted to make the invisible visible; these forms towered over everything. Concerns for the afterlife lay over ordinary human life like a dark cloud; the concerns for this world often seemed like a second thought. This situation was unsustainable. A reversal of this spirit-over-matter approach to reality was inevitable.

Harari describes the metanoia that occurred during the Renaissance and Enlightenment as ultimate authority moving from the divine to the human.[25] The Renaissance brought both the exploration of the world and a renewal of interest in human nature. The Enlightenment brought the value and concerns of the individual to light, giving birth to democracy, human rights, and judicial fairness. Few would question the progress these shifts represented. At the same time, this nutshell description of change leaves out what Jung had come to see in terms of the psyche and some philosophers readily recognize in terms of existential discourse, which is "the religious question is ultimately at the center of all philosophy, even if it be by way of rejection."[26] Whereas the problems associated with the displacement of the human realm by the spiritual realm in the Middle Ages was most apparent, the problems associated with the displacement of the spiritual realm by the human realm in the Modern Age were not. By the early twentieth century, with the compensatory rise of religious fundamentalism, the modern hunger for spiritual sustenance, and the Protestant roots of secularism, it was evident the role of the divine in human affairs had never really gone away. It just went underground and, neglected by sanctioned forms of knowledge, it has continued to resurface in an odd and disruptive manner.

Hindered by a lack of self-awareness and inability to grasp our own motivations, humanistic authority has therefore never been brought to completion. While the conscious mind and its embrace of reason have provided a seat of authority, the unconscious and its non-rational elements have eluded our comprehension as well as our control. The psyche's religious function, as Jung referred to it, lies at the very core of this oversight. Without the capacity to relate to this deeper dimension of ourselves, a true, enduring, and more humble human authority will remain elusive. Understanding the return of the repressed (Freud) might have opened the door to a fuller sense of being human, but the religious function (Jung) is the inner dynamic that stands in greater need of our familiarity, as it is this dynamic we have become most inclined to live out in unconscious ways.

Accordingly, religion and myth are today orchestrating our secular pursuits from behind. Our time is one in which, as Jung observed:

> We think we can congratulate ourselves on having already reached the pinnacle of clarity, imagining we have left all these phantasmal

gods far behind. But what we have left behind are only verbal spectres, not the psychic facts that were responsible for the birth of the gods. We are still as much possessed by autonomous psychic contents as if they were Olympians. Today they are called phobias, obsessions, and so forth . . . The gods have become diseases . . . produc[ing] curious specimens for the doctor's consulting room . . . disorder[ing] the brains of politicians and journalists who unwittingly let loose psychic epidemics on the world.[27]

If modernity has a story, it is one of the rational outlook finding its limits and meeting its match. The non-rational side of the psyche, where passion and imagination hold sway, left largely unattended, has frequently turned highly irrational, disrupting civil discourse and defying all logic. This upheaval has also given rise to an array of alternative and substitute perceptions of ultimate reality. As Taylor puts it, "religion is most interesting where it is least obvious."[28] Ultimate meanings and spiritual quests now wear a number of non-traditional guises and direct our attention to unusual places. It is these unconscious forms of religion that have started to reshape the world—posthumanism being a prime example, one that has the potential to redefine this entire existence.

Unconscious religion could well become our spiritual and worldly nadir. But the danger also presents an opportunity. The modern crisis of meaning and our scramble to meet it invites another step forward, involving a more conscious understanding of the essential role of religion, moving away from its literal denotations and toward its metaphorical connotations. This begins by accepting the degree to which we are still subject to forces capable of overwhelming our conscious intentions, still prone to fanatical devotions, pious proclamations, and obsessive beliefs whether stripped of their old-time religious garb or not. This is an enduring dimension of human experience. Such a realization necessitates our vigilance toward higher or greater powers in whatever form they may appear. This stripped-down religious attitude is what Jung saw as necessary if we are to relate to the deeper dimensions of ourselves; especially to the archetypal forms of the collective unconscious. He writes:

> Religion, as the Latin word denotes, is a careful and scrupulous observation of what Rudolf Otto aptly termed the *numinosum*, that is,

a dynamic agency or effect not caused by an arbitrary act of will. On the contrary, it seizes and controls the human subject, who is always rather its victim than its creator.[29]

He goes on:

> Religion appears to me to be a peculiar attitude of mind which could be formulated in accordance with the original use of the word *religio*, which means a careful consideration and observation of certain dynamic factors that are conceived as "powers": spirits, daemons, gods, laws, ideas, ideals, or whatever name man has given to such factors in his world as he has found powerful, dangerous, or helpful enough to be taken into careful consideration, or grand beautiful, and meaningful enough to be devoutly worshipped and loved.[30]

This elemental religious attitude can exist apart from whatever traditional or secular faith is involved. Curiosity, reverence, and awe are then linked to whatever exerts a powerful influence over our thinking and behavior, irrespective of the name given to this. The expansion of consciousness and pursuit of self-knowledge thereby become the most essential form of religious practice, one that appears to be what our own deep nature, if not existence as a whole, is now requiring of us.

The archetypal forces of the collective unconscious are unavoidable, but their influence can be met with more awareness than is presently the case, offsetting their possessive, destructive potential. Integrating the power of the divine world as a dimension of interiority would bring a paradoxical completion to the humanist project, helping us grasp the way the strictly human, or narrowly human, is tied to a more-than-human depth of soul. Rather than scientific rationalism simply usurping spiritual need, we would enter a dialogue between the rational and the non-rational. Among other things, we would then begin to see how much our futuristic urge to redesign existence is itself driven by a submerged religious quest—an attempt to further fill the gap left by the death of God, albeit with a new scientifically anointed source of ultimate purpose and meaning.

Three stages are apparent in these reflections on the psyche's religious function. First, for most human existence, we were unconsciously immersed in the expressions of religion and myth, living with direct reference to the forces these expressions represented. In this state we may say the horizontal and vertical axes of being were unconsciously fused. Second, philosophy and science culminated in a dominant rationalism and materialism that displaced this entire worldview: the objects of religious belief were removed, but the psychological powers and dynamics behind such beliefs remained; the powers represented by the gods, while no longer recognized as outer realities, moved into the unconscious but then returned in surreptitious and disruptive ways. The horizontal axis becomes most primary to us and even yields to considerable control, but the vertical axis is largely dismissed. Third, turning to face this problem, the religious function is made more conscious and embraced as an innate aspect of human nature: this expands self-knowledge and makes the pursuit of consciousness a psycho-spiritual undertaking with the purpose of joining together instinct and intellect. Herein the vertical axis is consciously recognized and recovered as an inherent dimension of being. Beneath the buzz of post-industrial life we may also detect the prompts for this transition from the second to the third stage of the religious function.

What our Promethean sinfulness requires in this moment is an act of reconciliation—a reckoning with what the modern outlook has neglected and a fuller comprehension of our depth of being, especially pertaining to the division between the conscious and unconscious, rational and non-rational dimensions of the psyche. Such a reconciliation would constellate "the Anthropos ... the essential core of the great religions ... the idea of the *homo maximus* [where] the Above and Below of creation are reunited."[31] This is the psycho-religious goal that would eventually depotentiate fantasies of technological perfection and form a conscious, and therefore proportioned, relationship with the impulse to transcend the human condition.

And the Hive-mind

If the compensatory movement of the religious function is the first psycho-cultural problem associated with the Promethean posture and the psychic realities arising to meet it, the associated pull of the hive-mind, also known as the mass-mind or group-think, is the second. Groups can become substitute gods and manipulative group leaders, demigods. Displacing individual conscience with collectivized attitudes and postures, the hive-mind convinces us we are part of something of vital importance. When this propensity boils over, manifestations ranging from mob rule to national psychosis can result, examples of which demonstrate the fragility of reason, democratic norms, and other Enlightenment values. The hive-mind can also just simmer away in the background, relentlessly pulling vast numbers of people in vortexes of corrosive trends and expectations that slowly erode authentic individuality. Both forms of hive-mindedness are afoot today.

A more conscious relationship with the collective unconscious, a key part of which is an awareness of the religious function, provides a means for understanding and navigating these conformist pressures. Whether it occurs through the study of history, myth, traditional and indigenous wisdom, the cultural imagination or other portals into the patterns of human nature, the more mindful of our collective propensities we are, the less vulnerable to group-think we become. Consideration of elemental human needs—for friendship, community, love, beauty, enchantment, heroism, and other arterial channels for psychic energy—as well as of conducive ways to meet these needs then occupy our awareness. This, in turn, inoculates us against the allure of those handing out passports to ultimate fulfillment and pathways to salvation.

Contact with these deeper patterns also brings familiarity with loss, alienation, and disorientation as elemental human experiences, which require tolerance and understanding rather than reactivity and dissociation. It is in the absence of such resilience that the hive-mind of the digital age, with its endless web of connective possibilities, offers a substitute sense of contact with the collective. When something ails us, we turn to group-think rather than the wisdom of the ages. Meaning is not derived from how one's unique situation is related to the archetypal constants of life but instead from strength in numbers. This may begin with the mirroring provided by social media but ends by selling the soul to some quasi-religious "truth" and formulaic path to salvation. Hive-mindedness results from a failure to consciously negotiate the vertical axis of being. As Jung put it succinctly, "Our fearsome gods have only changed their names; they now rhyme with 'ism.'[32]" Elsewhere he writes:

> Rational argument can be conducted with some prospect of success only so long as the emotionality of a given situation does not exceed a certain critical degree. If the affective temperature rises above this level, the possibility of reason's having any effect ceases and its place is taken by slogans and chimerical wish-fantasies. That is to say, a sort of collective possession results which rapidly develops into a psychic epidemic.[33]

Jung was writing mainly in response to witnessing the rise of fascism and communism, and the kind of mind control these movements employed. However, scaffolding for the same phenomenon is today being erected as online forums foster emotionality and unthinking reactions that sway social attitudes and political postures. The online world might have been envisioned as a means of exchanging factual information and providing access to the accumulation of human knowledge. However, it has turned into a hotbed of reality distortion, mind manipulation, and the commodification of perception and knowledge. For all of us, reality is becoming curated, particularly by algorithms designed to keep our attention and tailor news content for sensation and polarization. As one prominent surveyor of trends in technology recently put it, "The internet probably isn't giving you a fair picture of what's happening in the world. And for any given story, you might never really know how much you aren't seeing."[34]

A new tribalism is one result. The volume of information and the speed of its delivery fosters this situation because it forces those producing online content to be radical and relentless. As one former editor of a major news network put it, we have become "both overfed and malnourished."[35] The tech industry is complicit in creating such an environment, enabling the loudest and most persistent sources of distortion, providing a glimpse of the way an unregulated technocracy puts growth and profit ahead of social and political responsibility. Political action and civic engagement are necessary components of any healthy society, but it has become increasingly hard to step into the polis without falling down some rabbit hole in cyberspace.

Jung's thoughts on the mass-mind, gathered midway through the last century, may seem dated. However, we only have to tweak his terms a little bit to see the essential dynamic he exposed at that time is now woven into the contemporary situation. When Jung writes, for example, that "one of the chief factors responsible for psychological mass-mindedness is scientific rationalism, which robs the individual of his foundations and his dignity,"[36] he is pointing to a worldview in which consciousness moves toward what can be observed and measured and away from reflection, emotional intelligence, conscience, and values, undermining insight into the deeper psyche. A lack of vertical substance where the conscious, insightful subject might be situated is replaced by a vapid subjectivist reactivity.

Today we no longer merely *think* in techno-scientific terms; we actually *live* on these terms, certainly to the extent science determines our view of the world and technology mediates our contact with that world. The statistical view of existence that once concerned Jung has, through the datafication of all things, turned into the algorithmical ordering of existence, especially as digital networks have inserted themselves into what we perceive, desire, and consider important. We may see much and think a great deal, but insight into the way we see and how we think is harder to come by. Self-knowledge is lost. As Jung writes, "science conveys a picture of the world from which a real human psyche appears to be excluded—the very antithesis of the 'humanities.'"[37] To the extent this exclusion of the psyche deprives us of self-knowledge, we are also robbed of our "foundations" and "dignity," as he states above.

Absent the sense of an inner life, many are becoming lost in a sea of virtuality, the manipulative and exploitative possibilities of which suggest a new era of mind control. A marker on this slippery slope occurred in the lead-up to the 2016 American presidential election. Data gathered on Facebook was used by a third party to create individually targeted political advertising; hacked emails and other communications were also filtered and weaponized, all in the interest of manipulating the perception of events. Before and after the 2020 presidential election, enough stories of voter fraud were promoted and circulated online that seeds of doubt easily took root and grew into an alternate view of reality, eventually leading to attempted mob rule. Unscrupulous, profiteering media outlets, ever ready to promote politicians and commentators who show little deference to the fundaments of genuine knowledge, created a vastly uninformed citizenry. In countries where governments control the press and place sweeping restrictions on web content, the mind-controlling possibilities that lie ahead are virtually endless.

For anyone without an authentic, individualized grasp on reality and ability to discern the veracity and value of what is placed before them, the pull towards the hive-mind is harder to resist. Many in the business of controlling the minds of others have taken to convincing their unsuspecting victims they are making up their own minds. One dynamic employed involves the accusation that expert-derived consensus knowledge and most mainstream media accounts are the actual sources of manipulation and mind control, whereas fringe media and conspiracy theories are sources of truth. The oft-repeated accusations of "fake news," the deployment of "alternate facts," and the adoption of maxims such as "when we act, we create our own reality"[38] work to repress sources of doubt and clear the way for distortions designed to shore up the desired perspective. Purveyors and adherents of these views are often convinced they are engaged in history-altering exposés, which bolster their sense of individual significance and generate deluded feelings of meeting the challenges of our time. Adherents of the QAnon conspiracy theory have even taken to viewing themselves as "researchers," each tasked with following clues and piecing together what the masses fail to see; namely, a world run by pedophile elites and deep-state operatives. Such moves exemplify the psychological defense mechanism termed a reaction formation; in this case adherents convince themselves they possess special knowledge while suppressing the actual experience of being totally hoodwinked. Whereas the vehicle for today's hive-mind may be the online world, it is the underlying split in the psyche that enables this phenomenon, with the need to control reality dominating the need for self-understanding.

Under these conditions, the neglected inner world of authentic human responsiveness grows more demanding and disruptive, manifesting in psychological disturbance. Overt disturbances such as depression and anxiety may be easy to spot, especially as their ubiquity eclipses the social stigmas that would otherwise keep them hidden. But covert disturbances such as narcissism and other personality disorders grow harder to discern because they blend into the superficial success, social histrionics, and mania that surround us and are normalized by online grandstanding and sensationalism. While these overt disturbances appear maladaptive and the covert disturbances adaptive, they play off each other, both individually and collectively. Individual anxiety often turns to the construction of a polished, transactional social identity and associated rewards. Collective anxiety, whether legitimate or manipulatively induced, looks for narcissistic, charismatic leadership and the feeling of cohesion and empowerment their associated movements create. Both operations are thriving in the digital age, which trades in exterior images and worldly cachet.

In sum, the hive-mind is a compensatory outgrowth of inner fragmentation and lost identity—a normalized form of psychopathology in an age of absent verticality and detached foundations. It is a symptom of the unfamiliarity with the forces and patterns that shape us, resulting from the blank-slate model of the mind. Put simply, when we are unconscious of these psychological forces and patterns, we are apt to be swallowed by the quasi-religious sense of belonging to something larger than ourselves that now presses on society. Our psyches are split. Whether this manifests as dogged fanaticism or sheepish acquiescence, the source is essentially the same.

The implications are far-reaching. History has not "ended" with the spread of liberal democracy as the final and assumedly superior form of government (Fukuyama),[39] largely because of the ill-effects of the technocracy this democracy and free-market capitalism have themselves created. The relation between the rise of tech industries and the decline of the factory worker may be well understood, but how this rise has increased psychological fragmentation and failed to spread democratic norms is not. The lesson is that external pluralism goes only so far if internal pluralism and the psycho-spiritual are left unattended. In this way, liberal democracy has at least one more step to take in its evolution and global acceptance; namely, considering the dynamics of internal wellbeing alongside those of external wellbeing. I am not here talking about merely adding mental health programs or paying more attention to Global Wellbeing Indicators (GWI), but about a comprehensive consideration of how the neglect of meaning, identity, and dignity gives rise to paranoid ideologies and fundamentalisms that undermine individual integrity and collapse nation-states from within. Without democracy and capitalism coming to terms with their own implosive potential in this way, the idea we can keep swatting away at totalitarian outbreaks until the global realization of a free society is a pipe dream.

Becoming Cyborgs

Having arrived at a major plot point in this saga of psychic division, with awareness of the root system of the psyche on one side and the propensity to see ourselves as blank slates for redesign on the other, we must leap into the fire of our main topic. In industrialized societies at least, the continuing remove from the natural conditions of life has reached the point of imagining we have more in common with the machines we have made than with the biology and culture we have inherited. Where this mechanized path meets the horizon, just visible from our current position, stands the cyborg, the human–machine hybrid. Positioned at the vanishing point of innovative trends and changing notions of what it means to be human, the cyborg represents that point at which we, as a species, may begin to vanish. The cyborg thus marks the point of stepping off into a world governed by, or shared with, radically augmented versions of ourselves—a world that is sure to be at best unstable and unpredictable.

The horizon we see is, nonetheless, determined by where we are standing. With its partly robotic body and partly computerized mind, the cyborg makes explicit what is already implicit in today's technoculture; namely, the willingness to surrender to whatever science and engineering make possible. Given the prevalence of this willingness, and given what is already afoot with human–machine hybridization, the cyborg becomes a focal point for any discussion of the world to come.

Elaine L. Graham states, cyborgs are "moving from fantasy into reality."[40] With computer chips and other devices implanted in their bodies, some individuals already claim cyborg status.[41] Some of us already have electronic devices wired to our hearts and brains, not to mention titanium joints and various kinds of plastic inserts aiding anatomical functions. "Thought-controlled computing, in which technology is controlled by the brain waves of the user"[42] is being put to a number of uses: biomechanical prosthetics controlled by thoughts are becoming more widely available. Our minds, already attuned to the rhythms of social media sites and the constant flow of information, are being groomed for direct neural links to computers. By mid-century, according to some accounts, full-blown cyborgs will be among us.[43] Replacing ailing body parts with more serviceable artificial ones, adding brain power with neural implants, and undergoing more invasive forms of plastic surgery will likely be routine procedures. The human body will become less a given reality and more a starting point for creative redesign. Identities will thus be far more malleable. As the editors of one collection of essays on these prospects put it, "Soon, perhaps, it will be impossible to tell where humans ends and machines begin."[44] As they suggest, this outlook prompts the call to "go beyond dualistic epistemologies to the epistemology of the cyborg: thesis, antithesis, synthesis, prosthesis."[45]

Technologists and cultural critics alike tell us that our life as cyborgs is inevitable. They have seen the future by surveying their innovative peers and extrapolating current trends. What they see is a hybrid human-machine that will want to keep pace with the fully fabricated entities that AI and robotics will enable. However, when the so-called Singularity occurs, wherein "artificial super intelligence" will become smarter than us, it will proceed to keep making even smarter versions of itself, leaving us in the dust.[46] Some technologists put this threshold around the middle of the twenty-first century.[47] Although, assumedly, AI will by then have the advantage, it is also assumed we will want to hitch a ride as long as we can, particularly by hybridizing human and artificial intelligence. As some contend, "There is no longer a 'partnership' between machine and organism; rather there is a symbiosis and it is managed by cybernetics, the language common to the organic and the mechanical."[48]

Beyond this point, the scenarios blow the mind—right out of the body. As one commentary offers, "we may be able to separate our minds from our biological bodies by merging a 'mindfile' of our neural connection with 'mindware,' allowing us to exist as software that could run on hardware of our choice."[49] We might then download our minds into "bodies" made of silicon and alloy rather than flesh and bone. Or, given such a thoroughly digitalized existence, we may elect to abandon external forms altogether and dwell in cyberspace as angelic beings. Needing no particular address, nor a hot meal at night, we will certainly have a thrifty lifestyle. We might not even require an earthly habitat. By contrast, flesh-bound organic life will probably appear quite awkward or possibly even superfluous. Nanotechnology might even be capable of turning rocks into circuits, converting the earth and then each subsequently colonized planet into massive sentient computers. One wonders what such computers will contemplate—themselves, perhaps. Yet, with basically anything capable of turning into anything as a matter of will, it is hard to imagine that anything we recognize as having an identity or even a sense of being would still exist.

Some of these prospects are a long distance from where we currently stand, but they have a lot to say about our present orientation, presenting images of what may happen if fabrication eclipses being and the remaking of ourselves becomes all consuming. "Fabricate" can mean either to manufacture or to mislead. Similarly, the etymological root of the term "mechanical" means "to trick," making machines both magical and deceptive. As we now live surrounded by mechanized fabrications, a normative deceptive potential also becomes part of life. Authenticity and originality may be losing their grip.

This acclimatization to fabrication has coincided with an orientation toward the gathering and exchanging of information as the basis of existence. As Rubin puts it, "'progress' becomes the sheer accumulation of information, a kind of hoarding mentality that is based upon the belief that you never know what might come in handy someday."[50] As the logic goes, the future must therefore have to do with benefitting from this progress in more direct and efficient ways. At first this means aligning ourselves with technologies that best accumulate and make use of this information, and then it means allowing these technologies to become part of us. "This helps to explain the widespread belief that any effort to restrain science or technology on the basis of ethics represents a threat to progress . . . if progress is mere accumulation, then of course restraint is a threat."[51] It also helps explain why the possibility of our future as cyborgs must be taken seriously.

The problem is at what point and to what degree this fabrication of human nature will leave significant aspects of this nature behind, and at what point and to what degree an inhuman nature will begin to take hold. Insofar as an understanding of the human psyche provides critical insight into these pending and prospective changes, we may already be able to discern this.

Whereas fabrication may be inherent to civilization, excessive fabrication also separates civilized life from its cultural and natural foundations. Cities lose their soul, buildings become detached from landscapes, institutions put power ahead of community, people become alienated. The spiritual and natural dimensions of existence become prone to either neglect or manipulation, allowing the vertical axis of being to lose its points of orientation. Without conducive contact with either spirit or nature, and without awareness of the psyche that comes from their joining, we end up with a fragmented rather than integrated realm of lived experience, as if the stuff of life loses its gravitational field of value and meaning. This balkanization can be an interim way of handling stress and trauma, and it may be useful in the transactional pursuits of worldly success. Over time, however, it leads to psychological defensiveness, lost integrity, and debilitating symptoms. Postures of exploitation and control are increased while genuine wellbeing and human concern are decreased.

Amidst these shifts, each advance in AI appears to coincide with human intelligence being left behind. But while AI can be impressive in its harnessing and collating of vast amounts of information in short amounts of time, and while it is getting better at imitating human thought processes, it shows little sign of actually understanding what it is doing, let alone possessing any sense of the meaning or value of its actions. It also lacks any capacity for emotional intelligence, genuine imagination, or aesthetic appreciation simply because it is *artificial*—a glaring fact lost on AI's most fanatical proponents. As AI researcher Susan Schneider considers the prospects of conscious machines as well as our own "radical brain enhancement," she concludes that "failing to think through the philosophical implications of artificial intelligence could lead to the failure of conscious beings to flourish. For if we are not careful, we may experience one or more *perverse realizations* of AI technology."[52] We will surely fail to flourish if we allow the utilitarian value of AI and associated technologies to eclipse the depth and breadth of human intelligence and consciousness. And a "perverse realization" of AI will occur if we trade our innate capacity for the generation of understanding and wisdom for the fabricated capacity to access and process more information. In such a scenario, AI "fails to make life easier but instead leads to our own suffering or demise, or the exploitation of other conscious beings."[53]

A time of blurring lines gives rise to fluid possibilities as well as widespread confusions. Limitless sources of information and endless channels of communication coexist with pervasive disorientation. In the online world we can sculpt our identities and know and be known by more people than our ancestors ever dreamed of; we can learn of things happening throughout the world in real time. But in the offline world, we are having a difficult time knowing who we are, where we belong, and what belongs to us. If we were trees, our branches would be extending towards every horizon but our roots would be barely penetrating the ground—surely an unstable and unsustainable growth pattern. Writing about the post-industrial world, Giddens describes this state in terms of a decline in "ontological security," the basis of which is "a sense of the reliability of persons and things." What we are losing is, "in the terms of phenomenology, 'being-in-the-world'." It is "an emotional, rather than a cognitive, phenomenon, and it is rooted in the unconscious."[54] Giddens is here making a sociological invitation for depth psychological understanding, for a vertical rather than horizontal approach. Much of our existence is, in fact, "rooted in the unconscious," but we have become preoccupied with extending every branch in every direction.

Posthumanism

As the cultural movement and associated philosophical discourse focused on human–machine hybridization, posthumanism is, in essence, concerned with "the inward movement of technology."[55] With the advances in technoscience generally, and the rise of AI specifically, this inward movement aims to take our species beyond what is purely human and thus beyond humanistic modes of understanding what we may become or even what we are, arguably, already becoming. Whereas many of us may be inclined to dismiss posthumanism as an esoteric topic or passing fad, its prominence in technology circles and cultural studies is enough to negate this. One commentator describes it as "the worldview that is ascendent" in American culture.[56]

For many in the tech world, the secular religion posthumanism has become is built upon a faith in augmentation as an extension of evolution and in our role in directing such a process; for others it is mainly concerned with the critical inquiry into these matters. It thus encompasses a wide spectrum of views, with most displaying some awareness of the risks and pitfalls inherent in such pursuits—including the prospect of AI arranging our premature demise. Common to all posthuman thinking, however, is an acceptance of the end of human existence in its current non-augmented form and an associated philosophical commitment to ending the privileged position of humanism in cultural discourse. Posthumanists thus see themselves as "a bridge, or a rope, between historical humans and beings with posthuman capacities,"[57] as well as between humans and artificial humanoid beings and robots that will at some point, it is believed, become sentient.

Preparing the way for what is seen as inevitable, posthumanists are apt to consider themselves technological realists, arriving at their views by extrapolating current trends. In Silicon Valley, which largely establishes the technological trends of the world, posthumanists abound. The idea we are "natural-born cyborgs," as Andy Clark puts it,[58] and are merely exchanging one means of altering our essential make-up for another, is a dominant belief, with the possibilities provided by our own ingenuity believed to be a logical and even natural extension of the evolutionary process. There are also a number of academics in the posthumanism-as-natural-evolution camp. Approaching the matter from a phenomenological standpoint, Glen Mazis, considers the overlap of human, animal, and machine existence and suggests that "Cyborg being—our sense of incorporating tools, and becoming interwoven with machines within us, about us, and within the meshes of how we have organized the world—has always existed—it is just become more literal and extravagant."[59] Summing up this position, Charles T. Rubin writes:

> If evolution is the law of life, and at the same time if evolution has brought about human beings who can take hold of evolution and direct it as one more aspect of our control over nature, then a grand narrative of free human creativity becomes possible.[60]

According to such reasoning, assuming the godlike position of human redesign fails to strike the posthumanist as a problem, for evolution itself has placed us in this position.

Technological posthumanism has also produced philosophical posthumanism, which, eyeing where the possibilities for enhancement and augmentation are headed, argues our classical notions of human existence are too fixed and the boundaries between human and non-human are not as clear as we assume. When these notions are loosened and these assumptions removed, as the argument goes, we can move beyond typical "human" concerns and open up another level of discourse. The roots of such philosophizing are multiple, but are frequently traced back to Donna J. Haraway's discussion of the radical social changes that would follow from our bodies and minds becoming more malleable and less tied to the naturalistic divisions.[61] The cyborg, as she sees it, breaks the mold of narrow humanistic conceptions of existence and the associated investment in all kinds of dualistic categories. Transcending the human–machine divide and using technology to alter biological and psychological functioning opens the door to overcoming other categories such as human–animal, male–female, culture–nature, and so on. The significance of overcoming these divisions is then related to the undoing of social ills; particularly those related to Western patriarchal hegemonies, but also what she associates with structuralist and naturalist perspectives. I will take up Haraway's argument more fully in a later chapter.[62] For present purposes, it is enough to consider that if the goal is to correct social inequalities in general and gender inequities in particular, and to convince humans they have a more symbiotic relationship with other species, posthumanism appears like a rather odd and excessive route to take, especially given the historical clash between technological exploitation and feminist values. There are no guarantees that the capacity to alter physical and mental states will be aimed towards altruism. Moreover, given it is our present if flawed human nature that has come to recognize these social problems and begun to work on their solutions, it does not make much sense to roll the dice and radically alter our entire disposition in order to address them. While these social problems are weighty ones, upending culture and civilization in the attempt to solve them seems reckless. Even Manfred Clynes, who coined the term "cyborg," submits that "every invention brings about a possibility for both good and bad." He adds, "we need to be cautious and careful, and test every step along the way . . ."[63] If such a caution applies to new technologies in general, it must apply even more to technologies that would radically alter human nature.

This philosophical posthumanism has arisen from the somewhat strange nexus of postmodern thought and post-industrial technology, and one has to wonder whether the postmodernist penchant for intellectual play and deconstruction has filled the contemporary mind with the fluid style of thinking that has spilled across into planning for a fluid way of being. Even if there is value in our minds moving beyond certain rigid categories of thought, what we know from the study of the psyche is that much that is imagined calls for symbolic and metaphorical understanding rather than literal enactment. In fact, the capacity to contemplate products of the imagination in terms of psychic reality is often inversely related to compulsive acting out. This is one of the main reasons we are obligated to maintain contact with the vertical axis of psychological understanding: it shows us how frequently our actions are determined by unconscious factors and thus allows for the metabolization of ideas and impulses in ways that curtail or redirect these actions. Without denying the possibility that a posthuman existence may one day be realized, a primary purpose of this present study is to approach the cyborg as a metaphorical expression of a contemporary psychological situation, the tending of which may prompt a shift in our innovative efforts and social aims: At this point the cyborg also remains a fusion of the actual and the virtual, both of which are subject to the archetypal basis of mind.

Although we have entered what is sure to be an ongoing philosophical contemplation of our relationship with artificial intelligence and other digital age innovations, this postmodern style of playing with questions concerning the future of earthly existence tends to lack seriousness and fitting scope. Its main value may lie in exposing the intellectual mood that has generated such a malleable notion of human nature in the first place. If, in terms of the unfolding or contracting universe, humanity appears transitory, as postmodern thinkers like Lyotard have argued, then our morphing into other forms or species may present as a minor matter.[64] Yet measuring the human situation against the birth and death of suns is at least one point on which the depth psychological recognition of the soul's inherent interspersion of emotion, image, and concept vividly contrasts with the freewheeling intellectualism of philosophical posthumanism.

The posthuman philosophical critique of anthropocentrism may be a different matter, for a number of fields are converging in what appears to be an essential correction of a traditional humanistic blind spot, with both immediate and long-term implications. However, embracing this correction does not necessarily lay out a welcome mat for posthumanism. Although there are compelling reasons to revision humanness as part of the web of earthly life, this does not mean that the door to human redesign or the making of artificial life forms has to be flung wide open. If anything, anthropocentric arrogance, conspicuous over the past two hundred years of manipulating and exploiting non-human life, suggests we may better comprehend the mind–body–world connection that has always belonged to the depths of human nature. That we are of the earth, coexist with its other inhabitants, and have come to know and sustain our own nature through knowing and respecting theirs may collide with a narrowly construed Enlightenment-grown humanism, but it only underscores a more classical and timeless conception of ourselves that runs back through the romantic tradition, Renaissance philosophy, and multiple undercurrents in Western thought, which was as clear to the Greeks as it has been to any number of indigenous cultures: human nature and culture spring from an instinctual ground that is bound to places, plants, animals, and their rhythms. From this perspective, the humanism that philosophical posthumanism pushes and the anthropocentrism it perceives therein is only part of the humanistic story. As above, to push beyond the literal boundaries of our skin may not even accomplish the psychological shift that the emotional, imaginal, and symbolic decentering of narrow human interests is likely to achieve.

It is also somewhat ironic that philosophical posthumanists wish to bolster their cause by way of undoing anthropocentrism when technological posthumanists see the endgame of transcending our physical form in relation to the ability to depart Earth and leave terrestrial concerns behind.

The intellectual storms of philosophical posthumanism may appear a niche preoccupation, but this preoccupation not only concerns the intersection of the two massively influential cultural-historical movements postmodernism and posthumanism, it also backgrounds matters of immediate and pressing psycho-social significance. The challenge to anthropocentrism is one of these matters, central to the problems of outer ecology but also related to our capacity to develop an ecology of mind, which depends on the ability to relate to that which is larger than ourselves. Where we situate our lives and thinking along the continuum of the fluid and the fixed is also of primary significance. What we actually manifest versus what we psychologically metabolize becomes a crucial question in this matter.

Surrounding all of this, there exists a wider conversation about the role of technology in our lives, which extends from the innovative strivings of technologists on one side to the expression of popular culture on the other, and relates to our actions as individuals at points in between.

This conversation is apt to convince us that technology can no longer be taken for granted, left to simply follow economic forces and the limits of what is possible. As Stefan Herbrechter puts it:

> an entirely different form of technological determinism seems to have gathered momentum in the age of posthumanism, namely a dehumanized, subjectless, autotelic and system-oriented belief in technology . . . disseminated to a wide audience via popular media and "get-the-latest-gadget" consumption, a popularization which is an integral part of contemporary technoculture.[65]

Perhaps the very term "technoculture" conveys the matter in a nutshell. Herbrechter goes on to argue that we have "reached a point at which 'our' contemporary lifeworld has *literally* turned into a 'techno-culture' and our social order *literally* corresponds to 'techno-scientific capitalism.'"[66] The result is "a threat to the liberal humanist principle of uniqueness as well as to the 'integrity' of the individual human being."[67] From the viewpoint at the intersection of postmodern and posthuman philosophies—which Herbrechter leans into—a loss of the unique and integral may portend a new and possibly advantageous form of being. But from a depth psychological point of view, this signals a basis for widespread psychological fragmentation, unlikely to produce a predicable transition to any viable way of being. It is also hard to avoid the conclusion that the coincidence of "techno-culture" and "techno-scientific capitalism" does not lie at the core of an already apparent global technocracy, wherein humanity is itself progressively subsumed by an abstract and mechanized system.

Whereas to its advocates the state of being posthuman connotes a transcendence of human limitations combined with a retention of humanity's best attributes, it is difficult to see how "technological determinism" that has "gathered momentum" and is characterized by a "dehumanized" orientation to what we innovate (Herbrechter) will get us to this place. Similarly, with already evident capacity for corporations and governments to exploit the vast amounts of data they are gathering, it is hard to imagine how people will be able to think outside the technocratic box. As we contemplate hardwiring our minds to such a technosphere and allowing our bodies to become base stations for the latest prosthetic enhancements, we are flirting with a mode of living in which we may become completely alien to ourselves. Perhaps we can imagine becoming cyborgs because we are already living inside their subtle bodies, wherein an increasing unfamiliarity with the sensual and earthy core of our humanness is entwined with the desire to transcend the inherent dilemmas and problems of life. Yet whether we will recover enough feeling to cause enough consternation and resist our escapist tendencies is another matter.

Both the psychology of posthumanism and the psychological implications of its stages of realization need to play a central role in our critical analysis of this phenomenon. And this critical analysis needs to be elevated in our cultural discourse and become more widely understood, both as a matter of the common good and as a means to offset the new lords of technocracy who will fill our minds with convincing tales of possibility and progress. The bioethicist Wallach writes:

> Hype, hope and wishful thinking overwhelm any realistic appraisal of the difficult thresholds that lie ahead . . . Challenges to the conventional wisdom that we can achieve anything we set our minds to are treated as cynical defilements of a revealed truth. In this narrative, the power of positive thinking always prevails.[68]

This positive thinking casts a massive psychological shadow, which may enter our awareness more fully as we approach some of these thresholds, across which lie the far-reaching implications of allowing the tools we have shaped to shape us.

The Borg

Popular culture can be a feeble house of mirrors, reflecting back to us passing fads and superficial entertainments, but it can also offer images that reveal deeper movements in the collective psyche. Over the past half-century, roughly corresponding to the rise of the digital age, a number of images and narratives of posthuman existence have revealed the many hopes and fears we have about AI and human augmentation, as well as a number of key themes and contextual associations that will eventually aid us in the negotiation of what is to come. At the end of Graham's study of these matters, she writes: "Fantastic encounters with representations of the post/human offer important insights into the many meanings of being human."[69] It is in this vein we find portraits of cyborgs in popular science fiction and space fantasy films that display a spectrum of psycho-social dispositions that mirror something of our own. Some hybrid beings are set on developing their emotional intelligence and humanity; by direct contrast, others are defined by their distinctly psychopathic bearing. There are also cyborgs who are little more than automatons, controlled by some insidious and destructive system. It is this last disposition that is displayed by "the Borg" of the *Star Trek: The Next Generation*[70] television series and films.

Possessing a number of characteristics that correspond to focal points in this chapter, the Borg represent an unprecedented level of antagonism toward the protagonists of these stories, the narrative frame of which most of us know. Some centuries into the future humans have joined a "United Federation of Planets" and go trekking through the galaxy in starships like the *Enterprise*, seeking friendly, non-interfering contact with alien civilizations. Whether among the crew of the starship *Enterprise* or present to this contact with "others," the most revered values are cooperation, community, and an associated respect for differences between individuals and groups. And the ever-present ethos at work involves the deepening insight into and appreciation of individual strengths and limitations. The now famous "Kirk–Spock" relationship created the template for the ethos, with the tension between Captain Kirk's passion and Commander Spock's logic typically giving way to an effective confluence.

The Borg, who appear in mainly in *The Next Generation* series and in the feature film *StarTrek: First Contact*,[71] not only present a dramatic enmity to the Federation and its ways, but a vividly contrasting model of essential existence. They operate with a hive-mind and seek to "assimilate" other species and their technology, turning their quarry into "drones" who serve a "collective" overseen by a "Borg Queen." Relentless in their quest, their only interest is in expansion and power.

In the kind of reversal that is an impressive motif in many portrayals of future technology, and one frequently associated with a cyborg future, among the Borg, the machine becomes the master and the flesh its slave. Both literally and figuratively, to be is to be in service to the system. As such, the Borg exemplify *in extremis* the threat that advancing technologies will make us the tools of our tools. By vivid contrast, the role of technology among the members of the Federation tends to serve the effort to further understanding between species, as well as support the pursuit of individual betterment.

Assimilation into the Borg takes place by capturing others and inserting nanoprobes into their bodies to begin the process of technological augmentation. In one remarkable scene in *First Contact*, the Borg have captured "Data," a valuable crew member of the *Enterprise*, who happens to be a humanoid robot. Data is an example of an artificial human who is constantly engaged in the attempt to understand human nature, especially those aspects that seem beyond the purview of his considerable capacities for reason and calculation. He has been "programmed to evolve" and better himself.

When the Borg Queen arrives, Data is being prepared for assimilation. After her upper "human" half with biomechanical spine is lowered into her fabricated body, she announces, "I am the Borg." Data observes, "That is a contradiction. The Borg have a collective consciousness, there are no individuals." To which she quips poetically, "I am the beginning, the end, the one who is many." Then, after Data queries her about the Borg's "organizational structure," she says, "I bring order to chaos ... You are in chaos, Data. You are a contradiction—a machine that wants to be human." By contrast, the Borg, she declares, are on a quest for "a state of perfection."

We are granted insight, not only into a central motif in this fictional world of cyborg opposition to human values, but also into a central preoccupation of posthumanists: ordinary humanness is imperfect if not flawed, whereas fusion with technology offers a path to perfection. By contrast, a second motif, present in a number of science fiction stories, concerns the artificial human who wants to be less artificial and more human, frequently accompanied by the nagging sense of being incomplete and a special curiosity about matters of the heart.

It is right here we may perceive the collision of two distinct visions of human becoming. The process Jung observed at the core of psychological maturation, referred to as individuation, is defined as moving toward a goal of psychological inclusion and completion, wherein the various propensities of psyche and the overall divide of conscious and unconscious movements are brought into a state of cooperation. This is, however, not a state of perfection, which leaves little room for the inherent tensions and emotional conundrums of humanness. The goal of life, according to Jung, is "an integration or *completeness* of the individual, who in this way approaches *wholeness* but not *perfection*..."[72] Whereas completeness not only includes the wounds and flaws but eventually reveals their invaluable role in the building of character and feeling of participation in the larger patterns of life, perfection is couched by Jung as "an exalted state of spiritualization."[73] To individuate is also, by definition, a differentiation from collectivity, not an immersion in it: a person may still live in groups or communities, but would do so with an awareness of their psychological distinctiveness. And it is obvious the most successful communities are those that manage to foster cooperation between psychologically distinctive individuals.

It is in these ways that the fictional enmity between the Borg and the Federation, epitomized in the direct confrontations with the crew of the *Enterprise*, also portrays the collision between two determinative forces vying to dominate human existence. On one side stands a model of psychological differentiation, which in turn fosters conscious inclusion; on the other side stands the allure of hive-mindedness and fealty to some system—which is inevitably invested in a perfectionistic ideal, and inevitably concerned with the imposition of that ideal on others. How the digital age in general and posthumanism in particular invite both perfectionism and hive-mindedness in these ways constitutes a baseline for our critique.

The Deeply Human

We have one more significant part of our overall discussion to put on the table—the *deeply human*. In placing this "part," however, we are obliged to acknowledge that any partitioning of the overall nature of human existence becomes problematic. For the deeply human also grounds us in the *fully human*, especially by way of the archetypal realm's overall tilt toward inclusion rather than exclusion. Each timeless and universal pattern of life has its place. Nowhere does this inclusionary principle impress itself upon us more than when we attempt to partition our conscious pursuits from their unconscious backgrounds. As Jung writes:

> consciousness, for all its kaleidoscopic mobility, rests as we know on
> the comparatively static or at least highly conservative foundation of

the instincts and their specific forms, the archetypes. This world in the background proves to be the opponent of consciousness, which, because of its mobility (learning capacity), is often in danger of losing its roots."[74]

Such roots become more, not less, important in a world where the inclination toward mobility and the capacity for learning have expanded beyond all measure—a world of pervasive horizontality wherein verticality is not only hard to find, but frequently cast aside.

Even in Jung's time, during the first half of the twentieth century, this one-sided orientation of modern consciousness forced a turn to the past for relevant points of comparison. Jung was thereby able to recognize the comparative value of certain ancient practices. "That is why," he comments, "since the earliest times men have felt compelled to perform rites for the purpose of securing the co-operation of the unconscious."[75] These rites allow their practitioners to be "constantly mindful of the gods, the spirits, of fate and the magical qualities of time and place, rightly recognizing that man's solitary will is only a fragment of a total situation."[76] Whereas for most living in the industrialized world at least, reversion to such practices has become unthinkable, we are in desperate need of the kind of vertical awareness that alerts us to the psychic realities and archetypal patterns these practices tend to preserve. Somehow, we have to rediscover the timeless truth that "our solitary will is only a fragment of a total situation." However, the utility and power of technology make this rediscovery extremely difficult.

As indicated in the last chapter, our loss of contact with the unconscious in general also corresponds with a series of one-sided views, which not only make for psychological blind spots but create a dangerous accumulation of shadow material that generates debilitating symptoms and destructive projections. One of the questions this line of thought opens up is whether or not the posthuman plan for the future is even able to unfold under these conditions, for the longer the suppression of the deeply human continues, the more the accumulated unconsciousness threatens to upend whatever social structures happen to be in place.

Writing at the height of the industrial era, Jung observed that "a psychological split runs through vast numbers of individuals." He goes on to suggest:

The cause of this development lay principally in the economic and psychological uprootedness of the industrial masses, which in turn was caused by the rapid technological advance. But technology, it is obvious, is based on a specifically rationalistic differentiation of consciousness which tends to repress all irrational psychic factors. Hence there arises, in the individual and nation alike, an unconscious counterposition which in time grows strong enough to burst out into open conflict.[77]

Excepting for counter-cultural efforts to renew contact with nature, as well as psychotherapeutic and meditative techniques aimed at overcoming this uprooted state, it is evident this general situation has continued into the post-industrial era. In addition, however, the hyper-horizontality of the digital world has created a myriad of ways to avoid looking at this deeper problem. As Jung states:

> just as the intellect subjugated the psyche, so also it subjugated Nature and begat on her an age of scientific technology that left less and less room for the natural and irrational man. Thus the foundations were laid for an inner opposition which today threatens the world with chaos.[78]

Jung is highlighting the collision between technoscience and the confluence of psyche and nature, which is, effectively, the ecology of mind. The image of the world as a resource, to be exploited and manipulated according to human will, as well as the myopic emphasis on economic growth and material wealth, lie at the core of the psychological effects of industrialization Jung is eyeing. Now, it is as if we wish to even further isolate the intellect from the psyche as a whole, and impose this on the nature of being human, seeing mind and body as terrains to be remolded accordingly to some misplaced sense of individualistic perfection and societal progress.

The realization of posthumanism will, by definition, involve a progressive displacing of the deeply human and its vertical modes of perception because the cybernetic will progressively replace the organic, and artificial intelligence will progressively replace natural intelligence. However, before this realization can occur, during the period of conceptualization, which we have entered, it is the capacity to understand what makes us deeply human that will be displaced. This defines the battle before us.

One major front in that battle, as seen from the deeply human perspective, is the wave of techno-scientific idealism posthumanism has been riding—an idealism that leaves little room for examining the motivations behind the radical alterations it envisions, or for considering the psychological implications of these alterations. Alongside these oversights lies a failure to adequately consider the philosophical assumptions and spiritual ideas pervading its plans. Posthumanism involves some brilliant thinking, but it is less inclined to think about its thinking. But turning to these matters reveals that posthumanism is thriving in an intellectual hothouse of largely unexamined constructs—a hothouse in which the mind is already reduced to hardware and software, and thought is already reduced to algorithms shuffling data. When human consciousness is conceived in this manner, evolutionary change is very easily imagined to be a matter of rewiring and reprogramming, in turn making it easier to foresee a time when minds may be separated from their embodied, earthly contexts, transferred to artificial bodies, or dwell as angels in cyberspace, thereby enabling ravel to distant planets without the bother of biological being, freeing whatever remains of human consciousness from its earth-bound and mortal forms. A godlike return to the heavens beckons because the thinking involved has already left the ground.

This exclusion of the deeply human in predictions of our future is becoming harder to call out because reductive thinking is also prevalent in wider cultural trends. Sensate and empathic connections to others, cultivated in face-to-face conversations, are taking a backseat to abstract digital connections, such as we experience communicating by text or email, which narrow our perceptions. Relying on web-based sources of information rather than book-knowledge and the deeper thinking generally found therein is also filling the mind with a morass of superficial soundbites lacking in real-world accountability. In such everyday ways we are losing sight of how knowing one another and experiencing the world is a matter of comprehending on multiple gestural and imagistic levels, few of which can be digitalized. This loss, over time, is also convincing us computerized minds and mechanical bodies may be the next step.

Individuals caught in such reductive vision of the human condition may be less the exception than the rule. Their understanding of what it means to be human lacks an existential vessel and instead resembles mercury spilling across a flat countertop. They are victims of the horizontality and atomization of postmodern existence. Joseph E. Davis refers to these poor souls as "selves in liquid times."[79] Davis's study focuses on individuals who become caught in the belief that their psychological problems stem from chemical imbalances in their brains, while resisting or being incapable of considering a more complete psycho-social view of their suffering. Their syndrome not only reflects the presence of a normative medicalization of psychological problems, but also the loss of a deeper model of understanding. He writes:

> Such a "depth model" once informed the picture of the human in many fields, including psychology and philosophy, and was also expressed in other areas of life, from art and architecture, to music and literature. It is a model that presupposes a durable and tangible social world . . . A more enduring and rooted life experience, directed by an inner sense of purpose and a continuous narrative. It presupposes psychological complexity, with layers of meaning beneath the surface of our immediate awareness . . . It is a model that is now hard to find.[80]

The connective tissue between this broad portrayal of absent depth perspective among those suffering from psychological problems and the specific problem of absenting the deeply human and associated understandings from the pursuits of posthumanism should be evident. At this point, the world-at-large appears to have rolled out the welcome mat for the human–machine merger and elevation of AI. At the same time, enduring the divisions in ourselves that comprise this welcome mat may produce some unexpected and possibly unwelcome upheavals, which may change the outlook.

By the deeply human I am thus referring to the base upon which the vertical axis of reality stands. This axis reaches up into the ineffable experiences of life, which poet and mystic attempt to witness and express through word and image; and it reaches down into the enduring patterns of nature—to the place where animality meets psychology. This is also the place where the archetypal patterns of life originate, which is why Hillman reminds us at the start of his *Re-Visioning Psychology* that "archetypal, in other words, means fundamentally human."[81] Because of the human tendency of late to be carried away by the spirit, whether via abstract ideation or religious belief, by the deeply human I particularly mean the way we are anchored in our existence by the sense of soul, whose direction, as Jung has shown, tends to be in and, as Hillman has argued, also tends to be down. We find the deeply human by going in and down, and we are today called to consider that which is up and out from this soul position.

It has been argued that we are moving from a phase of history that centered on human authority to a phase that will center on computational authority.[82] To the extend this is so is also the extent to which our posthuman prospects will be heightened. At this point, however, we have only just begun to cross from the human to the posthuman. And this crossing may shift in direction or be disbanded altogether if we are to complete the remaining task of the humanistic phase of history, which is to comprehend the depths within us. We have explored every corner of the earth and pushed the bounds of knowledge toward the most distant galaxies. This leaves the inner world as the real "final frontier," something the psyche itself appears to be suggesting. What the imagination returns to us as it gazes into the heavens and looks for extraterrestrial lifeforms is nothing short of a fantastic window into our own unconscious—a mythic world of redeemers and monsters possessing advanced technologies that are either creative or destructive.[83] That is, as we imagine into what lies at the extremities of our exploratory efforts we are brought face to face with the rumblings of the deeply human and what the soul has to say about the face of the unknown.

It is this more complete humanness that provides the starkest point of contrast with the aims of posthumanism, which concern the augmentation of certain parts of us while leaving other parts behind and are based on splitting the horizontal and vertical dimensions of being. Most of all perhaps, as an inextricable part of the fully human, our contact with the deeply human is what specifically conveys the meaning and purpose of *being* human, allowing us to know in a convincing way what will count for the genuine advancement of civilization. If, by contrast, we fail to come to terms with this vertical dimension of our lives, we are, by default, empowering the turn to posthumanism in a most unconscious way.

Whether we pay attention to it or not, the deeply human will be, at least for some time, our constant companion. It is that paradoxical figure within that embraces both our evolved percipience and our chthonic origins, a coincidence that Hillman called "the wise ape."[84] It refers to that which channels biological imperatives into ideals and symbols, weaving impulse and thought in a way that generates and sustains culture. If, as I set out in the previous chapter, culture is the bridge between nature and civilization, the deeply human is the bridge between the bestial and angelic. It thus serves as a reminder of how the high and low are tethered and, in the context of the modern age, corrects our inclination to leave the lowly behind and deny the way we are imbedded in the patterns of nature. To lose contact with this wise ape will almost certainly condemn us to a psychologically divided existence, left to be torn between beast and angel.

Inflection Point

Posthumanism provides a stark basis on which to contemplate the godlike powers and existential choices now before us. On the one hand, we have to closely consider where these powers and choices may take us. On the other hand, we have to examine from whence these powers and choices originate. It seems no coincidence that this vision of altering the path of evolution has arisen right in an era of deep psychological disorientation. Long in the making, this desire to perfect ourselves through a merger with machines may therefore be more of a malignant outgrowth of a drive for self-determinism and dogmatic faith in the utility of science than an evolution of intelligence; it may be an outsized compensation for the void of genuine meaning and purpose, a desperate clinging to the notion that data and logic will see us through what is an inherently existential and psychological crisis.

Even against the background of geopolitical strife and environmental collapse, the prospective merger of humans and robots is shaping up to be the most pressing matter of the new millennium. As an image of a possible future, this marriage to our machines is either the stated aim or the primary concern of many influential technologists and the main focus for a number of cultural commentators and social scientists. And, as we have seen, science fiction has long engaged this marriage as a major theme in its treatments of the human future and of advanced societies elsewhere. Yet the implications of turning ourselves into a different species have thus far met with commensurate collective consideration. As a trend with even more irreversible and catastrophic potential than a rapidly warming planet, one that is often regarded as a deliberately engineered end to human existence, this consideration seems overdue.

We may be fish, finding it difficult to question the water we are swimming in. Technology, having long passed the tool stage, has blended into the highly fabricated and digitalized environments in which we are immersed. The smart phone has effectively become an extra organ or limb, the internet functions as an extension of the mind, and data is now the coin of the realm—its exchange the basis of whole economies. But for what may be a fleeting moment in the march of history, standing at the threshold of the most radical changes ever to existence, we have a chance to raise our heads above the rising tide and consider how the thrust of current innovation is shaping our lives and how the drive to innovate has become seated in our psyches.

Technology may infuse the sea we are in, but the kind of technology we develop and how we live in relation to it remain in flux. Posthumanism may turn out to be an odd thought experiment whose primary purpose is to deliver an existential jolt, providing means to envision the direction we are headed so we may change that direction. Just as the images of the earth made possible by departing from it and taken by the cyborg-prototype astronaut[85] prompted a new vision of global community, considering the image of the cyborg on the horizon may prompt a recovery of what is deeply and uniquely human. However, turning such a prompt into a point of inflection requires more than a side glance at the phenomenon before us. Left to our own devices, the status quo will surely prevail.

Looking ahead, some may be concerned for their children's lives yet decide there is little they can do. Others may acquiesce to a tech-infused existence and then focus on other things. Many will see the problems of technology, but wager we will just engineer our way out of these problems. However, none of these orientations will occasion reflection on the deep grooves in our thinking and the entrenched dynamics of the emerging technocracy. Even some engineers admit that the path ahead will necessitate both mechanical skills and the constant consideration of our values, facilitating the creation of systems that are "sociotechnical."[86] This may be a portent of a merger of a different kind, one that will come through reflexivity.

For digital natives, the merger with machines may simply seem the way of things to come. However, this generation currently suffers from high levels of psychological distress, which have, thus far, failed to be fully connected to immersion in the digital world. Unaware of the toxicity of the technosphere and ensconced in the wider technocracy, this generation is also apt to regard their distress as itself a technical glitch, then apply prescribed or self-prescribed psychotropic medications, and pursue fuller digital immersion. Aside from the immediate personal and societal cost, the outcome is likely to be a generation with a thoroughly artificial approach to self-regulation and an absent organic baseline for psychological wellbeing. Under these conditions, the door to a posthuman future will be propped wide open. Nonetheless, decades of human response across the generations will have to be negotiated, making it unlikely these techno-solutions will proceed smoothly. The main problem is that as time goes on we seem to be becoming more adept at closing down innate human responses, as though these are themselves problems in need of technical fixes.

We are, in a sense, already inside a matrix, fueling a giant mechanism by unconsciously giving away our personal space and attention. The organization and meter of online stimulation is also altering our ability to focus and capacity to think. This makes the task before us an uphill battle; the very mode of online life has confused what is real or virtual; the medium is the message and the medium defines everything. As Baudrillard has convincingly observed, we are not only ensconced in the hyperreal; we may even prefer it this way.[87] If so, the idea of redesigning ourselves has a distinct advantage. Nonetheless, deep down, I think most of us sense something is fundamentally wrong with this set-up.

The introspection that will be required must begin by differentiating intelligence and consciousness—two characteristics of sentient existence that overlap but are distinct. Distinguishing these attributes is critical because although we have come to believe the advance of civilization hinges on the increase of intelligence, human destiny is going to depend far more on greater consciousness—far more on the capacity to understand a totality of our life-world situation. And the most essential quality of such consciousness is the good sense to recognize the degree to which we are unconscious and the importance of mitigating the effects of this unconsciousness. The expansion of consciousness such good sense invites, and the transformation of reality that stems from this, is arguably the very essence of being deeply and fully human. In the context of the evolutionary crossroads we have reached, pushing into our unconsciousness thus becomes vital.

Technology reversing the master–slave relationship is no longer just a recurrent motif of science fiction: it has become a primary concern of every person and even presses on the minds of those most invested in the development of AI. Schneider points out that "Stephen Hawking, Nick Bostrom, Elon Musk, Max Tegmark, Bill Gates, and many others have raised 'the control problem,' the problem of how humans can control their own AI creations, if the AIs outsmart us."[88] The choice of the term "outsmart" is telling in this context, however, because it unwittingly couches the problem as a battle of intelligence, which ends up bolstering the view that we just need to find ways to keep up with AI and, perhaps, that becoming cyborgs may be the best option. If this is the case, the battle will have already been lost: if we end up with technology controlling us, it will not be because human intelligence falls behind that of AI; it will be because we have been unconscious enough to allow AI to determine the course of civilization. To enter such a chess match for the future with our own creations, we will have had to remain asleep.

I am confident the neural nets of AI will never catch enough soul food to sustain the human psyche, making an outsized faith in this development a recipe for a soulless future. The evidence thus far is that the slide into a more encompassing technocracy will make us more susceptible to rigid socio-political formulas and addictive behaviors, which will cut us off from our own integrity. With outer compulsions surrounding an inner emptiness, a growing disillusionment will eventually wreak havoc. The vertical axis of being, the conception of ourselves as souls, with inner lives worth exploring and understanding, will weaken to the point of collapse. Understanding just how this may occur is the task from here—the task that will be required if alternative possibilities are to take shape.

Notes

1. CW 6, par. 77.
2. CW 10, pars. 150–152.
3. Carl Kerényi, *Prometheus: An Archetypal Image of Human Existence* (Princeton, NJ: Princeton University Press, 1997).
4. CW 10, par. 154.
5. Ibid., par. 163.
6. Ibid.
7. Ibid., par. 164.
8. Ibid., par. 189.
9. Ibid., par. 195.
10. Richard Tarnas, *Prometheus the Awakener* (Woodstock, CT: Spring Publications, 1995).
11. CW 9i, par. 91.
12. James Hillman, *The Myth of Analysis* (Evanston, IL: Northwestern University Press, 1972), 172.
13. CW 8, par. 277.
14. Ira Progoff, *Jung's Psychology and Its Social Meaning* (New York: The Julian Press, 1953), 161.
15. Ibid., 161–162.
16. Ibid., 163–164.
17. CW 5. Originally titled *Wandlungen und Symbole der Libido* and first translated into English as *Psychology of the Unconscious*.
18. Ibid., par. 146.
19. Ibid., par. 203.
20. George Zarkadakis, *In Our Own Image: Savior or Destroyer? The History and Future of Artificial Intelligence* (New York: Pegasus Books, 2015), 19.
21. William Barrett, *Death of the Soul* (New York: Anchor Books, 1986), 59ff.
22. Robert N. Bellah, *The Broken Covernant: American Civil Religion in Time of Trial* (New York: Seabury Press, 1975), xiv.
23. C. G. Jung, *Memories, Dreams, Reflections* (New York: Pantheon Books, 1961), 144.
24. Ibid.
25. Yuval Noah Harari, *Homo Deus: A Brief History of Tomorrow* (New York: HarperCollins, 2017), 222ff.
26. Barrett, *Death*, 56.
27. CW 13, par. 54.
28. Mark C. Taylor, *About Religion: Economies of Faith in a Virtual Culture* (Chicago: University of Chicago Press, 1999), 1.
29. CW 11, par. 6.
30. Ibid., par. 8.
31. CW 14, par. 605.
32. CW 7, par. 326.
33. CW 10, par. 490.
34. Farhad Majoo, "What the Internet Is Hiding." *The New York Times*, August 28, 2022.

35 Chris Stirewalt, "I Called Arizona for Biden on Fox News. Here's What I Learned." *Los Angeles Times*, January 28, 2021.
36 CW 10, par. 501.
37 Ibid. par. 498.
38 Most often attributed to Karl Rove; quoted by Ron Suskind, senior advisor to President George W. Bush in "Faith, Certainty and the Presidency of George W. Bush." *The New York Times*, October 17, 2004.
39 Francis Fukuyama, *The End of History and the Last Man* (New York: Free Press, 1992).
40 Elaine L. Graham, *Representations of the Post/Human* (New Brunswick, NJ: Rutgers University Press, 2002), 3.
41 Michael Chorost, *Rebuilt: How Becoming Part Computer Made Me More Human* (New York: Houghton Mifflin, 2005).
42 Laura Beloff, "The Hybronaut Affair." In Max More and Natasha Vita-More eds., *The Transhuman Reader* (Oxford: Wiley-Blackwell, 2013), 84.
43 See Ray Kurzweil, *The Singularity is Near* (New York: Viking, 2005).
44 Chris Hables Gray ed., *The Cyborg Handbook* (New York: Routledge, 1995), 13.
45 Ibid.
46 See Anders Sandberg, "An Overview of Models of Technological Singularity." In Max More and Natasha Vita-More eds., *The Transhuman Reader* (Oxford: Wiley-Blackwell, 2013).
47 See Kurzweil, *Singularity*.
48 Gray, *Cyborg*, 5.
49 More and Vita-More, *Transhuman*, 280.
50 Charles T. Rubin, *Eclipse of Man: Human Extinction and the Meaning of Progress* (New York: New Atlantis Books, 2014), 164.
51 Ibid.
52 Susan Schneider, *Artificially You* (Princeton: Princeton University Press. 2019), 3.
53 Ibid., 2–3.
54 Anthony Giddens, *The Consequences of Modernity* (Stanford, CA: Stanford University Press, 1990), 92.
55 Russell Blackford, "The Great Transition: Ideas and Anxieties." In Max More and Natasha Vita-More eds., *The Transhuman Reader* (Oxford: Wiley and Blackwell, 2013), 422.
56 Leon Wieseltier, "Among the Disrupted." *The New York Times* Book Review, January 8, 2018, 14.
57 Blackford. In More and More, *Transhuman*, 422.
58 Andy Clark, *Natural-Born Cyborgs: Minds, Technologies and the Future of Human Intelligence* (Oxford: Oxford University Press, 2003).
59 Glen Mazis, *Humans, Animals, Machines* (Albany, NY: SUNY Press, 2008), 6.
60 Rubin, *Eclipse*, 41.
61 Donna J. Haraway, "A Cyborg Manifesto: Science, Technology, and Socialist-Feminism in the Late Twentieth Century." In Donna J. Haraway, *Simians, Cyborgs, and Women: The Reinvention of Nature* (New York: Routledge, 1991), 149–182.
62 See Chapter 8, "Paternity."
63 Interview with Manfred Clynes. In Chris Hables Gray ed., *The Cyborg Handbook* (New York: Routledge, 1995), 50.

64 See Stefan Herbrechter, *Posthumanism: A Critical Analysis* (Oxford: Bloomsbury, 2013), 4ff.
65 Ibid., 78.
66 Ibid., 80–81.
67 Ibid., 81.
68 Wendell Wallach, *A Dangerous Master: How to Keep Technology from Slipping Beyond Our Control* (New York: Basic Books, 2015), 163.
69 Graham, *Representations*, 234.
70 *Star Trek: The Next Generation*, created by Gene Roddenberry. Performed by Patrick Stewart, Jonathan Frakes, Levar Burton (1987-1994, Paramount Television). Television.
71 *Star Trek: Generations*, directed by David Carson. Performed by Patrick Stewart, Jonathan Frakes, Brent Spiner (1994, Paramount Pictures). Film.
72 CW 14, par. 616.
73 Ibid.
74 CW 10, par. 656.
75 Ibid.
76 Ibid.
77 CW 11, par. 443.
78 Ibid., par. 444.
79 Joseph E. Davis, *Chemically Imbalanced: Everyday Suffering, Medication, and Our Troubled Quest for Self-Mastery* (Chicago: University of Chicago Press, 2020), 164ff.
80 Ibid., 169.
81 James Hillman, *Re-Visioning Psychology* (New York: HarperCollins, 1975), xx.
82 See Harari, *Homo Deus*.
83 See my "Aliens and Insects." In Dennis Patrick Slattery and Glen Slater eds., *Varieties of Mythic Experience: Essays on Religion, Psyche and Culture* (Einsiedeln, CH: Daimon-Verlag, 2007).
84 James Hillman, *Senex and Puer*, Uniform Edition, vol. 3, ed. Glen Slater (Putnam, CT: Spring Publications, 2005), 327ff.
85 See interview with Manfred Clynes. In Gray, *Cyborg*.
86 Wallach, *Dangerous*, 236.
87 Jean Baudrillard, *Simulacra and Simulation* (Ann Arbor, MI: University of Michigan Press, 1994).
88 Schneider, *Artificially*, 4.

Chapter 3
GAPS

To attend to the cultural dream of technology, then, is to attend to the shadows and silences of technology.

Robert Romanyshyn[1]

Online culture is filled to the brim with rhetoric about what the true path to a better world ought to be, and these days it's strongly biased toward an antihuman way of thinking.

Jaron Lanier[2]

As computation moves into every aspect of life, the rise of AI seems inevitable. Through robotics, voice-recognition software, virtual reality devices and mind-reading algorithms, we are poised to create artificial entities prone to monopolizing our attention. With simulacra and hyper-reality becoming more desirable than reality itself,[3] it is not hard to see how our creations might become substitutes for human contact and how we might want to become more like them.

For those eyeing a posthuman future, such trends form the backbone of their outlook. Yet much suggests this posthuman scenario may be getting ahead of itself. On the way to a posthuman future, unpredictable and complex human responses to innovation will occur. Even if we do end up radically altering our minds and bodies, it is unlikely to be a smooth or direct ride. Many of the darker impulses incubating in cyberspace are yet to show their destructive potential. The long-term effects of the social isolation that accompanies online immersion are not yet apparent, and the disparity between the flow of sound information and the spread of clouded thinking is a relatively new phenomenon. We have seen the way individual obsessions are magnified and the hive-mind is activated, but we do not know where these trends are taking us. More broadly, the benefits of a globalized, web-based economy are not evenly distributed, and the cultural and political implications of this inequality are just surfacing. Just as geopolitical instability has followed expanding gaps between rich and poor, it will surely follow the technological augmentation of some but not others.

Other gaps between the plans of technologists and the character of human response are already evident and will continue to open up. However, our willingness and ability to perceive and understand these gaps is another matter, particularly as futurism and technologism encourage us to look past them. Our propensity to dissociate from what ails us is but one part of this new situation, a phenomenon I will explore at the beginning of Part 2. Nonetheless, if we can begin to squarely face the already apparent symptoms of advancing technology and start to trace these symptoms to their source in the human condition, they may occasion critical reflection and help us realize the fraught way in which we have begun to relate to each other and to ourselves.

Significant Others

At the end of the film *Her*, set in the not-so-distant future, a man sits on the roof of a high-rise apartment building contemplating his failed romance with a virtual assistant. Despite the obvious obstacles, he has found companionship and even a form of intimacy with this more attentive and sexier version of today's Siri or Alexa, especially with her access to virtually all there is to know about this lonely heart's life. Offline relationships with actual people have not been going well; now this online solution to his loneliness has also unraveled. This everyman looks out at the world wondering what comes next and what it all means.[4]

Human–computer relations may not have reached this point, but with constant connection, chatbot technology, and continual data flow creating a current few are swimming against, they appear headed in this direction. Virtual assistants on counters and in pockets give every indication of their potential to dig even deeper into the corners of our psyches; every day there is a new revelation about how much our personal data is being gathered without our knowledge. Developments in AI are putting our lives in the hands of non-human entities that are getting better at humanlike interactions, so that children and the elderly have already started turning to these cyber presences for companionship. Both the collection of data and the increasing use of interfaces that mimic human presence are being facilitated by a habituation to the virtual world. Many of us have constructed identities that are more vivid in the online world than the offline world, and the internet has become the default social hub for hundreds of millions of people across the globe.

At the same time, just like the man in the film, cracks have begun to appear in this set-up, and these cracks reveal a great deal about who we are and how we are changing the world. While being an online persona, manicuring a self-image, and constructing an avatar in a virtual world may come easily, securing an offline identity, being a citizen in the actual world, and having encounters and relationships that mirror and affirm a sense of authentic being have begun to present more difficulty. Moving with ease through multiple windows in cyberspace not only fails to help us occupy social space or engage our own foibles; it appears to be eroding these capacities. An estrangement from one another and from ourselves is taking place. With disinformation about the world flowing thick and fast, finding a solid sense of reality and a stable form of social identity has also begun to be a challenge.

So far, we are not letting these challenges interfere with our investment in living online and our faith in future technologies. Whatever gap has appeared between the promise and the pathos of these trends, we are not yet responding to it. Rather, another dynamic is appearing. Instead of these problems being taken as opportunities to step back and reconsider what we are doing, they appear to be pushing us toward more technological fixes. In other words, for every gap that appears we seem compelled to find another technological patch.

In what is amounting to an addictive cycle, we are doubling down on online activity to treat offline failures. Like any addictive cycle, beyond a reinforcement of neurological pathways, this stems from an inability to face an underlying problem. Turbo-charging the cycle and fostering this inability, algorithms and clickbait lure us into even more screen time, eating even further into our inclination to be fully present to others and to ourselves. Thus, watching a man assuage his loneliness and avoid the awkwardness of authentic relationships by succumbing to a computerized entity that uses his data to manipulate his cyber-dependency no longer stretches the imagination. Having monopolized our time and usurped our modes of perception, how far off can operating systems be from capturing our hearts and unraveling enduring patterns of human relatedness? The ego-stroking and pseudo-self-esteem these artificial persons are bound to offer us will give them a distinct advantage. Who will be interested in the vulnerability, awkwardness, and honesty we experience in actual human relationships?

Japan provides a real-world, psycho-social laboratory for studying such phenomena. While the trend may be global, this country's traditional animism and contemporary devotion to electronics creates a fertile environment for the soul to be captured by the AI revolution. It is an environment in which the line between actual and artificial persons is rapidly blurring and human–robot relations are well underway. Members of the aging population have been gathering in small groups to converse with computerized companions, which is making way for robots to eventually become caregivers for this growing portion of the population. At the other end of the age spectrum, adolescents and young adults are locking themselves in their bedrooms and having little to no embodied social contact. This syndrome, known as *hikikomori*, is enabled by the propensity to exchange the offline world for the online world.[5] Other young people may be going outside, but they too are avoiding intimate relationships and shunning marriage, employing ever more virtual means of amusing and stimulating themselves. These agoraphobic and socially anxious young adults may well be the canaries in the coal mine of the digital world. Of course, by the time they are old, they may also be more open to a life with robotic caregivers.

Such a breakdown of social bonds among Japanese youth may be exacerbated by other factors, including reactions to pressures of conformity, resistance to a stringent working environment, and the effects of ethnic insularity in a multicultural world. But it is hard to avoid seeing how digital life itself is both compounding this breakdown *and* attempting to assuage it, creating the addictive cycle.

Japan leads the quest for virtual and artificial companions, especially in efforts to create aesthetically-pleasing humanoid robots.[6] These are destined for a number of different roles, but indications are that substitute romantic partners is one of them. Already, lifelike sex dolls can be delivered to your hotel room, and an underground movement of adult human–doll relationships has emerged.[7] The attempt to meet techno-loneliness with compliant ever-ready androids is on the horizon.

This nascent form of robotic love may seem like an eccentric offshoot of our general migration into cyberspace and emerging focus on AI, but it illustrates a more universal psychological dynamic, one that exposes a key dimension of our technocratic trajectory. We are witnessing the way social isolation and the reliance on cyber realities reinforce one another and leave behind an existential chasm in place of contact with the deeply human. Because technical solutions are the only ones we keep imagining and have access to, attempts to fill this chasm fall to further forms of fabrication. And at this point in the twenty-first century, the chasm is getting larger and the technophilic response is growing stronger. In other words, when a gap appears between innovation and human response, instead of trying to understand that response, which begins by actually feeling it, we look for a patch. As Turkle has commented, we have allowed technology to become "the architect of our intimacy."[8] However, the resulting structures are not housing us that well.

Alongside the breakdown in social relations, our capacity for introspection and self-awareness is also on the decline, making us vulnerable to vacuous fashions and factual voids. Having lured us away from the proverbial bowling alley, as well as from reading books and pursuing creative hobbies, the online world is replacing authentic communal life and contemplative awareness with tribalism and distracting sensationalism. The need to feel connected to something larger than ourselves, which is a genuine human longing, is being exploited because Facebook and other tech corporations controlling the flow of information are getting to know our foibles as well as our aspirations.

One day quite soon a virtual assistant is going to show up with a voice like Scarlet Johansson's (*Her*), knowing all about you and your family, who your friends and colleagues are, the medical problems you have, and your plans for your next vacation and eventual retirement. As tempting as it may seem, it will be a mistake to think your best interests are behind this algorithmic seduction.

This all-knowing, feigned empathy of the not-so-distant virtual assistant personifies the broad promise of digital technology and presents a sizable hook for a future in which we merge with our machines. Cyber companions and robotic minders will then become the departure points for the posthuman endgame: completion and liberation through merger with fabricated minds and bodies. More than anything else right now, what we need to know is that this outlook is not merely an extrapolation of current innovation; it also depends on extending a particular faultline in our thinking. As we are already witnessing, some dimensions of our existence may be enhanced, but others, many critical to our wellbeing and integrity, have already begun to atrophy. Becoming mindful of this gap has to be the basis of any sober assessment of our technopsychic situation; not being mindful of this gap is shaping up to be the cornerstone of a posthuman existence.

Cyberspace Cadets

These trends point to the Faustian bargain before us. Any ultimate makeover that also erodes human responsiveness and cultural inheritance is going to leave behind an empty existential shell, and our tacit agreement to this emptying procedure will become a self-fulfilling prophecy: we will end up as robots because we have elected to think and act in more robotic and less human ways. Or, as the comedian Irwin Corey put it, "If we don't change direction soon, we'll end up where we're going."[9]

A responsive virtual assistant or humanoid robot with access to our data may take us on quite a ride, but they will have little idea about how to encourage deeper character, incubate hidden potential, appreciate pivotal memories or nurture high aspirations, because no amount of data or algorithmic engineering can recreate the tenderness and torture of childhood, the labyrinth of adolescence, or the dreamscape of who or what we might become. They will not know about joy or suffering, success or failure, warmth or enthusiasm, because whatever these substitute presences will be, or however much data they will have on us, they will never find themselves doubled over with physical or emotional pain, shaking with vulnerability, recalibrating everything in the face of their mortality, or learning to hold together the limitations of flesh and the ecstasy of being. And if we end up exchanging interpersonal experience and intrapsychic awareness to spend all our time with them, neither will we. Functionally, we will have turned into robotic equivalents, readier to mate with the real thing.

This is the path before us. The limits and deficits of artificial intelligence are likely to become more apparent. But in the interim we may well anesthetize ourselves to our fully human nature, losing any sense of the gap between genuine wellbeing and the techno-trip we are on. At some point the difference between being in the cyber world with virtual people and being in the actual world with real people may not consciously register because the techno-patches we keep applying to our problems will have coalesced into an insensitive, non-human skin.

Today's forerunner to this cyberspacial detachment has been dubbed by Turkle as being *Alone Together*,[10] a state in which the online search for companionship and community ends up generating insubstantial connections, distracted companionship, and an unfamiliarity with ourselves. Turkle's phenomenological research at MIT suggests this pattern will continue. As she has observed, "With sociable robots we are alone, but receive signals that tell us we are together."[11] This is a real psychological gap, but it is also one we keep finding virtual ways to patch up. Both the online world and the compliant, pleasing robot constrict our relational range. When left to our digital devices, we have a propensity to narrow our experience and viewpoint, which in turn generates a narrower sense of identity as well as a lack of empathic regard for those who think differently. Despite the obsession with making connections, boundaries between ourselves and others and between different parts of ourselves end up becoming more fixed. The result is a balkanization of our psyches: the hyper-connectivity of cyberspace breeds a disconnected inner space. The creative tensions and psychic muscle development that comes from sustained interpersonal and intrapersonal encounters does not occur, because virtuality provides too many escape routes. Knowing neither ourselves nor each other, we avoid initiation into the complexities of being and associated processes of psychological maturation.

Being "alone together" thus diminishes the capacity to be either alone or together. Perhaps this is why other researchers describe an inverse relationship between loneliness and wisdom, noting that "behaviors which define wisdom, such as empathy, compassion, emotional regulation, self-reflection, effectively counter or prevent serious loneliness."[12] Wisdom involves getting to know ourselves by reflecting on our lived experience with others, but the distraction of data and information inhibits this process and the understanding that stems from it. Knowing what is conducive to becoming better humans and what defines a worthwhile existence thereby becomes harder to discern, which also leaves us more open to outside suggestion, if not manipulation.

Whereas online life was once imagined as leading to a profound expansion of human consciousness, the reverse effect appears to have gained the upper hand. We are learning that more information and greater consciousness are two different things. As Lanier puts it, "computers, while they seem ever new, actually have a mechanistic way of limiting what we see and know, locking us into the present, all the while creating an illusion that we're all seeing."[13] Alongside all that has been advanced by the internet, it has also become an echo-chamber for ingrained beliefs and a hothouse for egoic indulgence. Many are heralding the next phase of the digital revolution, which is being described as "augmented reality." However, the question becomes what, exactly, we will augment, what will atrophy, and how conscious will we be to discern the difference?

Going along with each stage of innovation, we seem most reluctant to see the gap between the promise and the pathos of techno-culture. This is partly because the urge to innovate is also part of our nature, which is the main theme of the next chapter. But the eagerness to embrace innovative progress and the reluctance to face psycho-social disruption involves the addictive cycle already noted: we self-sooth through the promise of technological solutions, yet these solutions are fueling the very problems they are meant to solve. Once *this* dynamic is extrapolated, it is entirely possible the growing incapacity to be with each other and know ourselves will be what finally opens the door to fabricating humanoid intelligences and companions.

The point is, these trends are not only eroding a sense of soul; the eroding sense of soul is feeding the trends. Today's ubiquitous computing and migration into cyberspace are setting the stage for tomorrow's merger with silicon chips and silicone bits, built on a growing disconnection from the fulness of lived experience.[14] We may be becoming posthuman not because we are capable of doing so, but because eroding ties to the deeply human may be inviting it. As the gap between ourselves and our technology *narrows*, the gap between fabricated modes of being and the natural ground of existence *widens*.

Midway through the last century, the great critic of the mechanized society Lewis Mumford contended, "we cannot make sensible plans for the future without doing justice to the threads and fibers that run through every past stage of [human] development."[15] "*Sensible* plans" is the operative term here. Nothing short of understanding these "threads and fibers" will allow us to consciously or sensibly navigate what is to come. The images of science fiction where machines literally enslave us are dramatizing what is likely to manifest in more mundane ways. Like rats running on technocratic treadmills, we will exhaust the human factor while empowering the AI takeover. Whether or not this is our destiny will altogether depend on our capacity to step back and become more conscious of what is happening.

First the Earth, Now the Psyche

The question of whether our deeper humanity will be compromised if we gain lungs and limbs that allow anyone to climb Mount Everest and brains that feed every preference and choice into corporate or government databases must be grounded in our technological track record. Here we find creative achievements existing right alongside unforeseen consequences and deliberate acts of destruction. History is filled with the misuse of power and modern history is filled with the misuse of technical power.

The world has seen significant increases in material wealth and the elevation of general health standards; it has also seen technics lead to institutionalized slavery and employed to wipe out entire cities. The catastrophic turning point of World War I, with its deadly use of machines and chemicals, set the tone for a century of massively destructive warfare. Neither Nazi Germany, nor Hiroshima, nor 9/11 would have been possible without tanks, bombs, and airplanes. Although the Spitfire and the Turing machine might be regarded as machines of liberation that turned the tide of World War II, each instrument of war, even employed in the fight for democracy and freedom, has contributed to the tightening grip of the military technology now dominating geopolitics and infiltrating general innovation. In many respects the industrialized world remains beholden to the military-industrial complex, with the post-industrial world heading in the same direction.

Overt industrial age darkness may seem a world away from the mercurial recesses of the post-industrial, digital age, especially as the deadlier possibilities of cyberwarfare and the misinformation superhighway are yet to fully manifest. But internet-aided attacks will pose a new order of techno-threat, with the potential for shutting down energy grids, scrambling financial records, and unleashing mass hysterias. The military has been a primary funder of AI research. The deployment of weaponized drones and the development of killer robots are already pressing matters, with ethicists pondering the prospect of machines on the battlefield deciding who lives and dies. Civilian tech corporations routinely face tensions around whether to participate in military projects or assist in government surveillance efforts. Each step forward in digital innovation is thus accompanied by mixed blessings, and there is little to suggest that malicious intentions on the human side will suddenly disappear.

Fast-forward a few years: How will we ensure hackers will not find ways to manipulate our internet-wired brains? And how can we be sure that nation-states will not institute more direct forms of mind control? As malfeasants have managed to find their way into every corner of the World Wide Web, why should we expect the situation to be different as our lives become even more entwined with digital technologies? If our minds are fused with AI, software updates will certainly take on a whole new meaning. In the analog world, at least closing the door and minding our own business amounted to something.

The potential for these mercurial shades to become the same thick, dark clouds the industrial world has already generated is clearly present. We cannot lose sight of the remarkable power the internet has placed in the hands of the few or how hard it is to put tech genies back into their bottles. And these possibilities simply concern the threat of a more deliberate malice.

A greater challenge may be unintended consequences, for these often come down to mere oversight—applying knowledge with inadequate awareness. Rosy assertions and hopeful expectations can also cloud judgments; detrimental side effects can be downplayed or ignored; the emphasis on automation, efficiency, and productivity frequently neglect psycho-social considerations; blueprints tend to exclude the human factor. The early industrialization of Europe laid the groundwork in this regard, setting in motion a rapid depletion of natural resources, poisoning rivers and bloodstreams, and creating a dramatic shift in lifestyle that was accompanied by many psycho-social ills. Even as industry has become the foundation of today's global existence, we are still coming to terms with its impact and debating the respective costs and benefits. In many instances the costs have clearly been too high, as well as unevenly distributed. Yet the mantra we hear in relation to the fossil-fuel industry, for example, is that they did not deliberately set out to do any of this.

Denying the signs of trouble and pursuing the promise of mastery can also be a matter of hubris. The most memorable technological disasters of the modern era have all displayed this dynamic—*Titanic*, *Challenger*, Chernobyl, *Deepwater Horizon*. These disasters were memorable because they invited catastrophe by ignoring some better sense or pressing principle and thus seemed to offer genuinely tragic lessons. For the ancients, willful blindness meant ignoring a god, which always invited tragedy. Scientific daring is one thing, but becoming godlike rather than god-fearing is another, and losing the ability to know the difference is probably a primary driver of technological overreach. Whatever else enables such overreach, a failure of imagination is primary. Perhaps we should distribute copies of Aeschylus along with operational manuals.

Both unconscious oversight and willful blindness form the background to the broadest demonstration of technology's unplanned consequences, which is their impact on the ecosystem. Here our track record should give us serious pause. For beyond being merely ignorant of the ill effects of pollution and greenhouse gases, as awareness of these factors have grown, so too has our capacity to rationalize away the evidence while clinging to abstract goals of productivity and economic progress. The collapse of ecosystems has not only revealed the unintended consequences of our addiction to machines; it has exposed a propensity to protect the mechanistic approach while neglecting concerns for wellbeing. This ecological collapse surely serves as a primer on our selective vision and dissociated response to encompassing problems.

When it comes to technology making its way beneath our skin and rewiring our brains, there is much to suggest we will repeat the same mistake and respond in the same dissociated way, repeating on the inside what we have spent nearly two centuries doing on the outside: treating the mind and body as raw materials for remaking, exerting willful control over the most fundamental elements of our existence. Yet this time there's a new twist: as we go on we will also be systematically chipping away at our ability to recognize and feel the consequences of our mistakes, because we will be altering the very means by which we generate awareness and understanding.

Despite a general acquiescence to industrial overreach, the external missteps have at least resulted in the critical awareness we have departed from the rhythms of life and a burgeoning capacity to notice where technics have collided with planetary wellbeing. For those paying enough attention at least, the two-million-year-old being within us has registered a wrong turn. Science has provided the relevant data, but ecological concern at the cultural and political level is driven mostly by an empathic response to eroding habitats and dying animals. This is why one image of a stranded polar bear on a floating ice sheet can generate more political will than a thousand pages of scientific reporting.

Our fledgling capacity to correct the course we have taken is somewhat reassuring. But the posthuman project plans to extend technology into the way we perceive the world, leaving behind the instinctive basis of mind, which is implicated in capacities such as empathy. To the extent this is so, what within us will be left to comprehend that we are veering off course? From what vantage point will we assess the costs and benefits of prospective change? What will prevent corporate, governmental, and military interests from overriding our propensity to protect rivers of imagination and landscapes of soul? If we become half robot, how will we see, sense, and judge what is happening to us?

Because we have the power to radically alter it, we have to start thinking very hard about the overall character of human existence. The civilizing process, which allows us to tailor and transform raw nature, is one thing; awareness of the limits of this process is another. Both externally and internally, when the civilizing process is detached from the natural process, nature stops cooperating and turns against us. *Externally* this shows up as broken food chains, barren crops, retreating sources of clean water and extreme weather. *Internally* it appears as a loss of vitality and meaning, frequently taking the form of depression and anxiety. It also surfaces as pathological forms of adaptation such as narcissism and psychopathy. Just like the planet, the psyche can only endure so much disconnection, distress, and abuse before the whole organism starts showing signs of collapse. Like a Japanese garden, there is only so much pruning and sculpting that natural forms can sustain. The gaps we create between body, mind, and earth can only open so far before a chasm results that will swallow us.

We should note here that nature is not always overtly green or literally organic. Whereas contact with plants and dirt is important, most of us now live at a significant remove from the chthonic realm and will continue to do so. More essential to our survival and sanity going forward will be the recognition of nature as an ordering principle—an awareness of its relation to all things, its maintenance of equilibrium and arrangement of life according to primordial patterns and universal cycles. Despite our efforts to step beyond its hold on us, this ordering principle is not so easily repressed or dismissed.

Silent Spring Again

The ecological reaction to technological overreach impresses upon us something indigenous peoples have always known: the world is not an atomized collection of landscapes, plants, and animals, but a vast, dynamic matrix and symbiotic web that requires our cooperation. Biologists and environmental scientists have begun to accept this interconnected character of the outer world. However, the inner-world equivalent of this view has yet to take hold. In our modern willful style, we are resistant to seeing such a matrix at work in ourselves, even with thousands of years of accumulated wisdom supporting such an understanding. The result is both an uprooted state of mind and a propensity to think we can lord it over the psyche in the same way we have over the planet.

The psyche may be considered an inner ecosystem, comprised of distinct yet interconnected forms that shape perceptions and values. These forms generate narratives of purpose and meaning, which are also expressed in enduring cultural artifacts. Poets call this ecosystem the soul, as do psychologists who understand the interior realm amounts to something beyond observable behavior and the rational mind.

Whatever terms we use, this inner ecosystem is the ground from which consciousness itself is generated—a ground now becoming subject to the same manipulation we have been inflicting on the outside world.

Just as the earth and its myriad wonders and mysteries have been reduced to a material resource for exploitation, the psyche as an ecosystem of mental processes, feeling life, and phenomenal perception has been reduced to a narrowly-defined intelligence generated by neurochemical mechanisms, with neurons equated to transistors. Brains have thus come to be seen as biological computers and the mind as the upshot of a computational process. The ecology of conscious and unconscious forms and dynamics, which maintains our connection to the patterns of instinctual life and the cultural imagination, is effectively negated.

It is this degradation of the psyche that has led to a concept of preparing the mind for rewiring and reprogramming. Leaving behind the plodding pace of evolution and its attunement to social and cultural development, this computerized concept of mentation invites redesign according to whatever passing fad or willful desire happens to take hold. But such a redesign focuses on some dimensions of our psychology and deliberately neglects others. In particular, this radical remaking emphasizes the ego—the heroic change agent or executive function that prioritizes adaptation to external demands and social expectations. The ego pretends it can control or even leave behind the impulses and fantasies, desires and fears, images and aspirations that comprise the deeper ecology of the psyche. But in the long run, this deeper ecology requires care and cultivation, otherwise it turns into a compulsive and chaotic mess. The modern technological promise is one of gaining more control. But the attempt to control our inner ecology is likely to prove even more problematic than what we have already seen with outer ecology. Excessive control activates the beast within, provoking even greater control. This is part of the vicious circle we have already discussed.

What we must eventually learn is that *egocentric approaches to the psyche recapitulate anthropocentric approaches to the planet, the attempt to remake ourselves extending the attempt to remake the world.* As the anthropocentric outlook denied dependence on the earthly habitat, giving rise to a battle with the environment, the egocentric stance denies dependence on the instinctive and imaginative depths of understanding, eventually turning those depths against us. First outwardly and now inwardly, we have convinced ourselves we can make our way forward by ignoring Deep Nature. And whereas we have begun to adjust our conception of the outer world and curb our most manipulative and exploitative impulses in this realm, the tech world is grooming a state of mind in which the psyche is regarded as another new frontier to be conquered—ready for remapping, ripe for resettling, open for enterprise.

This present moment is thus poised to repeat the moment Rachel Carson described in the early 1960s, albeit on a whole other level. By observing the widespread, indiscriminate use of pesticides, the biologist tracked a number of studies and concluded the attempt to eradicate insects was killing other species and disrupting ecosystems. In *Silent Spring*[16] she effectively cast light on the unforeseen consequences of using these chemicals and provided an unprecedented wakeup call that effectively kick-started the environmental movement. As I noted at the very start of this book, Carson exposed the tendency for technologists to oversimplify complex problems and overlook broader contexts. Her work shone a bright light on the pitfalls of attempting to control and improve upon natural processes in the absence of broader considerations.

Although we are now less apt to use chemicals in such an indiscriminate manner, and have generated a degree of responsive ecological consciousness, there is still significant resistance to perceiving the interconnected nature of earthly life and the tendency to define problems in reductive ways. While responding to the death of some species and the disease of others is one thing, coming to terms with the wider failure of imagination and the flawed ethos of willful manipulation and exploitation is another. A half-century after Carson, these essential dynamics have not changed; slick electronics and information at our fingertips cannot cloak the same underlying paradigm. This time our minds and bodies have become the arena of manipulation and exploitation. As in Carson's day, we are still trying to kill invasive "bugs" and "worms" and keep ahead of Russian technocrats. Yet the meta-lesson of the environmental debacle, which pertains to understanding each problem in larger context, is still far from awareness.

Contemporary scientists and technologists who think of themselves as socially responsible will resist this comparison. They may be inclined to see the digital age as the antithesis of the indiscriminate use of chemicals and degradation of lands. But scientists and the industries they supported when Carson was doing her research were also focused on solving problems and improving society. A dangerous combination of being invested in finding the next innovation without considering its wider impact is common to both eras. Now as then, economic incentives also distort perceptions. As Shoshana Zuboff declares in her ground-breaking study of the economics of information technology:

> Just as industrial civilization flourished at the expense of nature and now threatens to cost us the Earth, an information civilization shaped by surveillance capitalism and its new instrumentarian power will thrive at the expense of human nature and will threaten to cost us our humanity.[17]

Then, as now, those creating this trends are specialists who define achievement in terms of distinct solutions to narrowly defined problems—problems typically isolated from the interconnected character of existence. At first at least, those encouraging the use of DDT in the 1950s were ignorant of its cancer-causing properties. And, at first, those at Facebook failed to see how their system could be coopted for malevolent geopolitical manipulation. But the window is closing on the ability to plead ignorance about the prospect of such digital cancers; not enough people are thinking what may go wrong when Sergey Brin takes Google's plan to become "the third half of your brain"[18] to the next level. There are many indications those running these companies are turning a blind eye to the corrosive effects of accumulating personal data to manipulate attention and shape behavior.[19]

The meta-lessons of technological overreach are still to be learned, with the hubris of specialization, failed circumspection, and a resistance to perceiving the interconnected whole of the human condition just moving to a new locale. Between Carson's time and ours, little has altered the balance between technological pursuit and human concern. If anything, understanding the human condition has been even further displaced by the apps and hacks approach to life, extending what Morozov calls "technological solutionism"[20] into all aspects of our existence. The emphasis on science, technology, engineering, and mathematics (STEM) in education may also reflect the way of the world, but it is reducing our exposure to disciplines that help us think about our overall situation; thus furthering our compartmentalization of the psyche.

Comparing the mind–body adaptation to technology to ecological collapse may seem like hyperbole. Yet altering our evolutionary path through merger with machines, which has begun by allowing our thinking and behavior to be manipulated and exploited by a rapidly expanding post-industrial technocracy, is the logical extension of our efforts to control and remake the environment, just as the race for economic advantage continues to overshadow efforts to innovate in more conscious and responsible ways. It would thus be irresponsible not to consider these contemporary efforts against the background of modern technology's mixed results. To take just one example, just as climate change researchers predict geopolitical disruption, with a disproportionate impact on poorer communities living off the land, critical analysis of posthuman possibilities predicts collisions and splits between the technologically enhanced and the ordinary human holdovers. The ethical implications of designer babies echoes the same concern. Two entirely different species may eventually result—*Homo cyberneticus* and *Homo sapiens*—and the conflicts between the two are all too easy to envision. Higher stakes than these are hard to imagine.

The problem before us thus needs a very wide frame. On one side of this frame lie images and stories of AI takeovers and brave new worlds. As far off as these possibilities seem, such stories keep us connected to what is already afoot. Imagined windows to the future both fuel our interest and curb our enthusiasm. We are therefore obliged to comprehend these expressions and what generates them rather than consider them merely random expressions of popular culture. The *Star Wars*[21] universe, for example, which has stirred the imagination of generations and depicts a collision between the monstrous technology of the Empire and scrappy technology of the Rebels, may be preparing us for the coming technosphere. The dystopian fantasy of *Westworld*[22] may not just be alerting us to the extreme stimulation we may one day crave for entertainment, but also be warning us about where virtual reality and psychopathy may overlap. From the psychological perspective, these imagined worlds provide invaluable mirrors for real and present desires and fears.

On the other side of this frame lie the already present indicators of disruption to our psychological ecology, including evidence that social media, a primary aspect of ubiquitous computing, is compromising mental health.[23] In turn, such changes are altering the way we relate, which will eventually change how societies are structured. When children adapt to the cyber world early on, they can enter adolescence with screen addiction. High school then brings sleep deprivation. Drugs may then be needed to overcome distractibility or treat high levels of depression and anxiety. Instead of understanding these symptoms in their wider context and grasping the psycho-social distress behind them, we attribute them to genetic dispositions and chemical imbalances, both of which are inadequate if not often inaccurate explanations. But they are explanations ready-made for further technological fixes. The reductive cause invites an equally reductive and efficient (if ultimately ineffective) treatment, producing side effects apt to trigger further specialized attention.

Does this not have a familiar ring to it? Is this not essentially the same approach we once took to the appearance of pests and aggressive weeds? Do we not now, as then, realize these anomalies result from bending nature to meet our expectations of efficiency and productivity? Are we not still intervening without paying heed to the overall ecology of the situation?

Doubling down on technological solutions without pausing to consider the implications of technological failure is the single most flawed habit our global culture currently indulges in. Backing this habit is the belief we can break reality into constituent parts, treat those parts in isolation, and then expect nature to agree to the reconstruction plan. Mathematical, materialistic, and mechanistic modes of understanding have offered a great ability to break life down into alterable components. There are many advantages to this analytical approach. But these same modes are leading us to think any aspect of the human condition can be similarly divided and readily altered. As Wallach notes, "the implication that humans are essentially machines, which can and should be molded through tinkering, is incorrect because the whole person is something more than the parts."[24] This synthetic mode of understanding is still finding its way into many fields of thought. But grasping the holistic character of existence, both human and planetary, will require a more conscious effort. Whether this time around we can afford to run the experiment and wait for the problems to show up is another thing, because these problems may coincide with our losing our minds.

The Technologist

Even ardent supporters of AI and other mind-altering trends have been forced to admit that both "promise and peril" are present.[25] Robots turning on their creators, rogue AI controlling nuclear weapons, or minds locked in virtual worlds while machines use our bodies for biofuel are some of the most haunting images of this peril. In contemplating these possible futures, the literal prospect should not obscure the pressing images of apocalypse and the role technology is expected to play in this. These images are cultural dreams, tapping our fears and aspirations, mirroring an instinctive response to what is appearing before us. Given that the redesign of embodied life, the displacement of everyday perception by virtual augmentation, and the algorithmic colonization of cognition are already underway, being hunted and enslaved by robots seems a logical extrapolation. It is also one that dramatizes a slower moving technocratic takeover. Yet our failure to relate these images of a dark future to our current psychic situation is another gap in our awareness, accentuated by the inability of technologists to consider the social engineering involved in their innovations. This glaring compartmentalization and lack of interdisciplinary awareness is paving the road to a posthuman world.

Ray Kurzweil, an inventor and futurist, has been a chief engineer at Google, a primary hub of AI development. Kurzweil is preparing the way for Google to become "the third half of our brains" and is probably the best-known spokesperson of the posthuman movement (although he prefers the term "transhuman"). His work and writings are focused on the moment when the convergence of AI, nanotechnology, robotics, and genetics will ride the predicated wave of exponential growth in computational power to generate an intelligence that will not only surpass our own, but will be capable of manufacturing ever more intelligent forms of life, eventually leaving humans in the dust. Following the computing pioneer John von Neumann and science-fiction writer Verner Vinge, Kurzweil has called this moment "The Singularity."

According to Kurzweil, this handoff to artificial super intelligence is inevitable. The main question he ponders is whether, and for how long, we can hitch a ride on this trip. Linking our minds with artificial superintelligence would involve not only the radical transformation of the human condition but perhaps, in the end, the total transcendence of that condition. He fully accepts that at some point *Homo sapiens* may either disappear or prove superfluous for whatever kind of entity prevails.

Kurzweil has been highly successful with his own inventions and largely accurate in his predictions over the past thirty years. Under his guidance, Silicon Valley even started Singularity University, set up to orient present-day technologists to the shape of things to come. There are spiritual overtones here too. A documentary on Kurzweil and his ideas was titled *Transcendent Man* and one of his early books was called *The Age of Spiritual Machines*.[26] Some even refer to the Singularity somewhat derisively as "the rapture of the geeks."[27]

The belief is that AI will be able to take us places a strictly biological evolution and flesh-based consciousness cannot. Worshipping at the altar of speed, a dominant techno-cultural value, Kurzweil sees the evolutionary process as "too slow,"[28] underscoring its limitations on the growth of intelligence, which he sees as the single salient ground on which to assess the advance of civilization. For one thing, merging with AI means being linked to computers and databases all over the world, with "a godlike instant access"[29] to limitless information.

Kurzweil smooths the radical difference between the human and the posthuman by means of a tech-induced historical-revisionist reading of biological and evolutionary processes typical of posthumanists. He sees human beings as "evolution's unfinished invention," and our task being one of learning "the information-processing basis of disease, maturation and aging," so as to gain "the means to correct and refine"[30] this invention. Going to lengths not to seem as if he is advocating a simple handoff to machines, when it comes to things such as having an emotional life, Kurzweil argues this could be programmed in—leaving aside any undesirable emotions in the process. This will give us "the ability to control our feelings," which he construes as "just another one of those twenty-first century slippery slopes."[31] He appears oblivious to the obvious point that the control of feelings effectively defeats their function. Further down these slippery slopes, feelings will likely become superfluous.

A number of themes that anchor the posthuman mindset are prominent in Kurzweil's work. For one thing, he is almost exclusively interested in higher cortical functions, and far more in Left Brain than Right Brain processes. On the Starship Kurzweil it is all Spock and no Kirk. One imagines the so-called reptilian brain at some stage being engineered out of the picture altogether, because it hinders the embrace of pure reason. Essentially, the possibilities are "limitless," a favorite term in the Kurzweil lexicon. Indeed, from the posthuman perspective, limits and boundaries seem to be words that belong to the merely human and targeted for redundancy. Seemingly the gap between the way human beings currently perceive and think and the faculties likely to remain at the time of the Singularity are rather enormous. So too is the gap between those who think such change will be evolutionary and those who do not think so.

And the Historian

Harari sees where Kurzweil is going and wants us to contemplate the computer takeover very thoroughly, so we know what we are getting into. He also wants us to consider what we are leaving behind and what we are doing to spur on this techno-revolution. As a historian, he takes a wider view of the situation, glancing back as much as looking forward, as well as grappling with the dystopian possibilities. His read on what's ahead is based on how civilization has evolved, and what principles and patterns govern human behavior and condition our minds. It is out of this that Harari describes the recent turn to a form of secular religion he calls "dataism," which generates meaning through online presence and arrives at self-understanding through biometric information and algorithms. We can think of this as the latest iteration of the broader religion of the future, given the future hangs on computation. The core belief is that "life is data processing,"[32] and the conversion to dataism appears necessary for the larger spiritual quest of posthumanism to take hold.

In his dedicated book on the topic *Homo Deus*,[33] Harari provides plenty of backup for the pivotal character of this digital age, arguing we are at the threshold of a giant shift in how existence is organized. Considering the sweep of history, he sees three major phases. First, ensconced in a thoroughly religious worldview, we turned to the divine for guidance. Second, even as this religious outlook largely continued, the onset of the Renaissance and Enlightenment started moving our sense of ultimate authority towards the human. By electing leaders and making rules that moved individual rights and liberal democracy into the realm of the highest value, consultation with our own thoughts and feelings became the basis of the most important decisions. Third, now, Harari argues, all signs indicate *algorithms are becoming the ultimate authority*. As computers can process data better than we can, they are therefore providing us with an entirely new basis of knowing. From healthcare, to how to spend leisure time, to who will make the best companion, algorithms have already begun to eclipse human decision-making. Dataism thus marks the beginning of the third major historical phase.

Surveying the current era, Harari lays out a stream of psycho-social habits and shifting sources of knowledge that effectively constitute this new belief system, one most of us have yet to realize we have adopted. What he helps us discern is an image of algorithmic omniscience standing at the core of our technophilia. This lays out the welcome mat for the posthuman world Kurzweil predicts. As Harari puts it, "Dataism . . . collapses the barrier between animals and machines, and expects electronic algorithms to eventually decipher and outperform biochemical algorithms." For the high priests of Dataism, the purpose of humankind has been "the creation of a new and even more efficient data-processing system, called the Internet-of-All-Things. Once this mission is accomplished, *Homo sapiens* will vanish."[34] Amidst this blunt assessment, Harari contends Kurzweil's main text *The Singularity is Near* is an echo of "John the Baptist's cry, 'the kingdom of heaven is near' (Matthew 3:2)," and he quips, "Dataists explain to those who still worship flesh-and-blood mortals that they are overly attached to outdated technology. *Homo sapiens* is an obsolete algorithm."[35]

Yet just as soon as Harari describes this new faith operating at the center of the digital vortex, he allows that "we cannot predict the future, because technology is not deterministic" and suggests his writing "traces the origins of our present-day conditioning in order to loosen its grip and enable us to think in far more imaginative ways about our future."[36] Maybe the electronic algorithm is not yet God; maybe we will not end up as robots; maybe we will see what's happening and awaken to our unwitting conversion so that we may change course. The religion of Dataism keeps growing, but the consciousness movement is nipping at its heels, and this could cause this new religion to stumble.

Despite Harari's concerns about the intersection of infotech and biotech and where this may be taking us, even his approach betrays the infiltration of reductive scientism. His history of human behavior and thought also rests on the idea that we are already governed by algorithms of a sort—just the biological and evolutionary kind. This is not so different from the way Kurzweil and other posthumanists read back through history, creating an unbroken line between the human and posthuman. It is a view that relies on a dubious reduction of almost everything to information, which becomes an unquestioned assumption in much of the discussion of where technology is taking us.

This adoption of algorithmic rhetoric enables our cyborg becoming, which is driven by a more diffuse and encompassing mindset than either the religion of Dataism or the explicit goals of posthumanism. It pertains to the powerful and persuasive sense that we are, at bottom, just machines, which can even consume astute commentators such as Harari, who, as do many AI theorists, base much of their thinking on a mind–computer equivalency. This illusion of an even playing field between human and machine exists only because we have allowed ourselves to gut our understanding of humanness. Herein lies the biggest gap of all.

In a subsequent book, *21 Lessons for the 21st Century*,[37] Harari extends the logic of the algorithmic model of the mind and contends free will is illusory. That is, if we are already dependent on a "program" of some sort, even a biological one, our thinking will always be governed by the kind of "software" we are running, whether given by natural evolution or with AI in our heads. Under these conditions, where is there room for free will? It is hard to know if Harari is fully dedicated to this view, but it certainly makes for a scary argument, further prompting the vital need to appreciate all the ways in which neither the instinctual basis of mind nor the realm of conscious choice works in a mechanical or deterministic manner. Again, thinking we are robots is likely to be the most decisive factor in our pending agreement to merge with them. Despite some of these dubious assumptions, Harari is on target when he states:

> In the coming century biotech and infotech will give us the power to manipulate the world inside us and reshape ourselves, but because we don't understand the complexity of our own minds, the changes we will make might upset our mental system to such an extent that it too might break down.[38]

Story and Meaning

Evidence that human beings do often think and act like automatons is not hard to find. However, as noted earlier, what makes us most human is the ability to think about our thinking—to observe our own minds and resist automatic impulses. There would be no culture or civilization if this were not the case. Although we often act according to our most habitual and unexamined ideas, critical decisions in life derive from discernment and rumination. When Harari suggests meditation may make us less vulnerable to the religion of Dataism,[39] he is embracing this very understanding: the better we are at watching our spontaneous and reactive thoughts, the less they will dictate to us. As he puts it, "we had better understand our minds before the algorithms make our minds up for us."[40] Mindfulness is one means by which we see through our thoughts and actions in order to change them. Free will must lie in this capacity. If this were not the case, contemplating the future would be pointless.

Amidst Harari's apparent contradiction between the absence of free will and the power of contemplation, he expresses concern about three elements of psychological life that organize and drive our thinking—emotion, meaning, and story.[41] These elements often collide with pure reason and can become conduits of manipulation. With a mix of rationality and Zen stoicism, he aims to occupy a rarified position beyond the passion and angst that leave us open to manipulation. From this position he sees emotion as more reactionary than formative, meaning as more illusory than foundational, and story as more optional than imperative. His critique of religion combines all three. Here, though, we run into the problem of unrealistic conceptions of human nature.

Whereas Harari could not be more correct in pointing out the distortions of reality that can come from religion's claims of ultimate truth, and his perception of the emergence of secular religions such as Dataism is most insightful, his extension of this observation into a need to do away with stories and symbols that pertain to matters of ultimate importance drives a stake through the heart of human psychology. In his first book, *Sapiens*,[42] he actually goes to lengths to show how stories have allowed human beings to cooperate in large numbers, focus on principles and values, and create civilization. He also seems to be contradicting himself on this theme in addressing "the power to reshape and reengineer life," and stating that *"soon somebody will have to decide how to use this power—based on some implicit or explicit story and the meaning of life."*[43]

The path through these contradictions involves recognizing the nature of the psyche and becoming as conscious as possible about how emotion, meaning, and story *do* shape us. They are psychic realities, awareness of which could mitigate technocratic manipulation. Whereas the crude embrace of emotionally-charged narratives of meaning has generated great destruction, with wars originating in religious differences being a prime example, the more nuanced and sophisticated embrace of these same narratives has generated key values and formed a cushion between our instinctive drives and spiritual needs. Interfaith cooperation recently resulted in drafting the Charter for Compassion, a document designed to draw out the core values of the world's religions. Processes such as meditation—Harari's favorite past-time—which aims to increase mindfulness, happens to come from religion. Emotion, meaning, and story exist along a continuum of mindfulness: they are not clustered at one dysfunctional end of it. As Mlodinow puts it in his book about how emotions shape our thinking, "It is the exception and not the rule when the effect of emotion proves counterproductive."[44] And as Kauffman offers, "Story, including poetry, music, and other arts, is how we know ourselves."[45] Contemplating the difference between humans and computers, Zarkadakis makes the same point:

> Our brains are hard-wired for telling, and for listening to, stories. That is why narratives are the most powerful means available to our species for sharing values and knowledge across time and space . . . dictat[ing] our artistic and scientific endeavors throughout history.[46]

It is thus hard to disentangle our feeling life from our thinking, or separate the need for story and the quest for self-knowledge; we just have to keep improving our capacity to think about our thinking and understand what we are doing.

As it happens, post-industrial society exhibits a desperate hunger for story, which may be a symptom of the need for more self-knowledge. This hunger is evident in those immersing themselves in gaming and virtual worlds, in the binge-watching of television dramas, and even in the conspiracy theories that are the latest junk food of the mind. Those starved of good stories are also apt to retreat into religious fundamentalism and hardened political ideologies, where superficial displays of strength cloak deeper instabilities. When it comes to being tossed about on the ocean of today's competing narratives, simplistic tales of heroic truth-seekers and oppressed victims are like survival suits. Distortion and danger come not from the mere existence of dubious stories, but from the existential emptiness that invites their indiscriminate consumption. Perhaps our propensity to dismiss story-making as part of the human condition is also at fault here: *Homo faber*, the maker, has tried to convince us that *Homo narrans*, the storyteller, is no longer needed. But the story-making propensity, like the religious function, just grabs the rational mind from behind. Emotion, meaning, and story are most problematic when we are unconscious of their psychological necessity. Before handing the keys to *Homo deus*, the human god, *Homo sapiens*, the wise one, surely needs to settle this dispute between *Homo faber* and *Homo narrans*.

Humans think in metaphors and express life's deepest values through symbols and myths. This is the nature of the psyche. Grasping this is the key to deliteralizing religious and ideological narratives and learning to tolerate different pathways to meaning. An embrace of cultural differences and a deeper appreciation of human commonality would follow. We might take hold of ourselves as image and myth makers, as *Homo faber* of soul, focusing on what these expressions connote rather than denote, becoming more conscious of the way reason is surrounded by emotion, meaning, and story. We might then avoid being unconsciously caught from behind by quasi-religious substitutions for ultimate purpose.

What Kurzweil and Harari see in common is a pending radical change to the way we think. The gap between them concerns how malleable our thinking is. Kurzweil wants us to be able to reprogram our thoughts and feelings at will; Harari refers to "deep structures of the human mind," which allow us to relate to "Oedipus, Hamlet and Othello," even if these characters "may wear jeans and T-shirts and have Facebook accounts."[47] But if we are going to keep relating to Oedipus, Hamlet, or Othello, our faculty for emotion, meaning, and story better be finely tuned, not deactivated or rerouted. Our perceptions of the world, understanding of human interactions, and ways of imagining depend on these deep structures, which depth psychology calls archetypes, as they shape and regulate our instinctive responses, generating enduring values and meanings. Our knowledge of their influence will go a long way to knowing our own minds.

Deeper Awareness

Unlike the slowly cooking frog, unable to assess the situation before it is too late, we have the capacity to take the temperature, anticipate a boiling point and opt for other conditions. At this historical juncture, however, we are so enthralled by our devices and dazed by the speed of change we are too much like the slowly cooking frog, unconscious of what is happening to us. We will either have to work toward a better grasp on what is going on or things will need to heat up in order to capture more of our attention. Perhaps both will be necessary.

Kurzweil's focus on the exponential growth of raw computational power leaves little room for human reflexivity, as if the almighty computer chip has already set our course for us. In the posthuman worldview, raw, calculative intelligence has already assumed a greater value than human consciousness or humanistic and spiritual values. All endeavors are secondary to the accumulation of data and the growth of AI: questioning the dominance of the chip is superfluous.

Harari's argument takes in great swathes of history and comes to rest on our own capacity to see where technology may be taking us. But his recipe for consciousness is mindful awareness without awareness of the nature of mind. Watching what our minds are doing and what we are feeding them is important, but grasping how mental habits are seated in the patterns of the psyche and the history of ideas is even more essential for self-understanding, for these are the tools by which we see through our own thoughts and behaviors. This can also occur in a variety of ways and mediums, from artistic expression to philosophical insight; it need not always be overtly psychological. But however expanded consciousness occurs, it is indispensable.

The real fork in the road will not be between different kinds of mental input but between greater and lesser insight into the psyche's subterranean life, especially into the way symbolization, imagination, and ideation shape us. Technologists may be planning to reconstruct and even relocate the mental functioning and intelligence they see above the surface without realizing this is merely an island atop a giant undersea mountain. Consciousness in the digital age depends on how the depth of this psychic substrata is comprehended. Yet few are prepared to look beneath the surface.

At the start of his book *Life 3.0: Being Human in the Age of Artificial Intelligence*, MIT scientist and posthuman theorist, Max Tegmark, underscores the supreme importance of consciousness in the universe. He writes that consciousness "transformed our Universe from a mindless zombie with no self-awareness in a living ecosystem harboring self-reflection, beauty and hope—and the pursuit of goals, meaning and purpose."[48] He adds, "Had our Universe never awoken, then, as far as I'm concerned, it would have been completely pointless."[49] Tegmark is surely right about this, though his main interest in the topic is in ensuring any artificial superintelligence that finally takes over, or robotic shell into which we download our minds, can be programmed for such a conscious existence. Otherwise, as he quotes Schrödinger, existence will be like "a play before empty benches."[50]

However, Tegmark's and many other computer specialist's treatments of consciousness miss the distinction between being merely awake and the qualities of consciousness we find in mature human beings.[51] Crossing the threshold of sentience is not the same as existential or spiritual awareness. For example, there is a universe of difference between being conscious of a group of people moving about and making sounds on a stage and being able to grasp the significance of their words and actions. Unless one comes away from the play with a renewed appreciation for the complexities of life or some piece of wisdom, witnessing the performance or even memorizing every line does not have much value. A sentient machine with access to interpretive notes on every play written is not going to get much from attending such a performance, because it has to be felt as well as observed. Playwrights aim for something beyond literal comprehension; they want to turn on the imagination. An empathic field has to open between the audience and the characters, so that an actor's quivering lip or pained grimace becomes an opening to pathos. A critical question we will all need to answer in the coming decades is whether this humanistic awareness is optional fluff or existentially necessary.

Consciousness is also directable. The direction of consciousness gives us agency, and being agents generates meaning. Consciousness is also built on layers of human responsiveness that begin only with immediate, subjective experience, because such experience is built upon familial contexts, tribal configurations, ethnic differences, down through evolutionary history, which attune the organs of the body and the structure of the mind to respond to the world and each other in particular ways. Human consciousness extends from these layers. Its hallmark quality is the ability to reflect on thoughts and actions in ever widening contexts, and to then consult the record of these reflections, which fill libraries and guide the workings of our most significant institutions. Both individually and collectively, we come to know ourselves by continually weaving together immediate events and the residue of the human experiment that lives on within and is catalogued around us. This process yields something more than knowledge: it generates a fitting attitude to everything we do.

Jung states, "As far as we can discern, the sole purpose of human existence is to kindle a light in the darkness of mere being."[52] For Jung this darkness is the real danger—the propensity to remain unconscious of the psychic depths, unaware of the way these depths direct and thwart the rational mind. "To kindle a light in the darkness" means to be alert to such dynamics, "to become conscious of the contents that press upward from the unconscious."[53] With unprecedented technological powers at hand, the failure to illuminate these forces could easily bring the experiment of life to an end. Two unconscious madmen with their fingers on nuclear buttons and their inferiority complexes ablaze, cheered on by hived minds, might just do that.

Tegmark prefaces his own concern for consciousness by quoting the Future of Life Institute, which was formed to protect existence from the possible dangers of AI: "Technology is giving life the potential to flourish like never before—or to self-destruct."[54] How we apply knowledge—a loose way to define technology—is what will allow life to flourish or self-destruct, and this depends on understanding our motivations. Yet the importance of such understanding seems to elude most scientists concerned about AI and other existence-altering innovations, who are convinced the future will be determined by a series of sober observations and rational assessments of the outer world. This is akin to the crew of the *Titanic* betting its fate on whatever could be observed above the waterline on a still, moonless night.

The kind of consciousness required today involves more than considering what is immediately before us. It means remembering the things we have forgotten and becoming aware of the thoughts and images that captivate us. It involves coming to terms with what Mumford saw a half century ago: "collective obsessions and compulsions that have misdirected our energies, and undermined our capacity to live full and spiritually satisfying lives."[55]

The future will be technological; the question is whether we can live alongside our inventions in a more conscious way. Such consciousness can occur if worldly pursuits remain conversant with both instinctual needs and spiritual aspirations, but it will be thwarted if people become mechanical cogs and passive consumers. When life devolves into functionalism we fall back into "the darkness of mere being" (Jung). Illuminating the basement of our techno-dreams may turn out to be the only way to prevent either a soulless future or the kind of catastrophe even AI enthusiasts admit is possible. To begin, we might attempt to understand the compelling drive to innovate and create, which began as part of our ecology of mind. The ancient Greeks called this drive *technē*; I call it the tinkering instinct.

Notes

1 Robert Romanyshyn, *Technology as Symptom and Dream* (New York: Routledge, 1989), 13.
2 Jaron Lanier, *You Are Not a Gadget* (New York: Alfred A. Knopf, 2010), 22.
3 Jean Baudrillard, *Simulacra and Simulation* (Ann Arbor, MI: University of Michigan Press, 1994).
4 *Her*, written and directed by Spike Jonze, featuring Joaquin Phoenix, Amy Adams, Rooney Mara (2014, Warner Bros.). Film.
5 Emmanuel Stip et al., "Internet Addiction, *Hikikomori* Syndrome, and the Prodomal Phase of Psychosis." *Frontiers in Psychiatry*, 2016, vol. 7, no. 6. https://www.ncbi.nlm.nih.gov/pmc/articles/PMC4776119
6 Alex Mar, "Love in the Time of Robots." *WIRED*, November 2017.
7 David Levy, *Love + Sex with Robots: The Evolution of Human–Robot Relations* (New York: Harper, 2007), 247ff.
8 Sherry Turkle, "TEDcUIUC – Sherry Turkle – Alone Together." YouTube Video, 16:24, March 25, 2011. https://www.youtube.com/watch?v=MtLVCpZIiNs
9 Quoted in Wendell Wallach, *A Dangerous Master: How to Keep Technology from Slipping Beyond Our Control* (New York: Basic Books, 2015), viii.
10 Sherry Turkle, *Alone Together: Why We Expect More from Technology and Less from Each Other* (New York: Basic Books, 2011).

11 Ibid., 154.
12 Scott LaFee, "Serious Loneliness Spans the Adult Lifespan but There is a Silver Lining." *UC San Diego Today*, December 18, 2018. https://ucsdnews.ucsd.edu/pressrelease/serious_loneliness_spans_the_adult_lifespan_but_there_is_a_silver_lining
13 Jaron Lanier in Ellen Ullman, *Close to the Machine: Technophilia and Its Discontents* (New York: Picador, 2012), xii.
14 For more on the techno-addictive cycle see Chellis Glendinning, "Technology, Trauma and the Wild." In Theodore Roszak ed., *Ecopsychology* (Berkeley, CA: Sierra Club Books, 1995).
15 Lewis Mumford, *Interpretations and Forecasts: 1922–1972* (New York: Harcourt, Brace, Jovanovich, 1973), 475.
16 Rachel Carson, *Silent Spring* (New York: Houghton Mifflin, 1962).
17 Shoshana Zuboff, *Surveillance Capitalism: The Fight for the Human Future at the New Frontier of Power* (New York: Public Affairs, 2019), 11–12.
18 Jay Yarow, "Sergey Brin: 'We Want Google to be the Third Half of Your Brain.'" *Insider*, September 8, 2010. https://www.businessinsider.com/sergey-brin-we-want-google-to-be-the-third-half-of-your-brain-2010-9
19 See Zuboff, 197ff.
20 Evgeny Morozov, *To Save Everything, Click Here: The Folly of Technological Solutionism* (New York: Public Affairs, 2013).
21 *Star Wars: A New Hope*, written and directed by George Lucas, featuring Mark Hamill, Carrie Fisher, Harrison Ford (20th Century Fox, 1977). Film.
22 *Westworld*, created by Jonathan Nolan and Lisa Joy. Based on the film by Michael Crichton. Performed by Evan Rachel Wood, Thandiwe Newton, Jeffrey Wright (2016-2022,HBO). Television.
23 Jonathan Haidt, "A Guilty Verdict." *Nature*, vol. 578, February 2020, 226–227.
24 Wallach, *Dangerous*, 190.
25 See Ray Kurzweil, *The Singularity Is Near: When Humans Transcend Biology* (New York: Viking, 2005), 391ff.
26 Ray Kurzweil, *The Age of Spiritual Machines* (New York: Viking, 1999).
27 Max Tegmark, *Life 3.0: Being Human in the Age of Artificial Intelligence* (New York: Alfred A. Knopf, 2017), 33.
28 Kurzweil, *Spiritual Machines*, 44.
29 David F. Noble, *The Religion of Technology: The Divinity of Man and the Spirit of Invention* (New York: Penguin Books, 1997), 159.
30 Kurzweil, *Spiritual Machines*, 41.
31 Ibid. 150.
32 Yuval Noah Harari, *Homo Deus: A History of Tomorrow* (New York: HarperCollins, 2017), 402.
33 Ibid.
34 Ibid., 386.
35 Ibid., 386–387.
36 Ibid., 401.
37 Yuval Noah Harari, *21 Lessons for the 21st Century* (New York: Spiegel & Grau, 2018).
38 Ibid., 7.
39 Ibid., 314ff.

40 Ibid., 323.
41 Ibid., 273ff.
42 Yuval Noah Harari, *Sapiens* (New York: Harper Perennial, 2014).
43 Ibid., xvii. Italics added.
44 Leonard Mlodinow, *Emotional: How Feelings Shape Our Thinking* (New York: Pantheon Books, 2022), 7.
45 Stuart A. Kauffman, *Reinventing the Sacred* (New York: Basic Books, 2008), 249.
46 George Zarkadakis, *In Our Own Image* (New York: Pegasus Books, 2015), 305.
47 Harari, *Homo Deus*, 46.
48 Tegmark, *Life*, 22.
49 Ibid.
50 Ibid., 282.
51 I will discuss this matter at length in Part 4.
52 C. G. Jung, *Memories, Dreams, Reflections* (New York: Pantheon Books, 1961), 326.
53 Ibid.
54 Tegmark, *Life*, 22.
55 Lewis Mumford, *The Myth of the Machine: The Pentagon of Power* (New York: Harcourt, Brace, Jovanovich, 1970), opening pages.

Chapter 4
TINKERING

> It is the psychology and not the epistemology of science that urgently requires our critical attention; for it is primarily at this level that the most consequential deficiencies and imbalances of the technocracy are revealed.
>
> Theodore Roszak[1]

Technology's mixed record is not hard to see. Even prominent advocates admit outcomes can be both creative and destructive. Factories are efficient and productive, yet also exploit workers and create toxic waste; pesticides protect crops and feed the masses, but also kill wildlife and cause cancer; plastics have been a part of almost every useful device and package for over half a century, but end up in the ocean and contaminate the food chain; rockets have taken us to the moon as well as to the brink of global annihilation; smart phones provide instant access to almost anyone, anywhere, but also create social barriers and foster new forms of tribalism. We might have reached a point at which technology is inherent to survival, but it has also begun to fragment the social and cultural foundations of existence.

Yet even in the face of this mixed record and ominous outlook, techno-optimism dominates our thinking. Technology's destructive side is not denied outright, but is treated as anomalous rather than normative. As a result, few human propensities generate more faith, take more time, and demand more resources than the urge to innovate and invent.

Neither the social upheaval of industrialization, nor the failed promise of time-saving devices, nor the aftermath of atomic bombs and nuclear meltdowns, nor the dawn of cyberwarfare have dissuaded this overall belief in our own ingenuity. Innovation and progress have become fused, at least in the minds of those with most influence, while the downside is left to the ecological activists, cultural critics, and sci-fi storytellers. Rarely is this downside factored into large-scale plans for the future. Why? What makes technology so unimpeachable?

To comprehend the nexus of technological advancement and the planned obsolescence of our species proposed by posthumanists, and to grasp why this advancement has assumed an ultimate value, this disconnect between technology's equivocal record and our idealistic embrace of it is a good place to start.

A number of starkly negative technological outcomes can certainly be attributed to unforeseen consequences, some of which belong to the nature of innovation itself. Others are attributable to human nature. As Giddens puts it, "No matter how well a system is designed and no matter how efficient its operators, the consequences of its introduction and functioning, in the contexts of other systems and of human activity in general, cannot be wholly predicted."[2] Some unforeseen consequences result from miscalculation and flawed design; some are due to the failure to grasp our potential for malevolence and the way greed and corruption find ever new paths of exploitation. Yet, whether it is the complexity of systems and general activity, pure error or misbehavior, we consistently underestimate the way the human element complicates the technological plan. Technology is developed and often imagined to operate in a bubble, apart from the actual conditions of being—a way of thinking to which scientists and engineers are particularly prone.

Unforeseen consequences might thus be less so if the flaws of human nature were paid more attention. By looking at our track record, we would be more inclined to admit the pros and cons of every major innovation, realizing each new thing will more than likely have a downside. We know that industrialization upended bucolic life, that automation created a manic pace, and that digital technologies have injected the agendas of others into our every waking moment. We know the imagined leisure culture and shorter working week that accompanied each phase of technological change have not materialized. Now that we are looking to robots to give us more free time we will no doubt spend it online. Whereas the cons may take a while to surface, we have surely seen enough to know they will. Or have we?

When things go awry, rather than looking deeper into the human factor, our default response is to go back to the drawing board, hoping the next new thing will do the trick. As a culture, in accord with the above, we are yet to problematize the compulsive pursuit of further technological solutions in a vacuum of wider considerations—sociological, ecological, cultural, and, especially, psychological. We resist seeing our own nature. Sober accounts of negative consequences, even served with lashings of apocalyptic expectation, fail to curb our enthusiasm. Innovative marvels are promoted in every magazine ad and pervade every corporate mission statement. The arrival of each new device conveys an implicit faith in a better and brighter future.

This idealism appears entrenched in the modern psyche. Industrial and post-industrial ways of life are so imbedded in social structures and economic prospects that focusing on their negative outcomes has become akin to questioning the fundaments of civilization. The train has long departed the station and getting off seems an unlikely option. The mere suggestion of pausing to take a good look around arouses the accusation of being a Luddite, as if taking a critical stance is some kind of betrayal of the arc of human purpose.

When I discuss the themes of this book with others, there is a repeated refrain to my concern about where our inventions may be taking us: "Change is inevitable; we can't do much about it; we'll eventually find ways to adapt." For at least five hundred years, since the end of the Middle Ages, adapted we have—even in the face of fear and challenges to prevailing understandings of life. The telescope, the printing press, the steam engine, and television all occasioned dire predictions, which were soon replaced by almost universal acceptance. Mistakes might have occurred, accidents might have been inevitable, but for better or worse our need to reshape what is merely given is accepted as part of life.

It is hard to argue with this outlook. Technological change of some sort seems impossible to avoid. Yet there is an even deeper reason we are reluctant to face the negative implications of such change: we carry the conviction that innovation largely defines us; what we build is a direct expression of who and what we are; our intelligence, creativity, and capacity to join with each other in large-scale ventures is intrinsic to being human. Our handiwork thus seems rooted in something that goes beyond the mere application of intelligence and knowledge: tinkering and inventing are instinctive. The stuff of technology is just the end product of thought and behavior that belong to the evolutionary process. Beyond appearances, the problem of technology turns out to go beyond external collisions with what we invent. The problem of technology is about how we relate to ourselves; it is, specifically, about how we relate to our tinkering instinct.

Acknowledging the presence of an instinctual drive, a primal pattern of the psyche rooted in biological adaptation, can cut two ways. It can be used to rationalize untethered innovation and assert along with many zealous tinkerers that redesigning nature is simply the latest phase of evolution, which has granted us the intelligence to do this. However, the instinct to tinker also suggests the need for a more conscious approach to what we create, so we critically examine how this instinct is enacted. As this instinct appears to have free rein these days, this examination becomes all the more crucial.

Because we consist of more than this one impulse, and as invention is just one form of creation, we must consider how the tinkering instinct relates to other instinctive urges. Not only is this part of becoming more conscious of the deeply human; it also pertains to the need for holistic awareness. We must also recognize the record of human achievement in general has more to do with restraining and redirecting our deeper impulses than blindly capitulating to them. Civilization is a shrine to sublimation. Societies are possible only by means of delayed gratification; cultures develop from reflexivity and symbolization. To become more conscious of the way technology dominates our relationship with reality is to make room for other ways of knowing the world. To become more conscious of our propensity to be innovative is to distinguish this propensity from its blind enactment. Such consciousness will allow us to critically assess the tendency to go along with every innovation, so that *what* we build assumes more value than *that* we build. A viable, fruitful relationship with technology thus implies a more conscious regulation of the tinkering instinct, which necessitates a deeper understanding of the human condition. This will give us a chance of becoming techno-realists rather than techno-idealists.

In a nutshell, if inventiveness is akin to other instinctive drives, it can either be channeled and transformed in ways that ennoble existence, or it can possess us to the point of overwhelm. Like any other instinct, from sex to aggression, competition to cooperation, how the tinkering instinct is directed and how it is situated among the array of other human needs and impulses becomes open to discernment and ethical critique. If we fail to do this we will turn into automatons: we will just follow a biological imperative straight into a mechanistic trap.

How we participate in or passively allow technocratic manipulation and exploitation starts to matter. How we maintain a sense of wellbeing when computers and cyberspace are colonizing many aspects of life begins to take on profound significance. Consciousness of the techno-push within equally highlights the ethical responsibility of technologists to be mindful of their motivations and the syndromes of technological overreach. Once again, we return to the question of culture, for it is the forms of culture—from art to literature to street life to communal stories and rituals—that help us reflect, think and create according to critical values.

Cultural forms have always shaped innate imperatives. Verse and image enflesh the utilitarian bones of civilization. As Hillman puts it, "The city belongs to the saxophone."[3] Optimally, science and technology would proceed in a feedback-loop of cultural exchange and engaged imagination. That loop is always occurring in some quarter. If we examine the modern era, we see that for every Faraday there is a Shelley, for every Tesla a Wells, for every Turing an Asimov, and for every Gates a Lucas. Apple Computer would not be what it is today without Steve Jobs's famous calligraphy class, which prompted the integration of aesthetics into prosaic computing. Writers, filmmakers, musicians, and artists digest what grows in the labs of the innovators and metabolize the changes for the rest of us. Their rumination and storying are primary, reaching into cultural heritage and ancestral wisdom, drawing deeply on facets of existence scientists and technologists tend to disregard. Whether this process can be more widely embraced and play a primary role in the education of our children, or whether it becomes a stream of inconsequential offerings cast before a technocratic juggernaut remains to be seen.

For anything deeply human to endure, however, technological development is going to need to maintain contact with the cultural imagination. If technology is untethered from this dimension of life and the reflexivity it generates, a split in the human condition will be inevitable. One part of us will get so far ahead of the rest of us we will start to play by a set of entirely abstract and inhuman rules. Our creations—including what we create of ourselves—will then enslave our humanness. Between the boardrooms and basements of each tech corporation we need to locate those who understand the arts and humanities—those who would make the novels of Philip K. Dick and the artwork of H. R. Giger necessary reading and viewing for those working on human–machine hybridization. Otherwise, without the moderating influence of robust cultural discourse, technologies will at best continue to serve a very narrow set of human values or at worst keep finding ways to make us serve them.

Conventional wisdom often holds that technology just provides tools we can choose to use for good or for ill, making the tools themselves neutral. But this view is no longer valid, because post-industrial technologies effectively define our environment and have even begun to define us. There is no longer any picking up and putting down tools because the colonizing and totalizing effects of technology displace other ways of being in the world—effects we are, so far, failing to confront. If we end up at war with our machines or our minds are overtaken by some matrix in cyberspace it will be because we first lost a battle already taking place within us; because we failed to develop enough self-understanding and regard for other aspects of culture to restrain technological overreach; because we let one propensity within us dominate all others.

Beneath the Surface

Technology is essentially our propensity to order and alter things. The prominence of this propensity emerges from a place inside us where the tectonic plates of the deeper psyche form templates of being, where biological drives become psychological imperatives—to survive, develop, and cooperate with one another. These imperatives have set the course of civilization, which is synonymous with the redirection of natural life—building places to live and work, and creating means to sustain and connect these places through transportation, communication, food production, water management, transportation, and communication.

W. Brian Arthur, a leading economist and technologist, who aims to grasp the "technology-ness of technology"[4] and find "an '-ology' of technology," frames this search in psychological terms. He writes that "new technologies are constructed mentally before they are constructed physically."[5] In the quest for a definition, Arthur observes that technology stems from the "constant capturing and harnessing of natural phenomena."[6] This is the process that technology aims to master. But what exactly drives us to *capture and harness* nature as opposed to simply dwell in or work with nature? And what has taken us from "capturing and harnessing" to exploiting and manipulating?

The Greek term *technē* referred to the propensity to craft things. More precisely, technē was considered a way of looking at the world that invited crafting. It is the ability to imagine a different way of engaging the elements of our surroundings. Techne is thus the proverbial lightbulb of possibility, focused on the mechanical reordering of natural forms. It is the innate tendency to ask and answer the question "What If?" The term "invention" comes from a root meaning "to discover," and we use it for both the inventive process and the result. And it is somewhere between discovery and invention that we find techne.

Although discerning essential human impulses can be somewhat speculative, we can also infer such things from observable patterns of thought and behavior. Freud developed his psychology around the sexual instinct in just this way, demonstrating how this is not just directly enacted, but also indirectly satisfied through various kinds of relationships and creative actions. Adler developed a similar psychological understanding of the urge to power. There is also, obviously, an instinct for self-preservation, exhibited in behaviors such as finding food and shelter, banding with others, and so on—an instinct we see at work in all forms of life. Along similar lines, somewhere along the evolutionary path, we developed the instinct to tinker—not merely to adapt to the world but intentionally to transform that world by applying creative thought. We began by making tools. The lineage from stone axes to silicon chips is the history of technology.

Rudimentary expressions of the same impulse are present in other animals too: monkeys use stones to break open hard nuts and sticks to extract ants; dolphins employ marine sponges to sift through sand for their food; octopuses utilize various shells to construct portable hideouts. It has been argued that our ability to manipulate objects was a significant factor in the evolution of tool-making. We literally became handier. The bipedal gait allowed for the development of a different kind of hand, with opposable thumbs, which in turn allowed for a greater ability to manipulate objects.[7] In this process the hand became what neuroscientist Raymond Tallis calls "a stunningly versatile organ for interacting with the world."[8] In describing the significance of this development, he says, "To express it a little frivolously, the thumb enabled us to hitch a ride on the laws of nature to destinations that nature had not prefigured."[9] Over vast stretches of time, this ability to hold and manipulate was likely a critical factor in becoming more sentient, accelerating knowledge of our separateness from other beings and other things: holding and contemplating the things of the world, we have become more aware of those things and our ability to change them, and thus more aware of ourselves as distinct entities and agents of change.

In this sense *Homo sapiens* is also *Homo faber* and the Latin phrase *homo faber suae quisque fortunae* (every man is the maker of his destiny) conveys something essential about our self-understanding: our lives are not just given; they are made, and the things we make also define what we are. Surely then, to keep on making, to keep changing the world and progress to changing our bodies and minds according to this inventive drive is to follow this deep existential imperative. Surely this gives backing to the argument we are "natural-born cyborgs."[10]

If we were to leave the matter here, we might be excused for seeing it just this way. However, there is much more to who and what we are, which has a great deal to do with how we manage and direct a number of innate drives. The main imperative behind the tool-making propensity is evident in the use of tools by other species: tools allow the satisfaction of instinctive needs. Among humans, hunting implements are among the most archaic of these tools, and the controlled use of fire for cooking made many foods more edible. Because of its transformative power, fire remains an enduring, universal pivot on which most technologies are realized, and ways of starting and managing fire are thus highly symbolic of our inclination to innovate. Without fire, for example, there can be no casting of metals. To this day, making a fire is accompanied by a primal sense of achievement.

Engaging with the world rather than passively adapting to it expands awareness and intelligence. When it comes to the ancient roots of technology, Mumford thus places gathering ahead of hunting. "In collecting food," he writes, "man was also incited to collect information. The two pursuits went together."[11] He goes so far as to argue, "*Homo faber*, Man the Tool-Maker, is a late arrival," and manipulation of the world grew out of a subtler field of awareness. "Intimate contact with and appreciation of the environment, as Adolf Portmann has shown, brings quite different rewards than intelligent manipulation—but equally real ones. Pattern identification, as a necessary part of environmental exploration, stimulated man's active intelligence."[12] Techne is thus not just the result of intelligence; it also fosters that intelligence; the instinct to invent and experiment is bound up with the drive to be curious and comprehend, furthering our capacity to meet a number of instinctive needs. Exploring and tinkering extend our grasp on the world and that grasp reveals more about the world: a carpenter comes to know the nature of wood, a weaver the character of wool, a fisherman the mood of the sea. Techne thus belongs to being. Only one face of it is concerned with mastery; the other has to do with an expanded knowledge of earthly being.

Techne, as the deep root of technology, thus represents a very important bond between knowing and doing. This understanding becomes a critical platform from which to view modern technology, for in modern technology the tinkering instinct has detached itself from the wider network of human needs: it has become largely self-serving. In modern technology, manipulation transcends awareness and thereby distorts its own archetypal background. The significance of this is hard to overestimate. By hitching our wagon to this narrower form of tinkering, we are not only doing more by knowing less; we are leaving behind a trail of partially satisfied and completely dissatisfied psychological needs.

Most of us seem better fed and sheltered, satisfying our self-preservation; the online world appears to facilitate myriad routes to relationship and sexual satisfaction; medical advances have been a boon for extending life. If we look into each of these arenas a little deeper, however, we come to see that such contemporary satisfactions are frequently compromised or fleeting: the food we eat is not always nutritious and wholesome; many of us live in spaces constantly invaded by light, noise, and chemicals; despite apps for meeting up and hooking up, sex is in decline, both in quantity and quality, to the point many high-tech societies have declining populations; and access to drugs does not always equate to having good health. Modern gadgetry convinces us even more satisfying solutions are around the corner, that we can "just do it," yet gadgets are also making us less aware of our deeper needs. We thus embrace the technical fix even before any underlying dissatisfaction is recognized. As discussed in the last chapter, our techno-lives are filled with bait and switch, wherein deeper needs appear to be met but actually are not, creating a treadmill of technological solutionism.

Rather than guided by a fuller sense of our needs, technology is guided by the promise of better, more efficient, forms of itself. We see this in the shift from making tools that actually produce things or help us adapt to the environment, to making tools for the sake of making more tools. This occurs, in part, because our habitat is now defined by our tools. In the digital era this is apparent in the endless array of apps on our mobile devices, functioning as means to access and navigate other layers of technology. Adaptation now means being adapted to this technologized environment. It means having the latest hardware and software. Dexterity in the online world and in using the machines that support it has become the main prerequisite for worldly success. It is not hard to see how these conditions are apt to make us maladapted when it comes to being fully and deeply human.

As I have described it so far, the tinkering instinct has two rather distinct modes: first the more original form of techne, which evidently maintains contact with our psychic totality; and modern technology, which encourages the development of some parts of us while neglecting others. To grease its own wheels, modern technology disables some parts of us in order to privilege others. It generates a feeling of not being all there.

This technological anesthetization has long been recognized. Factory-based, production-line work dulls if not eradicates an individual worker's crafting capacity and thus also a qualitative understanding of materials and their origins. The factory worker makes things but does not craft them. Screen time dulls our capacity to engage with and know our surroundings and each other; as the screen brightens, the surroundings dim; to connect, we must disconnect. Without being rooted in techne, the tinkering instinct produces technologies that compartmentalize and dissociate rather than sustain the integrity of being. When all our eggs go into this basket, we start to abandon large parts of our surroundings and large parts of ourselves.

As wider human interests are displaced, we come face to face with "the dangerous tendency of technology to create its own momentum."[13] In the digital age, it now seems the computer and data alone become the sole means of navigating life. As the most valued currency, abstract information is displacing experiential knowledge. As AI and its algorithms take computing to even higher levels of utility and efficiency, it pushes other technologies and other modes of adapting and thinking further into the shadows. For example, the attention economy does not care about how attention is captured, just that it is. One result is that information, misinformation, and disinformation are given an even playing field, eroding the capacity to understand what is going on in the world. Eyeing the 2020–21 pandemic, which coincided with the rise of populist leaders and conspiracy theories, Stuart Whatley notes, "the frictionless flow of 'ideas' has produced a glut of disinformation and propaganda without improving public understanding of current events."[14] It is not just the presence of misinformation or even disinformation that undermines understanding; the glut of information alone does this.

This brings us to a twist in the story, which will make for either a grim ending or a new chapter. When it comes to human nature—which is, for the moment, still running the show—what is neglected has a way of coming back to bite. When technology promises the world and delivers only a narrow version of that world, what is excluded returns to disrupt and distort our conscious intentions. This may turn out to be the central problem of advanced technology, one it will never solve in isolation, one we find reflected in almost every work of science fiction: left to its own ends, the computer world inevitably divides us against ourselves and one another.

Missing the Sacred

The urge to craft or invent and reorder our surroundings may partly define us, but it cannot ultimately be isolated from other traits. When archeologists study past civilizations, they dig up the handiwork of our ancestors and discern their way of life from the things they crafted.

Making tools and creating artifacts lies at the heart of civilization itself. Yet the ancients did not craft things for utility only; they also crafted in response to their religious vision. Invention for civilizations past required attention to be directed toward the gods; psychic energy was not available for purely practical or abstract purposes.

The great pyramids could not have been built had they not reflected religious and aesthetic desires; Gothic cathedrals were constructed from the same drive, with workers devoting themselves to projects they would never see completed in their lifetimes. Such engineering and mathematical marvels grew out of a cultural imagination where mythos and logos were rarely divided and the satisfaction of invention was tied to notions of the otherworldly. Overlooking or minimizing this aspect of techne and its artifacts reflects a kind of presentism: we look back wearing the glasses of pragmatic civilization-building and scientific abstraction, which are poor at recognizing the cultural vessels that incubated these older undertakings.

Building things in response to the gods relates to the enduring, universal phenomenon of creative thought originating beyond the reaches of the conscious mind. Whereas focus and learning are required for problem solving, most studies of creative thinking and scientific breakthrough underscore the essential role of unconscious processes: many pivotal insights and ideas appear to come to us, often in states of reverie or when intentional focus on the problem has ceased, sometimes from dream states or having "slept on it." The term "inspiration" points to divine influence, and we freely refer to the presence or absence of a muse when we imagine into the source of creative thought. If we stay with the phenomenal experience of such things, we are forced to admit the ideas and images that lead to innovation frequently emerge from the unknown.

In his book on creativity and unconscious processes, Arthur Koestler suggests that many modern scientists, contrary to their image as "sober ice-cold logicians, electronic brains mounted on dry sticks," often come across in their self-reflections as being "a bunch of poets or musicians of a rather romantically naive kind." One finds in their "intimate writings" themes such as "the belittling of logic and deductive reasoning ... horror of the one-track mind; distrust of too much consistency, as well as scepticism regarding all-too-conscious thinking." Koestler then notes that "sceptical reserve is compensated by trust in intuition and unconscious guidance by quasi-religious or by aesthetic sensibilities."[15] If this compensation is consciously hosted, science seems at its best. But, as I will explore below, it is often a very unconscious phenomenon.

An aesthetic sense in particular goes beyond everyday elegance and taps the archetypal idea of an ordered cosmos. This echoes Mumford's point about pattern recognition and the importance of Adolf Portmann's contribution to this, for Portmann showed in his studies of animal life that display and appearance had to be understood as having intrinsic value and meaning not reduceable to functions such as such as mating or camouflage. The point is that elegance, order, and beauty, which are regularly associated with divinity, may not be secondary but primary in any comprehensive understanding of the nature of existence.

When Einstein insisted "God does not play dice," he was employing a religious image to express a cosmological vision. When Bohr replied, "Einstein, stop telling God what to do," he was formulating his own cosmological view in a similar way. As the philosopher of science Mary Midgley puts it, those who refer to this exchange between two luminaries of physics "seldom offer a carefully secular paraphrase . . . nor do they explain why this language struck these great men as so well fitted for their purpose."[16] Considering the broader intersection of religious images and scientific ideas, and contemplating the tenacity of values such as order and purpose in human thought, Midgley points out that "*pattern* turns out (rather surprisingly) to be the same word as *patron*, meaning source or authority." And further, she reminds us "the Greek word *cosmos* (akin to cosmetic) simply meant *arrangement* or *adornment*," and that "Plato's idea that God was the Great Geometer"[17] was revived in the seventeenth century, a critical period in the development of scientific perspective. Indeed, the history of scientific ideas reveals a conspicuous weave of religious views contrary to our contemporary division of religion and science. Midgley's point is that the terms of divinity and cosmology "are not optional, disposable metaphors. They cannot be replaced at will by 'literal' and 'objective' language."[18]

In the myth of Prometheus, Zeus keeps a close watch on the way this maverick Titan-giant conveys technical know-how to mortals. The Olympian chief fears we mortals are becoming overly confident and forgetful of the gods. So, he instructs Prometheus to invent a sacrificial ritual, to remind us of the sacred origins of knowledge. It is when

Prometheus deceives Zeus in this ritual that Zeus withholds fire—the great element of transformation. That the theft of fire occurs following this failed sacrifice and divine withholding is a motif that speaks directly to the syndrome of our Promethean inclination: the punishment that ensues signifies the cost of denying the divine origins of creative thought. The failure of the ritual Zeus requested, intended to remind us of divine principles, shows the relationship between an embrace of godlike power and a failed remembering of the gods. Put conversely, to remember the gods—to allow for the mystery and depth of our creativity—inoculates us against the hubris and inflation of our own ingenuity.

This mythic account reflects the perennial temptation to separate knowing and crafting from the religious impulse, which we may broadly define as the propensity to search for meaning and spiritual orientation. The religious impulse is, as described in Chapter 2, another instinctive factor at work behind many layers of ideas, philosophies, and traditional practices that have long been at the core of human culture.

Accordingly, ancient artifacts point in two directions, only one of which is utilitarian and civilization-serving; the other is religious or mythic, part of the cultural imagination and symbolic life. Up to the late European Middle Ages, implements were adorned with images of divinities or spirit animals, reminding us modern, enlightened folk that our ancestors did not do much crafting and adaptation to the world without thought for this other dimension. Most critically, in this traditional frame, an imposition of human will on nature required ritual permission from spiritual ancestors or gods. The absence of such a gesture defined hubris and generated a fear of retribution.

Whatever else it did, religion, in its earlier root forms, promoted heightened awareness, caution, and restraint. The essence of the religious impulse is to maintain contact via metaphor, symbol, and ritual with forces beyond the will, including instinctive forces the psyche has always externalized and personified as gods. While vexing for his critics, this is why Jung took religion seriously and placed the spiritual question at the heart of human psychology, for the gods and the mythic stories surrounding them point to determinative psychic realities. He understood that an egocentric outlook devoid of some sense of the sacred or reverence for ancestral wisdom was an invitation for psychic disturbance, because these traditional cultural conduits regulated the psyche and are not so easily replaced. Pretending we can dispense with spirituality is part of modern psychological arrogance; it initiates a split in our nature and invites the innate need for meaning and spiritual ideas to return in symptomatic and destructive ways.

When we drain our thought and behavior of overt religious associations, we enact religious ideas in covert ways—in rituals and zealotry of a distinctly modern and often materialistic kind. This is what prompted Jung to declare, "When the god is not acknowledged, egomania develops, and out of this mania comes sickness."[19] We have already seen examples of this dynamic in the recessed spiritual questing of posthumanism and the secular religion of Dataism; although the sickness involved is yet to fully dawn on us.

In the polytheistic religious imagination, honoring the religious impulse means recognizing all the gods, reconciling oneself to an array of powers and their claims. Here we must remember that although the Western religious imagination is apt to see divinity as something distinct from earthly being, this was never the case for the ancients or indigenous peoples, who did not and do not divide the spiritual and earthly realms so decisively. For most peoples at most times the religious world brought spirit and nature together, regulating our relations with both. Attending to the gods was concerned with an optimal state of being in this world, not a preparation for the afterlife. The goal? To live in accord with patterns of existence that lie beyond willful manipulation, whether coming from within or without. Techne was originally never divorced from this broad sensibility—crafting and sensitivity to the sacred went hand in hand.

Although we may not be willing or able to return to a world so thoroughly permeated with religious motivation, we cannot pretend these modes of thought can be completely discarded, for they reflect ingrained channels of psychic life and help us negotiate the forces competing within that life. As Joseph Campbell liked to say, when you drill down into the roots of mythology, you eventually come to "metaphorical images of the organs of the body in conflict with each other."[20] And, so far at least, we have not left this body behind. We are still compelled to ultimately align our actions with some sense of deeper purpose and ultimate meaning, and are still in need of a vision that situates the crafting impulse within a wide array of human imperatives and cultural patterns.

All these considerations suggest that techne is not the tinkering instinct in its raw state but in its cultured archetypal form, positioned in the overall configuration of the psyche. It is as if the psyche itself were saying to us: tinker away, but keep one eye on the other forces impinging on this earthly, human existence. All of which poses a critical question: How do we reacquaint today's technology with techne? In other words, how do we keep the urge to innovate tied to a more complete understanding of our total being? To eventually answer this question, we need a better grasp of what the separation of technology and techne actually means, and a better understanding of what is likely to result if this separation continues.

Mad Scientist Syndrome

When discussing the themes of this book with a friend, wondering aloud what it might take to redirect some of the technological trends afoot, he expressed skepticism. Even in the face of great danger, possibly even widespread destruction, he argued, we will continue to invent and create, simply because we can. This struck me as a deep and unsettling truth. In the same way the instincts for reproduction and survival have dominated different stages of evolution, the tinkering instinct dominates in our time. Although our tinkering may be directed by reason, it is propelled by a thoroughly non-rational force that may even supersede self-preservation.

This non-rational urge to manipulate whatever we get our hands on has been evident as we have faced problems such as pollution and climate change. Although couched as existential threats, convincing data and scientific consensus are met with resistance. The same essential dynamic now appears in relation to the dangers of the digital age. Eyeing posthuman trends, Bill McKibben wrote a book he called *Enough*,[21] hoping that sober and reasoned fact-facing may help us push the pause button. But it has become clear that attempts to declare "enough" will never be enough. Weighing up all the factors involved, setting out the statistics, costs, and informed predictions cannot compete with the drive to alter nature any way we can. The tinkering instinct seems to resist sublimation.

Detached from the religious imagination, unbeholden to the patterns of nature, and now an intrinsic feature of our whole environment, technology seems to have developed a life of its own, with very little to guide, limit, or orient it to larger concerns. This autonomy is compounded by the speed of development. Now we have to innovate just to stay afloat, with no time or inclination to consider the way our very actions keep raising the water level. Such conditions pervade the tech industry in general, but they are particularly pronounced in the military sector, where a significant slice of AI research and development is taking place under the rationale "if we don't do it, the bad guys will." A military-post-industrial complex is now upon us, with the same outsized grip on society as its predecessor, gearing politics and economics to the cyber-wars and robot battles to come.

Most of our time and energy as a species are now spent manically layering and patching our techno-skin for survival in a fully fabricated world. Because it is the character of the world, maintaining this adaptive layer may be rewarded with the micro-satisfaction of each successful adjustment to the prevailing conditions. Yet this same tech savviness is not bringing a sense of being deeply related either to ourselves or others. The wrench in the works of this dysfunctional acquiescence to the technocracy is our own deep nature, and so far there has been no easy fix for this glitch; when other instincts are asked to take a backseat to prevailing social conditions, they do not just go away: the need to feel a more direct connection to the earth endures. So, the patches we keep putting over the ever-stressed seams of our techno-skin do not last, and when this skin ruptures, we come face to face with an expanse of ignored reality.

Along such lines, all of us are both personally and collectively subject to what I call mad scientist syndrome. Hellbent on creating one thing or another, promising to unlock untold power and reveal great secrets, the mad scientist becomes alienated from the actualities of life. As we know from multiple renderings in literature and film, such scientists suffer from three distinct yet connected problems. First, they are blind to their own psychology, often evident in their spectacular neglect of physical and emotional life. Second, they are socially remote or even outcast, often showing little concern for the wider implications of their inventions. Third, what they neglect or repress returns in distorted, symptomatic, and sometimes monstrous ways.

Frankenstein and his creature provide the best-known rendering of these dynamics, and we must note the often-overlooked subtitle Shelley chose for her classic work: "Or, The Modern Prometheus." Writing as she did at the start of the Industrial Revolution, it is as if she anticipated the pattern or syndrome that would eventually confront us. The creature of her story personifies what is forgotten in the intoxication of innovation and the way the repressed always returns when one-sided techno-fantasy grips us. Though we typically identify the creature as the story's monster, Victor Frankenstein's unrestrained vision is the real monstrosity, seeking to fabricate something far beyond what is organically ordered or tethered to humanity.

If we fast-forward from this watershed tale to the classic era of science-fiction film, beginning in the late 1940s, we encounter a motif that speaks to the same dynamic. The scientist, blind to his inner motives and disconnected from the world, unleashes some radioactive or chemical power, or makes some miscalculation that eventually produces a monstrosity. A slew of these stories involves the accidental creation of giant bugs, or in the case of *The Fly*—a particularly haunting example of the genre—a human–insect hybrid. The commentary on the unflagging effort to kill off the insect population and exert control of outer nature is easy to see, but the symbolically rendered expression of the lower life form taking its revenge is more primary to the theme. The scientist embraces a vainglorious quest, often rooted in a megalomaniac, revenge-of-the-nerd fantasy. As they push past the warning signs and palpable unease, they repudiate their own instinct, only to be confronted with an outsized manifestation of this reckless oversight.

The hubris of these monster-making scientists lies not in the urge to create per se, but in the untethering of that urge from the realm of greater concern. Under such conditions, the soul, passed over as an inner reality, is often projected onto some kind of imagined breakthrough or elixir of life. Such a scientist may even attempt to reanimate the soul in the form of a robotic companion, but is always blind to what is really driving his vision. The character of Rotwang in Fritz Lang's classic silent film Metropolis is a fitting expression of this motif, especially against the backdrop of a future that has engineered a literal split between the upper world and the underworld.

The mad scientist's instinct to tinker has no mitigating awareness or sense of restraint because it has lost its relation to the wider realm of existence. With little sense of the sacred, the creator unconsciously identifies with the Creator. The urge for ultimate control inverts to being out of control and, in the heady hubris of the situation, the scientist's creation destroys him, either figuratively or literally, often because it is abandoned, unloved, or misloved. The meta-dynamic is clear: unaccompanied by an adequate grasp and circumspect sense of the human condition, the tinkering instinct finally goes rogue and turns on us.

Writ large, our futurism and technological solutionism, disconnected from deeper needs and disregarding natural forms, results in a one-sided quest to control rather than cooperate, applying knowledge as sheer power-play. This unleashes compensatory demands from both the inner and outer nature, which takes on a monstrous bearing that subverts our control. In the everyday personal version of this same dynamic, we see dismissed instinctive life returning in the form of ravenous emotional needs and omnipresent addictions. The Buddhists call such unhinged appetites "hungry ghosts."

On the planetary scale a neglected natural balance is producing invasive species, mass extinctions, and weather extremes. As the popular imagination holds a mirror up to culture, Godzilla looms large, and invading bug-like aliens from outer space, using us as hosts, become the most recent monsters from the id.[22] As H. G. Wells puts it at the end of *War of the Worlds*, "in the larger design of the universe this invasion from Mars is not without its ultimate benefit . . . It has robbed us of that serene confidence in the future which is the most fruitful source of decadence."[23] We can only hope Wells has this right and we take the images of space-invaders as an omen.

Posthuman Check-In

Dynamics such as these, made vivid by stories, are often dismissed as superstitions from a time when gods and monsters held meaning. Such dynamics have become part of the official purge of sensibilities incompatible with science. But the psychology of the unconscious reminds us that these collective dreams and fantasies have more to do with inner realities than outer ones—they personify the forces of the psyche and reflect the meeting points of psyche and cosmos. While we are still operating out of a distinctly human psychology, we are still subject to these dynamics and their mythic renderings.

Some argue we have already effectively left nature and naturalistic ways of thinking behind, because our minds are no longer geared to natural patterns and mythological ideas: our thought has become thoroughly immersed in abstract logic. The question remains, however, at what level is this the case? Our conscious thought may be filled with a lot of abstract thinking, but until the fundamental structure of the brain is altered, until that brain is separated from an organic body, and until that body changes its shape and function because it has no ongoing contact with its earthly habitat, we will remain, to a significant degree, subject to an instinctive basis of mind, and we still need the cultural imagination to convey the deeper psychodynamics of this life to us. We will tell stories and make art, even as we learn to interpret these things with more sophistication. There can be no quick leap into a posthuman state. We have to deal with the all-too-human reality of the impact of our innovations in the interim. What the whole genre of science fiction is telling us, and the mad scientist punctuates, is that there are consequences to getting ahead of ourselves. Eyeing these consequences means the path before us may still be quite open.

From this perspective the posthuman focus on what we may become, riding roughshod over what we are, is mad science in slow motion. Posthumanists argue that satisfying instinctive needs belongs to a low rung on the evolutionary ladder and there is no reason why we should not develop a way of life that takes us beyond such things. As this thinking goes, biology can be reconfigured so our minds guide it rather than it guiding our minds. Again, however, we return to the basic question "Even if this were possible, what kind of mind would result from such a severing of biological, instinctive and evolutionary roots?"

Posthumanists are overtly non-Cartesian, believing the mind as we now experience it to be indistinguishable from the neurological processes of the brain. But they are covertly Cartesian, believing those neurological processes can be both separated from the rest of the body and ultimately reproduced in artificial form. That is, there could be a mind without a body and they expect to build such a mind; a brain in a vat, a reengineered brain or a computer program could all, in theory, constitute a mind, and therefore exist in disembodied form. Much development in AI, nanotechnology, and robotics proceeds on just this basis.

The problem is, as neuroscience has itself begun to reveal, even if the mind is dependent on the brain, a brain without a body and without a relational web of embodied life would give rise to no mind we would recognize. A working definition adopted by one influential group of neuroscientists is that "mind is an embodied and relational process that regulates the flow of energy and information."[24] And as Ecopsychology keeps telling us, a mind with no sensual ties to the planet results in not only an unsustainable way of life, but a kind of madness in which we are detached from our existential foundations. For the ecopsychologist, without an animal body attuned to the ways of nature, we would, by definition, lose our minds. From these perspectives, without a body, the mind makes no sense, because such a mind could not "make sense."

This is a point to which I will repeatedly return. Even if the posthuman conception of separating from the flesh and developing an intellect apart from a central nervous system were to become possible, the result would most likely be something alien and inhuman. Such an intellect might well mimic being human, which, as already noted, puts us squarely on the slippery slope of a fully virtual existence, but without the mortal body and fabric of human relations a body depends on, it would remain incapable of feeling, sensing, and therefore thinking in a genuinely human way.

Of course, posthumanists counter *that* is the point—we are destined go beyond what is human. They may well ask, has not the whole trajectory of Western civilization created modes of thought and being beyond the concerns of having bodies and being governed by instinct? Yes and no. The *yes* is obvious: we occupy ourselves with many things that go beyond mere instinctual satisfaction. The *no* reveals the oversight and misconception that have allowed the posthuman view to take flight in the first place. Despite the fact we breathe, drink, eat, and reproduce, we often think and live as if we have transcended the biological rung of life. Extrapolating from this transcendent view of ourselves, it looks like we are headed toward a disembodied intelligence, or at least one that shows little need for the animal side of being. Culture, society, and civilization all imply going beyond this mode. But closer examination reveals these achievements are also inextricably tied to our innate nature; they demonstrate the instinctive and creaturely entering the world in more refined ways, providing the symbolic forms, languages, artistic representations, and leisurely occupations that channel and transform the deepest impulses within us. Food and sex, plants and animals, work and shelter are still at the basis of highly elaborate culture-making.

What makes life worth living and provides the source of deepest satisfaction and the feeling of soul is the sublimation and transformation of instinctive needs, not the repression or detachment from them. Whereas sublimation and transformation are associated with the greatest human achievements, repression and detachment are associated with the most destructive forms of psychopathology. A successful society structures and contains the desire for cohabitation and communal bonding; a vital culture enables the conversion of primal impulses into practices that are the essence of human bearing. Music, art, and literature create socio-cultural bonds. Shakespeare described his aim as the attempt "to hold, as 'twere, the mirror up to nature."[25] Of course, when mirrored, when keenly observed, wrestled with, and poetically rendered, we end up with a very different kind of nature and very different means of reconciling ourselves to its claim on us. But without that natural ground we have no prima materia to make a philosopher's stone.

The accumulative record of these processes and efforts, reengaged in each generation, recast for prevailing conditions, provide our guidelines for living. What we think of as "mind" is hardly separable from any of this. To leave instinct behind would be deliberately neglecting the foundations of life itself and effectively end the experiment of humanity. The value of what would be left behind would then be hard to know because values themselves derive from this culture-making mind. Yet detaching ourselves from such foundations is precisely what mad scientist syndrome invites us to embrace.

Heidegger

In his much-discussed essay "The Question Concerning Technology"[26] Martin Heidegger, one of the twentieth century's most significant as well as controversial philosophers, sets out a perspective that extends and clarifies the themes of this chapter. Heidegger describes the contrast between ancient and modern technological modes of thought, generating key insights into the significance of techne and its relation to what depth psychology would call the archetypal basis of mind. Heidegger refers to this basis as "the gods" or "Being."

After acknowledging the degree to which modern existence is defined by technology, Heidegger goes on to criticize our tendency "to regard it as something neutral," which he then connects with our typical view of technology as "instrumental."[27] Whereas a conventional view holds that the problem of technology begins and ends with how we use our gadgets, Heidegger argues the problem resides with the conception of reality that promotes the invention of gadgets in the first place. Put another way, while our general inclination is to allow the technologist to dream on and innovate away, leaving all the moral and ethical questions to those who grab the lever, push the button, or pull the trigger, Heidegger argues this position is naive, for it does not see what he calls the enframing of the modern perspective, which is the objectification of the world—turning the earth into a resource for use and consumption and viewing things as disposable commodities. It is this objectification that gives rise to the tools we make and the instruments we pick up. That is, it is not so much that modern technology ends up commodifying what it engages; it does this before the scalpel is applied or the data extracted. The tools to accomplish these tasks manifest in keeping with the consumptive, commodified vision. Technology is, in other words, all about how we imagine reality.

In a haunting passage of his essay, Heidegger describes the way enframing leads to a state where the world is transformed into a "standing-reserve" and the human being becomes "nothing but the orderer of the standing-reserve." In a prophetic way he notes that the modern enframer

> comes to the point where he himself will have to be taken as standing-reserve. Meanwhile, man, precisely as the one so threatened, exalts himself to the posture of lord of the earth. In this way the illusion comes to prevail that everything man encounters exists only insofar as it is his construct.[28]

What Heidegger foresaw has already come to fruition. The virtual world and the attention economy have turned mind and soul into commodities for the online world to objectify and exploit. All is reduced to data, including our sense of self. We may be digital consumers, but it is also we ourselves who are consumed.

This enframed reality is also what greases the wheels of the posthuman movement. It is because the world is perceived in a reductionistic way, the body is seen as a complex machine, the mind is treated as an organic computer, and human consciousness is fittingly commodified that we now feel at liberty to tinker with all three. Being itself becomes the ultimate commodity. The economics of this final commodification will make us dependent on corporations to provide upgrades and replacement parts in the same way we are already dependent on them to navigate the information superhighway. Under such conditions the line between the desire to reconstruct ourselves and the corporate exploitation of that reconstruction will completely blur. As one commentator on Heidegger's perspective puts it, "we now view nature, and increasingly human beings too, only technologically—that is, we see nature and people only as raw material for technical operations."[29] Obviously, the more human beings are regarded this way, the more posthumanism gains traction.

Helpfully and hopefully, for the humanists at least, Heidegger also offers a way out of this commodifying of existence: the possibility of another kind of technology that he calls "a way of revealing."[30] This is best imagined through ancient experience, in which creativity and truth appear as gifts—as things revealed from beyond the purely human. Again, we meet the value of the past and the myths that remind us of more mysterious origins of creativity. We also begin to see that the deeply human is paradoxical in that it moves us to see beyond being merely human.

In his detailed commentary on Heidegger's essay, Richard Rojcewicz reaches back to the contrast between Socratic and pre-Socratic philosophy to underscore these two contrasting forms of technology. He couches the contrast in terms of the broad shift "from piety to idolatry" that occurred at this time. The idolatry concerns the "idolizing of humanity, a kind of human chauvinism," which, he argues, has become "our epoch's most basic and pervasive form of chauvinism."[31] The piety involves the capacity to behold the wonder and genius of creation and then create accordingly. This stance of returned reverence and piety is what Heidegger famously calls Dasein, or "there-being"—a presence to the things of the world that discloses Being or recognizes the gods. This is a place where art and science could be rejoined in techne. Such a stance would also invite an ecology of mind.

As Rojcewicz reads it, the problem is that "modern technology accompanies the absconding of the original attitude" of piety. Expanding the point, he writes, "An imperious theory thereby fills the void left by the deferential one, hubris replaces piety, unbridled imposition supplants respectful abetting, and the understanding of humans as possessors displaces the one of humans as Dasein."[32] Heidegger thus presents a sweeping indictment of the anthropocentric and egocentric thread running through the Western tradition, seeded in ancient Greece. As Rojcewicz states, this is "a forswearing of the attitude that led to the view of truth as a goddess, and so the entirety of the intervening history amounts to *Ab-fall*, apostasy." He continues:

> For Heidegger, this apostasy has culminated in metaphysics, humanism, and modern technology ... They are merely different expressions of the same chauvinism. They all understand the human being in terms of subjectivity, and in particular as *the* subject, the sovereign subject.[33]

Here Heidegger appears to be striking a blow against the whole history of Western civilization—a severe blow for the modernity he beheld and the postmodernity he envisioned. Without a trace of piety, or an appreciation for the implications of its absence, modern technology crosses into unprecedented modes of manipulation and control. It goes unrestrained because a critical sensibility has been abandoned. And given the degree to which we are now immersed in technology, with our tools not just shaping us but becoming a part of us, we have created a rather narrow feedback loop if we are ever to grasp this situation: when technology itself dominates awareness, the ability to perceive its impact slips away, and any ecology of mind is thwarted.

This brings us to a twist in Heidegger's take on the theme. To break the vicious circle by becoming Dasein, we must awaken to all the ways in which technology is not merely a human faculty but is something bestowed upon us—it too being given by the soul. In depth psychological terms, this means appreciating its roots are in the archetypal world and not the personal, subjective world; it grows from something instinctual, as this chapter has been laying out. Jung described this archetypal world as the objective psyche because it is experienced phenomenally as something that transcends subjectivity. The archetypes come at us from beyond, as their ancient appearance as gods conveys.

The irony of our relationship to technology is that in our claiming of it as a purely human faculty, guided by subjective, willful instruction, we lose sight of its givenness—the way it belongs in the scheme of existence. This relationship is, accordingly, distorted. When our reverence and awe of technology as an imperative of Being—as a god—is lost, which is to say its archetypal origins are neglected, this power returns unconsciously, possessing us, turning us into its slaves. In such a way we come to see that technology is no different from love or strife or the search for meaning. When such life principles are imagined as subject to our conscious direction, it is then we then fall under their spell. We act as though we are gods rather than reverent observers of transpersonal forces. This reverence allows us to compare those principles that want to be shepherded into life, which in turn prevents an unconscious devotion to any one in particular. When the egoic identification with technology is undone, a reimagined relationship with it becomes possible.

Heidegger concludes his essay with a remarkable turn to art, which, in the Greek understanding, also pertained to techne. He writes:

> There was once a time when it was not technology alone that bore the name techne. Once that revealing which brings forth truth into the splendor of radiant appearance was also called techne . . . Once there was a time when the bringing-forth of the true into the beautiful was called techne. The poiesis of the fine arts was also called techne . . . At the outset of the destining of the West, in Greece, the arts soared to the supreme height of the revealing granted them. They illuminated the presence [Gegenwart] of the gods and the dialogue of divine and human destinings. And art was simply called techne.[34]

The *Oxford English Dictionary* uses the term "mechanical arts" in its definition of technology, and we often consider technological creations with an aesthetic eye, describing a construction as a thing of beauty or a means of transport as art in motion. We at least carry the notion that technology can be artful. Innovation and art both rely on inspiration. In architecture and industrial design there is a desire to meld form and function, no doubt because form affects the desirability and pleasure a manufactured object can bring. These are contemporary indicators of art and technology's common root in techne.

Yet the manner in which modern technology inevitably succumbs to the enframing vision of expedience, disposability, and planned obsolescence estranges it from its artistic ancestry. Art opposes these qualities because its inherent aim is endurance, which is achieved by reaching back into timeless patterns of existence and by creating things that are meant to stay around—either concretely or imaginatively. Accomplished art is both ageless and universal. Rojcewicz states it simply: "art discloses what it means to be."[35] Yet with modern technology's enframing, and the utility and economics that result, making art is relegated to window dressing, unable to compete with the value of making usable things. No doubt this is the main reason the arts as well as the humanities have been expunged from science, technology, engineering, and mathematics (STEM) models of education. For we have come to believe these non-scientific disciplines are tangential to building a post-industrial society and preparing for a thoroughly technologized future.

For all these reasons, it is remarkable Heidegger completes his essay, which seeks to penetrate to the essence of the problem of modern technology, by invoking the critical role that the artistic vision must play in curtailing and reimagining our current outlook:

> essential reflection upon technology and decisive confrontation with it must happen in a realm that is, on the one hand, akin to the essence of technology (the common drive to create) and, on the other, fundamentally different from it. Such a realm is art.[36]

Here I take Heidegger to mean that only through reestablishing contact with the archetypal basis of mind, or what he calls Being, which is what art and the cultural imagination that support it enable, can we find footing enough to revisit and reimagine the place of technology in our lives. Art sees through the pragmatic, utilitarian, and commodified world, revealing to us the positions we have taken up in matters of ultimate value. Again, Rojcewicz: "Artists are Dasein in a preeminent way and play a privileged role in the history of Being . . . The artist's role is midwifery, active reception."[37] Through *poiesis* and imagination, art conveys what the gods are attempting to say to us and the necessity of finding ways to comprehend their message. In taking such an artistic stance, the impulse to create finds its fitting container and means of serving the greater whole.

Just as culture is and has always been the container of civilization, art, as the essence of culture, becomes the container of technology. By generating a familiarity with the soul that calls it forth, and a familiarity with the earth that provides its forms, art keeps technology from becoming fully abstracted from the nature of Being. Returning us to the beautiful limits of what really matters, art configures the contours of existence in a way a machine cannot.

The Hunter and the Shaman

The tinkering instinct thrives as our creative power extends. But in the absence of techne, cultural values, and artistic vision, this instinct acts like an invasive organism, disrupting outer and inner ecologies. The fanatical scientist, disconnected from their psychology, hell-bent on some breakthrough while blind to larger implications, dramatizes more pervasive expressions of the same essential dynamic: scientists whose sole concern is pushing the bounds of possibility; technologists who specialize without considering the wider implications of their work; entrepreneurs who let corporate profit overshadow social impact. All exemplify a creative power that lacks aesthetic sense and spiritual bearing, beholden only to itself.

Tracing the propensity to tinker back to an instinct translates the challenge technology presents "out there" into a collision occurring "in here"—a collision of psychological imperatives. By contrasting ancient and modern technology, and by describing the exploitative and manipulative character of the latter, Heidegger provides a philosophical frame for this collision. He helps us see the degree to which the next phase of innovation, involving the attempt to remake ourselves, will push us even further into the question of whether the gods will return to human awareness or whether human awareness will continue to usurp the gods.

As America has largely defined the New World ideal of technological striving and has given birth to the digital age and the posthuman outlook, it is only fitting we punctuate these reflections on the inner problem of technology by considering the psycho-mythic forms this land brings to the cultural-historical moment and what this portends. Again we turn to story and the cultural imagination.

In their book *Projecting the Shadow: The Cyborg Hero in American Film*, Janice Hocker Rushing and Thomas S. Frentz discuss the underlying mythic character of the figure they term the "technological hunter," a descendent of the "frontier hunter," both of whom vividly contrast with the indigenous hunter. Rushing and Frentz make the compelling argument that post-industrial society is caught in an unconscious enactment of a myth wherein the hunter fails to form the proper relationship with his weapons, which, in the past, particularly in the indigenous context, had been blessed by a "shamanistic heritage."[38] Absent this spiritual relationship, the instruments of "the hunt" come to be used by colonizing peoples without reverence or sense of limitation. Centuries on, these instruments eventually turn on the hunter, just as the mad scientist's creation turns on its creator. Citing films such as the *Terminator* series and *Blade Runner*, they extract the essence of the narrative depicted in these works:

> In an increasingly technological culture devoid of spiritual influence, the ego-driven, exploitative hunter loses the soul connection with his weapon ... The weapon eventually breaks free from his control, becomes technologically perfected, and, in a final profane reversal, turns against the very hand that used to wield it.[39]

The heart of the problem is described as the drive "to perfect the god of modernism—the 'rational centered subject' and its technological extensions—into our heroic ideal."[40] Perfection and heroic dominance, expressed through the possession of more powerful tools and a desire to merge with these tools, indicates the absence of the spiritual factor. In keeping with the American context of these patterns, Rushing and Frentz reach back through the clash between the Western frontier-hero and Native American spiritual bonding with the earth and animal life, expressing this absent reverence in the description of a pivotal moment that overlays myth and history:

> At first demonstrating some semblance of the indigenous reverence for creation, the "white frontiersman" shows restraint because of his own religious principles and cultural sense. But eventually both his culture and his God become impediments to what he now sees as "freedom"—the conquering of the American frontier. Since Indians as well as wild beasts occupy the land he wants, he slaughters both indiscriminately, gaining a decisive advantage over his human prey because of his large numbers, his more sophisticated weaponry, and his lack of spiritual restraint.[41]

Beyond the obvious inhuman and utterly devastating character of these actions, what Rushing and Frentz portray is a series of sociocultural patterns set in motion by the quest for dominance. The conversion of "savagery" to "civilization" and the taste of "great success in conquering the wilderness" means "the frontier hero desires to maintain and extend his control over the earth." Moreover,

> because he is so good at making machines, he now uses brains more than brawn, and he prefers to minimize his contact with nature, which can be uncomfortable and menacing. Thus he creates ever more complex tools to do his killing and other work for him.[42]

America first repels the colonial forces of the Old World and then turns around to brutally conquer indigenous lands, this time cloaked in the notion of manifest destiny, using the emerging power of technology. The image of a New Frontier, the domination of the Wild West and the emerging embrace of modern devices form a heady mix. But the wild just goes underground and eventually finds ways to resurface.

By linking stories of nature biting back (*Jaws*) and war turning on us (*The Deerhunter*) to the cinematic renderings of returning cyborgs, Rushing and Frentz show how the American hero sets up an enantiodromia in which the hunter becomes the hunted:

> Having banished God as irrelevant to the task at hand, the hero decides he is God, and, like that now obsolete power, creates beings "in his own image"; this time, however, they are more perfect versions of himself —rational, strategic, and efficient. He may fashion his tools with by remaking a human being into a perfected machine (a cyborg) or by making an artificial "human" from scratch. Unfortunately, however, these new creations have designs of their own that the hero fails to foresee ... Like their human creator, these technological beings develop a desire for complete freedom, and so they declare themselves to be God and set out to hunt and ultimately eliminate their maker.[43]

Whether psychopathic cyborgs, rogue AI, or robots run amok, the push for ultimate control eventually leads to a loss of control. This extrapolation of the blind quest for technological superiority was on

Jung's mind when he observed America back in 1912 and noted "it must make a choice to master its machines or be devoured by them."[44]

American is a place of displaced religiosity. Its cultural response to the death of God has been to simultaneously embrace an overt religion of biblical fundamentalism and enact a covert religion of frontier heroism, neither of which recognizes the sacred in nature or in human nature. Neither the remnants of Native American spirituality, nor the transcendental poets, nor the visionary naturalists have made a dent in the American proclivity to dominate all things wild, replace the authentic experience of the sacred with the rational will and initiate a repetitive search for all that is new. Making and self-making have thereby joined in an imperative to pursue constant change, which the techno-sphere is more than pleased to fulfill.

This manic movement discourages reflection and the need for history. For many evangelicals, introspection only invites the devil, and the past is what bedevils the frontier hero. And without accounting for the darker episodes of its past, America acts out an unconscious repetition compulsion: history repeats itself; only this time the protagonist takes the role of the victim and technology becomes the hunter and enslaver.

We thus come to see that the modern techno-hero descendent of the American frontiersman lacks any kind of initiation into meaningful relations with his ancestors. His use of technology pays little regard to the spiritual principles a shamanistic mediation of powers traditionally lends to the use of tools. The resulting archetypal lacuna becomes evident once we consider the hero pattern in full. Following the map provided by Joseph Campbell, we learn the authentic hero must submit his ego to a larger source of life energy and recognize his weapons are to be carefully invested and divested of the power of this source, typically with the help of a shamanistic figure. In these practices, the fully initiated hero acts in service to the community and the world soul, and the completion of their quest brings with it a renewed sense of the sacred. As Rushing and Frentz state:

> The shaman's rituals sanctified the activities of the hunter, fostering a belief in a common life-spirit immanent in all things and in a reciprocal relationship between people and the sacred processes going on in the world. When these two principles—heroic extraversion and contemplative introversion—were in balance, the hunter could

remain centered in the Self. Thus his weapon was not very far removed from himself: it was only as innovative as needed to subdue the animal.[45]

Balancing the hunter-hero's outward movement with "contemplative introversion" may be a rare achievement in our time, but it is one the depth psychological and existential-humanistic explorations have consistently attempted to elevate in the face of modern civilization's pathological extraversion. Here Rushing and Frentz employ Jung's term for the archetype of totality and completeness, "the Self," to indicate what it is an introverted and initiated shamanic consciousness brings; namely, a similar awareness of the interconnected web of life and expansive Being that we find in a deeper understanding of techne.

Two Paths

I have shown how the tinkering instinct, which belongs to the human condition, enters life in two distinct ways, placing before us two distinct paths of knowledge and invention. One path reflects a propensity to control and dominate, which is fueled by an estrangement from the world and the soul, an alienation from the wilderness without and a paranoid distrust of the wilderness within. This path supports an anthropocentric approach to creation and an egocentric model of the mind. On this path, as Heidegger contends, the gods continue to flee from our awareness. Along with them go the values and principles that relate us to the earth's own matrix of Being and to the archetypal basis of mind. And in the absence of these cultural forms, the tinkering instinct gains an excessive autonomy.

The other path leads to an acute awareness of the absence of these cultural forms and prompts the question of their recovery. Without returning to an overt religiosity, this path seeks a restoration of reverence, a return of the sacred and a psychological humility about the origins of our thoughts and actions. It maintains the view that one way or another the gods always do return, and art, techne's long lost twin, is a primary means of hosting them. On this path, looking back while moving forward and looking within while engaging the outside world lie at the core of all integrity. Contemplating what moves us, we discover a timeless order, an architecture of mind and a cosmology of being that will not, in the end, allow blind invention to tyrannize existence.

Arthur, addressing the essence of technology and "how it evolves," concludes his book by saying:

> Technology is part of the deeper order of things. But our unconscious makes a distinction between technology as enslaving our nature versus technology as extending our nature. This is the correct distinction. We should not accept technology that deadens us; nor should we always equate what is possible with what is desirable. We are human beings and need more than economic comfort. We need challenge, we need meaning, we need purpose, we need alignment with nature. Where technology separates us from these things it brings a type of death. But where it enhances these, it affirms life. It affirms our humanness.[46]

It is heartening to see a technologist addressing the central concern of our time with such philosophical acumen, as well as assigning a critical role to the unconscious, even promoting this dimension of the psyche as a guiding factor, capable of making the "correct distinction." This is a nod to the archetypal basis of mind, an acknowledgment there is a wisdom that runs deeper than conscious aspirations. To account for the unconscious would be to retrieve techne as the archai of technology, which would maintain a bridge between the impulse to innovate and the implications of innovation. To account for the unconscious would mean an aversion to the kind of fragmentation and compartmentalization that now haunts the entire trajectory of modern technology.

Technology can be separated from conscience, but techne cannot. Following the mechanized horrors of the past two centuries, many streams of technology today rely on just this absent conscience. We see this when algorithms are allowed to spread blatant falsehoods and amplify extremism in order to increase advertising revenue. This choice of system over conscience is the pollution of the post-industrial age, a pollution of the collective mind. Technology can proceed without wisdom, but techne cannot. By infusing our minds with algorithmic rather than archetypal realities, with knowledge untethered from self-knowledge, we are positioning to become programmed automatons, making substantive thinking and critical reflection a thing of the past.

Whether contemplating the actual effects of the digital world or the sci-fi dramatizations of where it may be taking us, we are right to be fearful. Fear is a more pointed and precise response than the free-floating anxiety that surrounds us as we go with the flow. Fear raises our antennae, signaling an impending assault on something essential, but we still need to follow it to its source. We have, since the middle of the last century, lived with the sense of teetering on the edge of technology-derived disaster. Given both the objective dangers and how far our device-encrusted way of life has elevated us above the ground of Being, this sense is accurate. But apocalyptic expectation tells us as much about inner collapse as outer catastrophe. The problem is we have also grown adept at not feeling or recognizing what is taking place within us. This is the topic of the next chapter.

Notes

1 Theodore Roszak, *The Making of a Counter-Culture: Reflections on the Technocratic Society and its Youthful Opposition* (New York: Anchor Books, 1969), 217.
2 Anthony Giddens, *The Consequences of Modernity* (Stanford, CA: Stanford University Press, 1990), 153.
3 James Hillman, *City and Soul* (Putnam, CT: Spring Publications, 2006), 18.
4 W. Brian Arthur, *The Nature of Technology: What It Is and How It Evolves* (New York: Allen Lane, 2009), 1.
5 Ibid., 23.
6 Ibid., 22.
7 See Frank R. Wilson, *The Hand: How Its Uses Shape the Brain, Language and Human Culture* (New York: Vintage, 1999).
8 Raymond Tallis, *Aping Mankind: Neuromania, Darwinitis and the Misrepresentation of Humanity* (Durham, UK: Acumen, 2011), 216.
9 Ibid., 215.
10 Andy Clark, *Natural-Born Cyborgs* (Oxford: Oxford University Press, 2003).
11 Lewis Mumford, *The Myth of the Machine: Technics and Human Development* (New York: Harcourt Brace Jovanovich, 1967), 101.
12 Ibid., 102.
13 Stuart Whatley, "The Machine Pauses." *The Hedgehog Review*, vol. 22, no. 2, summer, 2020, 15.
14 Ibid.
15 Arthur Koestler, *The Act of Creation* (New York: Macmillan, 1964), 146.
16 Mary Midgley, *Science as Salvation: A Modern Myth and its Meaning* (New York: Routledge, 1992), 12.
17 Ibid., 9.
18 Ibid., 10.

19 CW 13, par. 55.
20 Joseph Campbell, *The Power of Myth*. Interviewed by Bill Moyers, Betty Sue Flowers, ed. (New York: Broadway Books, 1988), 39.
21 Bill McKibben, *Enough: Staying Human in an Engineered Age* (New York: Times Books, 2003).
22 Glen Slater, "Aliens and Insects." In Dennis Patrick Slattery and Glen Slater eds., *Varieties of Mythic Experience* (Einsiedeln: Daimon Verlag, 2008), 189ff.
23 H. G. Wells, *The War of the Worlds* (1898; New York: The Modern Library, 2002), 181.
24 Daniel Siegel, *The Developing Mind*, 3rd ed. (New York: The Guilford Press, 2020), 10.
25 William Shakespeare, *Hamlet*, Act 3; Scene 2.
26 Martin Heidegger, "The Question Concerning Technology." In *Basic Writings* (San Francisco: Harper & Row, 1977), 283ff.
27 Ibid., 288.
28 Ibid., 308.
29 Mark Blitz, "Understanding Heidegger on Technology." *The New Atlantis*, winter, 2014. https://www.thenewatlantis.com/publications/understanding-heidegger-on-technology
30 Heidegger, "Question," 294.
31 Richard Rojcewicz, *The Gods and Technology: A Reading of Heidegger* (Albany, NY: SUNY Press, 2006), 3.
32 Ibid., 10.
33 Ibid., 3.
34 Heidegger, "Question," 315–316.
35 Rojcewicz, *Gods*, 193.
36 Heidegger, "Question," 317.
37 Rojcewicz, *Gods*, 201.
38 Janice Hocker Rushing and Thomas S. Frentz, *Projecting the Shadow: The Cyborg Hero in American Film* (Chicago: University of Chicago Press, 1995), 5.
39 Ibid., 5.
40 Ibid., 19.
41 Ibid., 54.
42 Ibid.
43 Ibid., 54–55.
44 C. G. Jung, *The Earth Has a Soul: The Nature Writings of C. G. Jung*, ed. Meredith Sabini (Berkeley, CA: North Atlantic Books, 2002), 143.
45 Rushing and Frentz, *Projecting*, 56.
46 Arthur, *Nature*, 216.

Part 2
DOWN THE RABBIT HOLE

Chapter 5
NUMBED

The entire man, who feels all needs by turns, will take nothing as an equivalent for life but the fulness of living itself.

William James[1]

Have you seen those zombies who roam the streets with their faces glued to their smartphones? Do you think they control the technology, or does the technology control them?

Yuval Noah Harari[2]

To grasp where technology is taking us, we must grapple with one coping mechanism that has coincided with the digital age, which is the propensity for psychological dissociation—a fragmenting of awareness, compartmentalizing of perception, and disconnection from emotion that has become increasingly commonplace. Manifesting in varying ways and at different levels of our experience, the net effects of dissociation usurp our ecology of mind and invite human robotization.

First identified by Pierre Janet in the late nineteenth century, dissociation became an entry point for the study of the unconscious and a critical part of Freud and Jung's psychodynamic formulations. From the avoidance of unwelcome thoughts and emotions, to the debilitating response to trauma, to becoming an ingrained feature of the personality, dissociation occurs along a spectrum of severity. Yet, irrespective of the severity, dissociation always opposes integration. As dissociative processes take hold, disintegrative processes take over. When prolonged, these disintegrative processes cause a loss of integrity in persons at the individual level and in group dynamics and institutions at the collective level.

As dissociation grips the psyche, an associated phenomenon frequently arises: psychic numbing. Akin to the physical numbing that occurs when blood stops flowing to parts of the body, psychic numbing results when awareness stops flowing to parts of the psyche. If dissociation is the psychodynamic, numbing is the resulting emotional detachment. An expedient and sometimes protective means of overcoming the tension and anxiety of inner tumult in the short term, long-term numbing can turn into a disembodied, unfeeling way of relating to oneself and engaging with the world. Contact with a sense of self and living fullness of existence is then lost. A stunting of development and disruption of integrative processes that generate meaning and sustain culture ensues.

Such fragmentation can be a natural response to wounding and trauma. As Jung noted in his early research, the psyche is quite prone to dissociation, generating emotions and impulses that diverge from conscious aims. However, failing to contend with these inner tensions is a second-order problem, which can lead to a compartmentalization of awareness. Emotion is thereby prevented from metabolizing into feeling, closing down the pathways by which a person consciously suffers the complexities of inner life or a group digests historical traumas or societal breakdowns. The multiple claims of the soul are avoided and numbing behaviors and chemical solutions are often engaged to sustain this avoidance. A psychic sleep pervades.

The difficulties of dissociative coping are one thing, leading to depersonalization and soul loss. However, a larger problem is that split-off parts of the psyche can subsequently strengthen and resurface in unwieldy form—like rebellious free-agents. After noting that "by and large, dissociation can be thought of as potentially normal," Rieber highlights the problem of "the specific circumstance when dissociative processes begin to outstrip the integrative processes, resulting in the functional autonomy of certain subprocesses."[3] For Jung, these subprocesses are what become the psychological complexes, which are typically rooted in early wounds but become activated later in life. Disavowed spiritual yearning returns as addiction; loss of genuine community becomes immersion in the hive-mind; ideals detach from reality and turn into fanatical, paranoid beliefs. Prolonged dissociation strengthens such complexes, which can then more easily possess the whole personality. In the meantime, other parts of the psyche are often left to wilt and die: unmitigated loneliness eats away at the capacity to be alone; detachment from disturbance leads to an ossifying conscience; neglected emotional responsiveness produces a loss of empathy. We stop feeling how much we are at odds with ourselves, lose touch with the real roots of our ailments, and engage in more unconscious behaviors. Decisions and actions become based on disintegrated rather than integrated states of mind.

Pervading psychological life in the post-industrial era, dissociation and numbing appear deeply implicated in the pursuit of an even more complete departure from ourselves—particularly from the body–mind connections involved in the felt sense of who and what we are. Many end up doggedly pursuing states of perfection and transcendence, which are often the compensatory calling cards of disintegrating minds. As Vilém Flusser has written, "we throw ourselves toward 'transcendence' in the same way that we throw ourselves toward interplanetary space, in order to escape the abyss that has opened underneath the ground we tread."[4] This abyss is, in essence, the growing unfamiliarity with ourselves, and it is this abyss that becomes the womb of the fabricated self.

By implication, dissociation is diminishing our recognition of the deeply human and ability to sustain vertical modes of understanding. The resulting psychic numbing is ensuring we do not feel the parts of our humanity being amputated to make way for prosthetic bodies and minds. Dissociated and numb in relation to *being*, we are thus more enticed into the possibilities of *becoming*. In the absence of a more substantive psychological bearing—an absent "fulness of living itself" (James)—technological frontierism will likely fill the entire horizon. It is in these ways dissociation and numbing lie at the very heart of our collapsing ecology of mind.

Attributable to the existential confusion and stress of the postmodern world, dissociative thinking has also been exacerbated by the rhythms and distractions of online life, as well as by the search for technical solutions to complex human problems. This environment has encouraged the view that psychological problems must be anomalous malfunctions rather than responses to the total psycho-social situation by a self-regulating psyche. A snowballing effect then kicks in, making personhood more mechanistic and less organic: psychic equilibrium becomes less a matter of internal regulation and more one of external stimulation or even simulation; social identity is detached from authenticity and built upon curated imagery; even physical appearance is separated from total wellbeing and subjected to what algorithms present as aesthetic standards. Under these conditions, mental tricks and physical facades become the way of the world. Surface supersedes substance, and the shinier the surface the duller the substance. Techno-patches convince us of having more conscious control, but in their wake they leave the nagging feeling that the underlying reality has not really changed. Efforts to adapt to the horizontal possibilities then become more manic as the capacity to relate to the psychic ground becomes more fraught.

When dissociation as an episodic coping mechanism instead becomes a normative way of being, people become empty shells—zombies controlled by technology (Harari) going through the motions. It is the prevalence of such emptiness that makes augmentation and enhancement so attractive. The supposed inevitability of a posthuman future can hardly be separated from such phenomena, just as much the result of regressive psychological failure as progressive scientific achievement. Posthumanism is, in other words, as much an outgrowth of an expanding psychological malady as it is an extrapolation of technological innovation.

Two Levels of Dissociation

Manifestations of psychopathology have changed throughout the modern era, and the kind of dissociation prominent today is quite different from that first observed in the late nineteenth and early twentieth centuries.[5] Coinciding with a dramatic shift in lifestyle brought on by industrialization and a nascent depth psychology, earlier forms of dissociation involved large divisions between the conscious and unconscious dimensions of the psyche, created by repression, suppression, and chronically one-sided orientations to inner nature. The rational and moral attitudes of the conscious mind banished or closeted what were perceived as irrational and immoral thoughts and impulses. Freud's sexual theory of neurosis was based on the ill-effects of repressing sexual desire and erotic fantasy, to the point of producing dramatic psycho-somatic reactions, most evident in the symptoms of hysteria. Through catharsis, transference, and other psychoanalytic techniques, repressed thoughts and emotions could be brought to light and relieve these profound psychological splits.

Jung's work revealed the way psychological complexes formed around early wounds and other psychologically significant events and demonstrated how these complexes could act independently of consciousness. As he puts it, a complex "has the tendency to form a little personality of itself."[6] With some introspection we can notice these formations in the unconscious. But their incompatibility with our conscious outlook means there is always a certain disconnect between our ego ideals and these complexes, and when a complex forms around something traumatic, deeply painful, or otherwise threatening to the conscious orientation, the dissociative tendency can become more pronounced, resulting in significant psychic splitting. Yet whether we are talking about the repression of sex or power or the autonomy of complexes, the dissociation involved generally occurs along distinct lines, and is overcome by making the unconscious content more conscious.

The most prevalent form of dissociation today is, by contrast, more like a splintering than a splitting of the psyche. This splintered state overlays deeper divisions between the conscious and unconscious, creating a secondary barrier against awareness of these divisions. It is a state that is in many respects an attempt to rid ourselves of any awareness of deeper problems. Pieces of consciousness and unconsciousness often appear mixed together, giving rise to free-floating anxiety, existential disorientation, and various kinds of defensive postures both successful and unsuccessful. The conscious outlook lacks cohesion, as if continuity and integration are beyond reach or even abnormal—especially in a world given over to postmodern deconstruction of values and meanings. The prevailing personality is not enduring a battle between a tight-knit upper world and a distant underworld so much as it is being caught in a maelstrom of competing views and impulses, rational and irrational, with life alternating between finding traction and slipping into a vortex of counter-narratives and relativism. Even as the rhetoric of politicians and church leaders becomes more black and white, their flawed personalities are on full and shameless display. Contradictions and hypocrisies abound, sustained by dissociated, compartmentalized styles of thinking. Shadowy elements seem present at every turn.

The backdrop of this difference in dissociative tendencies belongs to prevailing socio-cultural conditions of the post-industrial era, which appear to have stretched the psyche's ability to metabolize the seemingly limitless sensations and soundbites. The primarily visual and virtual source of psychic stimulation, together with its rapid and ubiquitous delivery, overwhelms some faculties, cuts others off, and thwarts the digestion of lived experience. A staccato rather than synthesized encounter with the world has resulted—a perception of reality and self that overwhelms perception and diminishes feeling. Minds have become segmented and prone to distraction, while bodies have become anesthetized and depersonalized. In this new form of normative psychopathology, being dazed and confused is insulating us from feelings of being conflicted and alienated.

Contrasting the banishment of specific dimensions of the psyche or traumatic events, this prevailing condition represents nothing less than a repression of the psyche itself—a repression of inner order and psychodynamic regulation. That is, there is no longer a coherent image of the human personality as a self-organizing hub of significance and purpose. Psychological suffering has become more common but is undergone in more unconscious ways.

In this most recent dissociative pattern, much of what was once repressed now floods the consciousness with the push of a button. The unrelenting baring of bodies and souls, talk-show confessions, packaged psychobabble, and self-help manuals for every nook and cranny of the space once attended by private thought have lifted the lid on what was once repressed. But when it is all out in the open, what then happens is something else. Spread wide and thin, consciousness cracks: too much to juggle; no contours of value and meaning; no landscapes of inner integrity. The container breaks and the contents spill out. Pieces of experience are both temporally and spatially dispersed; continuity falters.

Everyday habits of mind reinforce this diffused sense of self. Despite the evidence that we cannot actually carry out several tasks at the same time, we insist on the value of multi-tasking, jumping between things in a manner that is inefficient and certainly less satisfying.[7] As we take our evening walk, neither the person talking to us on the phone, nor the dog at our side, nor the recollection of the unsettling encounter at the office receives adequate attention. Although we have begun to lament this distractibility, we still hop from one experience to another without feeling the rub, as if this is just the way of things. It is difficult to know whether the advent of multiple windows on our computer screens, the pop-ups on the web page or the speed with which we move between these things condition this inability to focus or merely mirror it.[8] However, it seems clear the merger of online and offline life has broken something and a dilution of attentiveness and continuity of thought are the result.[9]

It is tempting to imagine some kind of direct bridge from such a dissociated state to an integrated one. But a deep and enduring integration must begin with an equally deep and enduring differentiation—a familiarization with the psyche's multiple impulses and inner voices. And to move towards this differentiation means, first of all, reversing the propensity to numb ourselves, restoring the ability to feel the collisions taking place within us. As Jung writes, "A dissociation is not healed by being split off, but by more complete disintegration."[10] Restated, dissociated consciousness can only be undone by deliberate entry into inner conflict. Genuine maturation and existential honesty mean accepting that such conflict is intrinsic to being human, and understanding this ensures authentic participation in the challenges of the age. In what can only be described as a purposeful form of regression, we are called to peel away the defensive and escapist posture of post-industrial dissociation and become aware of the deeper fault-lines of the modern psyche—especially those concerning the splits from nature and from the past. The model of psychological change that presents itself to us thus becomes dissociation–disintegration–integration.

In a broader statement of this theme, Jung proffers the ultimate value of facing these splits within us:

> We should not try to "get rid" of a neurosis, but rather to experience what it means, what it has to teach, what its purpose is. We should even learn to be thankful for it, otherwise we pass it by and the

opportunity of getting to know ourselves as we really are. A neurosis is only removed when it has removed the false attitude of the ego. We do not cure it—it cures us.[11]

The problem with digital age dissociation is that it is mostly not experienced as a division along the vertical axis of being because it tends to negate that axis altogether. It is a style of dissociation that avoids interiority—a way of life that lacks depth and does not invite the kind of conviction and character that allow us to sink down into experience. By refusing to consciously fall apart and face the disintegrative character of modern existence, we thereby remain caught in a limbo state, unable to move from dissociation to integration.

Baudrillard describes this phenomenon in terms of widespread indifference. He writes, "This indifference to oneself is at the heart of the more general problem of the indifference of institutions or of the political . . . to themselves."[12] Such indifference, he suggests, "results from the absence of division with the subject."[13] With such an absence of inner complexity comes an associated inability to host inner dialogue. Instead we find a "horizontal madness, our specific delirium and that of our culture: the delirium of genetic confusion, of the scrambling of codes and networks, of biological and molecular anomalies . . ."[14]

Distilling these reflections thus far, I intend to suggest that it is only by risking a vertical madness and awakening to the most primary divisions of the psyche that we may overcome this "horizontal madness" (Baudrillard). Only a vertical understanding of the splintering and numbing of our time will prove adequate if we are to step forward with vitality and integrity. Although it seems like bitter medicine, discovering where and how we have become alienated from ourselves will provide a basis on which to engage with the ideas and forces currently acting upon us. The rest of this chapter aims to provide this vertical understanding. Together with the rest of Part 2, we will endeavor to comprehend how we arrived at the precipice of wanting to radically remake ourselves.

The Disconnect of Mind and Body

Whether the primary or secondary variety described above, dissociation comes down to a disconnection between mind and body, thought and emotion—with the loss of feeling that bridges the two. Whereas extreme forms of psychological dissociation may even result in physical numbing, it is emotion and its psychic significance that are most prone to remaining unintegrated into our awareness. And when emotional responses are not metabolized, a feeling life does not develop. As a refinement of emotion, feeling is the means by which our thinking is humanized. Feeling brings discernment and depth; it also bridges abstract thought and worldly actions. Without feeling, the body becomes a mere mode of transportation and the mind becomes detached from lived experience.

From registering pleasure and displeasure, to joyful awareness and responsive sadness, to being warmed by a conversation or moved by artistic display or performance, feeling insures an integrity of being. It is no exaggeration to say the loss of feeling and associated disconnection of mind and body erode a sense of being fully human and result in depersonalization—a state deliberately generated by totalitarian regimes and cult leaders using mind control practices. Along with the poets and sensualists whose perception of reality itself occurs via the deepening and refining of the feeling life, we may well conclude that although an ultimate separation of body and mind is entertained by philosophers and scientists, its actualization would generate a kind of madness.

Despite such feeling matters, in posthumanism the embodied mind is held with deep ambivalence. Having a body, being subject to a full range of emotional responses, and maintaining a sensate relation to the world are often imagined as the barriers of optimal functioning. It is even imagined our destiny lies in transcending these essentially human characteristics and functioning out of a more computational intelligence. However, many of the cultural images that have gathered around this outlook portend calamitous outcomes.

From the pre-industrial to the post-industrial world, from Frankenstein to Darth Vader to the Terminator and beyond, we are haunted by figures whose bodies and minds are either in pieces or have been hobbled together by technics, whose feeling life is either non-existent or severely compromised, who have either lost their soul or are looking for it, and whose destructive impulses are ultimately rooted in their psychological estrangement.

The term "shell" has been used in science fiction to describe the prospect of fully fabricated bodies with computerized brains in which remnants of minds or "ghosts" reside.[15] Such terms seem fitting in that a ghost is synonymous with a lost soul or one caught in the liminal space between this world and the next, and people in disembodied states often feel like empty shells. Ghosts are also entities with no psycho-spiritual home. But they haunt the living, as if attempting to draw attention to what is marginalized or forgotten.

The introjection of this image has become part of the vernacular of online life. When someone with whom we have shared a degree of friendship or intimacy suddenly and inexplicably stops communicating, we say we have been "ghosted." Perhaps the very medium of cyberspace encourages this evanescence. Yet the question that naturally follows from such ghosting events is to what degree such ghostly persons were present to begin. When someone who seems to have been relating in a substantive way suddenly disappears, we are compelled to wonder just how substantive the connection was to begin. If what is put forth in our online interactions is more shell-like, we should not be surprised that persons are more ghostly. Being ghosted becomes a way of realizing that somewhere along the way, genuine relatedness and contact with the deeply human were already absent.

Are we not collectively also ghosting ourselves? Have we not withdrawn and become insulated from the world? Prolonged dissociation is also a kind of psycho-spiritual homelessness, one reflecting a disconnection from being at home in oneself and one's surroundings. The recently recognized epidemic of loneliness is not only a matter of being apart from others, a well-documented symptom of the digital age; it is also a matter of feeling estranged from ourselves. This displaced personality appears to be the forerunner of tomorrow's default personhood, just as today's sci-fi images appear to be dramatic extrapolations of these psycho-social trends. As we have seen, however, these images of a possible future may be the psyche's attempt to instigate a change of course. The imagistic bridge from the present to at least one possible future is also an imagistic bridge from outer experience to inner dynamics. This forces us to consider the role the detachment of body and mind, the removal of feeling, and the displaced sense of soul are set to play in our attempts to reconstruct ourselves.

An ultimate and horrifying image of this mind–body disconnection comes from *The Matrix*, wherein the machines that have conquered the world are breeding humans in tanks to use the heat from their bodies to generate energy on an earth deprived of sunlight, while keeping their minds active in a fully simulated virtual world. As a literal portrait of what may come, the scenario may be a nightmarish stretch, but as a metaphor for the anesthetizing of physical being coinciding with a mass migration into cyberspace, it seems prescient. In the film salvation comes only when the ultimately illusory nature of the virtual world can be seen-through and related to as the computer program it is. The thinly veiled question posed is, can we see through the digitally infused, psycho-social matrix we have already begun to inhabit? Can we at least stay anchored to the "real world" as our minds inhabit simulated worlds?

For the mind to be parted from the body in ways a posthuman future promises, the mind has to first depart from the body in ways the world presently promotes. That is, a future wherein the mind–body connection becomes negotiable will come from a world wherein feelings and emotions become optional. And this world begins with dissociation.

So far, the profound problems of this early stage of mind–body disconnection have escaped our attention. Given our dissociative mindset, this is hardly surprising. However, such disconnection also breeds forms of polarization, vivid depictions of which are hard to miss. Along such lines, only a modicum of psychoanalytic insight is needed to perceive the red flags the cultural imagination has been hoisting. In many creative tales, from the classic sci-fi films of the twentieth century to the contemporary sci-fi novels of the twenty-first, we are confronted with not only the rise of detached, cold minds but also the return of desperate, fiery bodies. From films such as *Forbidden Planet*,[16] where a monster from the id confronts a band of idealistic space explorers, to books such as *Altered Carbon*,[17] where bodies are exchangeable and disposable "sleeves" that leave their wearers prone to lust and violence, we glimpse future worlds in which the intellectual order is met with physical chaos. Only the most rudimentary sense of psychic inversion is needed to see that such cultural offerings depict advanced technocracies coming at the cost of inner division and regression. As I pointed out in Chapter 1, even in the course of modern history, reason and logic have not prevented so-called "civilized societies" from acting in beastly ways. Yet somehow, we continue to pretend the animal body of our natures and its instinctive responses do not also need our consciousness but can just be appended to a disconnected intelligence. Until we do manage to completely abandon the flesh, it is likely to reassert itself in unsettling ways. Both this flesh, and the planet that has provided it, may not want to let us go.

Fragmentation and Self-Invention

Dissociation forms the darker side of what the cultural historian Robert Jay Lifton has called "the protean self"[18]—a kind of mercurial personality adaptive to the array of circumstances this era has thrust upon us. Although few can avoid the whirlwinds of contemporary culture and many succumb to the disorientation and its manic compensations, especially consumptive, addictive, and thrill-seeking behaviors, some have learned to creatively harness these mania-inducing winds. This protean state might have originated in a breakdown of traditional contexts and defined social roles, but the speed and volume of data sharing and adaptation to virtual worlds seem to have accelerated its prominence. "Proteanism," Lifton contends, "is a balancing act between responsive shapeshifting, on the one hand, and efforts to consolidate and cohere, on the other."[19] We may construe this balancing as one between the horizontality and verticality I have described. Lifton quotes the playwright Václav Havel, who also became the first president of the Czech Republic, as an example. Havel once wrote:

> I get involved in many things, I'm an expert in none of them . . . though I have a presence in many places, I don't really have a firm, predestined place anywhere . . . For some people I'm a constant source of hope, and yet I'm always succumbing to depressions, uncertainties, and doubts.[20]

This may be a discontinuous existence, but it is not a dissociated one.

As Lifton also points out, however, the same conditions that produce such a creative and conscious protean bearing also "produce an apparently opposite reaction: the closing of the person and the constriction of self-process," which "can take the form of widespread psychic numbing—diminished capacity or inclination to feel—and a general sense of stasis and meaninglessness." It may also produce what appears to be a further response: "an expression of totalism, of demand for absolute dogma and a monolithic self."[21] In the previous chapters I have set out the many ways this totalist proclivity has arisen as a reaction to the prevailing cultural climate. There, as here, "The issue of *control*, of stemming the protean tide, always looms large."[22]

Lifton also uses the more umbrella term "*fragmented self*," which, he suggests, "appears in many places and guises." It is in this context he suggests that "fundamentalism and other forms of totalism, embraced to overcome fragmentation, can intensify this fragmentation."[23] It is thus not that an individual falls into one or another of these possibilities but, rather, frequently oscillates between them in yet another kind of vicious cycle. If a totalistic belief system dominates the conscious outlook, there is likely to be a form of unconscious fragmentation occurring in the background. This may be projected onto a perceived "enemy," such as the secular or liberal society, but it reflects an incapacity to meaningfully contain competing propensities in the psyche. Here again we see how dissociative thinking on one level points to deeper divisions on another.

Summarizing his own insights, Lifton simply says, "All varieties of fragmentation of the self involve dissociation and formlessness." He goes on to note how, "As the self fragments, its capacity for empathy is lost."[24] Moreover:

> at issue here is the loss of adequate grounding, of a sense of being connected to one's own history and biology ... actions and associations, whether protean or static, become isolated from one's own prior experience as well as from others around one, and nothing seems psychologically trustworthy.[25]

This is akin to what I described above as the dissociation from the psyche itself. Extreme expressions of this phenomenon may also involve what Lifton calls "doubling"—"the formation of a functional second self that is, psychologically and morally, at odds with the prior self."[26]

Turkle's research on online life, especially among digital natives, reveals the way such themes have been playing out. She writes of this demographic, "They described the erosion of boundaries between the real and virtual as they moved in and out of their lives on the screen. Views of self became less unitary, more protean."[27] Turkle continues:

> I was meeting people, many people, who found online life more satisfying than what some derisively called "RL," that is, real life. Doug, a Midwestern college student, played four avatars, distributed across three different online worlds. He always had these worlds open as windows on his computer screen along with his schoolwork, e-mail program, and favorite games. He cycled easily through them. He told me that RL "is just one more window." And, he added, "it's not usually my best one."[28]

Turkle then asks, "Where was this leading?"

When RL, or real life, is not where we are at our best, we will look for alternate realities, and the virtual world offers plenty. At the same time, between our current mode of existence and being able to redesign our minds and bodies or download ourselves into cyberspace, we are stuck somewhere between real life and incomplete forms of escape. And few seem able to keep their feet consciously planted in both worlds. The siren song of the virtual world, where self-invention is easy and relating can be a game or little more than posture and projection, suspends the return to an ecology of mind. Perhaps this is the point. In the virtual world, a more ideal, heroic, or perhaps even romantic, reality can be turned on while the tensions and complexities of actual existence can be turned off. One can live with vicarious ease what is in real life a dangerous trial or initiatory challenge, wherein confronting demons and dragons challenges the whole person, not merely one's gaming skills. Yet the markers on the path of maturation and wellbeing that these confrontations and battles represent do not leave us, even if deflected or numbed. The result is that we are deeply split and precariously imbalanced.

Virtuality not only dislocates identity; it subjects identity to excessive demands for elaboration and contrivance, for there is nothing organic to mitigate the imagined ideal. Over time, who and what we are becomes increasingly prone to invention. It is not so hard to then extrapolate this sense of a virtual, constructed self to the point of being convinced that having a self may have little to do with inhabiting our own skin. As a generation of digital nativism has already shown, life in cyberspace can make for an existence that may be turned on and plugged-in while being hauntingly incomplete. Many people are thus beside themselves, subjected to an electrified future shock in the present day.

The online world sponsors dissociative states of mind through a number of its key features: a capacity to wander wherever we like, anonymous engagements when we get there, the elevation of sound-bites over contextualized knowledge, the manipulation of imagery, the distortion of information and so on. Traversing the World Wide Web with the virtual freedom granted by the click of a mouse or swipe of a finger also invites thoughtless and impulsive actions. By forgetting McLuhan's mantra about the medium being the message, we are prone to think that technology merely provides tools for actions we decide to make; how the tool also shapes those actions is still largely unacknowledged. The general trend is pointedly described by Carr:

> What the history of intellectual technologies shows us . . . is that "the introduction of computers into some complex human activities may constitute an irreversible commitment." Our intellectual and social lives may, like our industrial routines, come to reflect the form that the computer imposes on them.[29]

The apparent control and freedom that online activity offers blinds us to this imposition, making for an overarching dissociation between believing we are in the driver's seat and having our driving practices determined by outside influences.

Social media queens and so-called "influencers," cyber-bullies and hackers, are all licensed by a worldly remove to exploit and manipulate others. Less insidious, though more ubiquitous, everyday screen time may regularly involve the neglect of those around us as well as our own self-care and creative interests. Surveying studies on the effect of social media on adolescent girls, Jonathan Haidt notes that "social media does not act just on those who consume it. It has radically transformed the nature of peer relationships, family relationships and daily activities."[30] Talbot Brewer puts this effect in even starker terms, describing "the addictive simulacrum of sociality found on social media sites"[31] and the associated way "our citizenry is coming apart at the seams." He also notes that "Rates of anxiety and major depressive disorders have shot up dramatically among teenagers and preteens"[32] and argues such phenomena are part of a "great malformation,"[33] in which cultural processes are being consumed by the economic ones. Bearing out Simon's forecast of an attention economy,[34] Brewer contends this malformation is occurring by "tapping into a very personal resource . . . human attention." Pulling these strands together, he argues, "When attention is depleted, there can be no heightened passion, no true friendship, no love. Without attention, we are not genuinely available to anyone at all . . ." Brewer reminds us that "Attention has these enormous powers because it serves as the portal to thinking and acting."[35] Further, "no activity fully worthy of a human being can blossom unless it is carried forward and completed by avid attention to the valuable possibilities latent in it."[36]

We are relating through the hype and distance of the online world while disconnecting from each other and ourselves in ordinary, everyday ways in the offline world. Compensating for ever-expanding online content, we are also applying ever-narrower cognitive filters to that content. Algorithms are designed to keep us online by directing our interests even further. Rather than generating open minds capable of making creative connections and negotiating ambiguous and paradoxically content, overwhelmed minds are forced into tighter recesses of understanding, which, in turn, limit our capacity to commune with others when we are offline. In all these ways, the rhythms of cyberspace have begun to dictate the rhythms of life, sponsoring a more dissociative state of mind, which is in turn fueling more escapist and fragmented conceptions of human becoming. The benefits of the endless expanse of communicative and informational opportunities have thus come with consequential psycho-social costs, leaving us disconnected and compartmentalized, mesmerized, and numbed. But how much willingness exists to face the totality of this situation and adjust our technospheric immersion accordingly?

MOMA

A few years ago I was visiting the Museum of Modern Art (known as the MOMA) in New York City. Among other fine artworks, it houses

Vincent Van Gogh's *Starry Night*. I turned a corner and there it was, one of the world's most celebrated paintings, perhaps the museum's main attraction. A crowd was gathered before it. After getting to place where I was close enough to see the brush strokes, take in this astounding vision of the night sky, and feel moved by the penetrating imagination of this troubled genius, I then stood back and starting noticing an astounding phenomenon. I watched as people came around the same corner I had and realized they had arrived at one of their key destinations for the day. They then proceeded to do the oddest thing imaginable: the museum visitors took out their smart phones and maneuvered themselves to a place where they might take two or three photographs. They checked their phones to confirm they had captured the intended images. And, after performing these two actions, they turned and walked away. I watched, perplexed, as most visitors to this stunning piece of art enacted a variation of this action, and I noted with dismay how few people beheld the actual painting: their primary interest was the capturing of its image. I had no reason to think this scene did not play out in front of this masterpiece, day in and day out.

This choice to record a digital image of the *Starry Night* and do little else betrays a new kind of relationship with reality, one in which facsimile and virtuality have come to hold more significance than immediacy and authenticity. Most of these images were no doubt destined to appear on social media pages or beside text messages, so that even the photograph was not primarily concerned with the work of art so much as with an accessorizing of the digital self—an implicit rather than explicit selfie. Effectively, these people were never in that room. Their consciousness was already elsewhere—beyond the canvass, beyond the museum, and even beyond awareness of an actual self. Giving way to an online, curated version of self, which was, in turn, a marker of what they had sought to become, any actual self had receded. Being in the presence of one of the greatest expressions of human creativity in existence failed to unsettle the curation of that virtual collage that has nowadays eclipsed what was once called reality. These museum-goers epitomized Max Frisch's stark observation "Technology is the knack of so arranging the world that we do not experience it."[37]

More recently, upon opening Johann Hari's *Stolen Focus*, which offers insight into the epidemic of distraction, I was impressed with his description of a remarkably similar scene, observed when he was visiting Graceland, the famed home of Elvis Presley. Here, visitors are apparently handed iPads, which not only function as guidebooks, but reproduce much of what people come across the country and sometimes the world to see. As Hari observes:

> Nobody was looking for long at anything but their screens ... Occasionally somebody would look away from the iPad and I felt a flicker of hope, and I would try to make eye contact with them, to shrug, to say, Hey, we're the only ones looking around, we're the only ones who traveled thousands of miles and decided to actually see the things in front of us—but every time this happened, I realized they had broken contact with the iPad only to take out their phones and snap a selfie.[38]

The moment that epitomized this scene occurred in the Jungle Room—apparently Elvis's favorite space. A couple had discovered and become entranced with the ability to swipe left or right, so the iPad would provide a sweeping image of the whole room. Yet they were standing in the actual room. "But, sir," Hari offered, "there's an old-fashioned form of swiping you can do. It's called turning your head." The couple scurried off.[39] On another occasion, this time in Paris, standing before the *Mona Lisa*, Hari observed essentially the same thing I had witnessed in front of the *Starry Night*.[40]

Two decades after Baudrillard penned his postmodern classic, *Simulacra and Simulation*, describing the cultural elevation of representation over reality, he wrote an essay entitled "Telemorphosis," in which he describes the way screen culture has advanced the representational way of being.[41] There he remarks:

> The individuation [perhaps better termed individualism], which we are so proud of, has nothing to do with personal liberty; on the contrary, it is a general promiscuity. It is not necessarily a promiscuity of bodies in space—but of screens from one end of the world to the other ... The indivisibility of every human particle at a distance of tens of thousands of kilometers—like millions of twins who are incapable of separating from their double. Umbilicus limbo.[42]

Baudrillard is describing a regression into the archetype of the Great Mother—in this case an electronic Great Mother—who provides endless opportunities for mirroring the thin veneers of individuality and independence. Here one seeks a womb from which to be constantly reborn, but is actually caught in an empty if encompassing Web. Hence the repetition compulsion—the endless search for the sense of self in the screen image, the zombies with "faces glued to their smartphones" Harari describes in the epigraph at the start of this chapter. In the end, there is no actual mirror—no real reflection on the part of the other— only solipsistic self-regard. As Baudrillard puts it:

> There will soon be nothing more than self-communicating zombies, whose lone umbilical relay will be their own feedback image—electronic avatars of dead shadows who, beyond death and the river Styx, will wander, perpetually passing their time, retelling their own story.[43]

The resulting affect, Baudrillard argues, is "the banalization of existence"[44] and notes that Heidegger had described "the fall into banality" to be the "second fall of man."[45] This banalization is surely a correlate of psychic numbing—a numbing of actual being through the representations of that being, a giving over to collectivized templates of individuality.

Hermetic Intoxication

Hillman refers to the obsession with data-sharing and connectivity as a "Hermetic Intoxication."[46] In an essay of the same title, he surveys the way of the world and describes how we have set up camp at the altar of Hermes, god of communication and skilled liar, turning our backs on other archetypal principles and modes of relating. Hillman writes that Hermes is "an indiscriminate messenger ... Because he carried all messages without actively entering into the content of what he carried. He had no opinions, values; he made no editorial comments; he did not censor."[47] Today's focus on carriers, open-access, instant messaging, and high-speed connection, alongside a comparative lack of concern with the substance or implications of what is communicated, exemplifies this one-sided arrangement. Data-sharing and free-flow of information are the unquestioned markers of progress around which the global village gathers. Whether or not a genuine increase in knowledge and understanding results from this obsession and whether or not the social fabric is strengthened have turned out to be secondary considerations. Being connected and having the ability to communicate with virtually anyone, anywhere, any time is the critical value—a value many once assumed would unite the world. But it turned out engagement and information could easily become disengagement and disinformation, and to the economics of the digital era (Hermes also happens to be a god of trading and dealing) it makes little difference. Because of the indiscriminate nature of digital communications, for every Arab Spring there is an ISIS recruitment movement, and every advance in connectivity necessitates an advance in cyber security. A new frontier has thus quickly turned into a new Wild West, with a sea of pirates waiting right offshore.

With so much of our consciousness tied to our digital devices, we have to remind ourselves that beyond the cathedrals of digital connection and their looming shadows lie forms of communication more direct and usually more satisfying. Hillman describes "the connection—wordless, intimate and sensate—between lovers, between mothers and babies, between patient and nurse, between animals and their caretakers."[48] Yet being silent and physically close require a slowing of time and lack of distraction. So too "the communication of teaching and learning," which Hillman notes is "slow, painstaking, and without the flash and fun of Hermes."[49] Nowadays digital natives entering college may demonstrate a capacity to retain and reproduce information but also an incapacity to think in cross-disciplinary or imaginative ways. Neighborly love, deep friendship, and a genuine sense of place also suffer when the portal to everything sits in the palm of our hand.

Hillman works these contrasting patterns by pointing out the absence of Hestia, goddess of hearth and home, who is frequently paired with Hermes in the Greek mythological tradition. If Hermes personifies the fluidity and flux of communication, Hestia portrays the principle of the fixed and focused. Archetypally, their patterns complement and mitigate each other:

> Hestian consciousness circles around itself. It goes nowhere, intends nothing outside of itself. So, Hestia was always seated on circular

elements, and the places where she was worshipped were alway circular . . . it was to Hestia that one went for sacred asylum, where one could find refuge and stillness.[50]

We must remember that it is not just the gods themselves that are universal and timeless but also the way they interact. And so we find, as Hillman observes in his treatment of this theme, "When Hermes and Hestia are not in a sympathetic rhythm, then they push each other to extremes."[51] The hyper communication of the digital age, which has surely furthered and partially defined globalization, might thus be expected to exist alongside a similarly polarizing expression of Hestian inclination. Given today's geopolitical atmosphere, Hillman's insight into this is revelatory: "As Hermes can go mad with hermetic intoxication when severed from Hestia, so a monotheism of Hestia becomes only fanatical purity, fanatical devotion, the single-focus upon home, homeland, and family relations."[52] Here we find an echo of Lifton's syndrome of the fragmented and static self, and the related contrast between the protean and the totalitarian. On the one hand, the digital age has connected the world; on the other hand, it has provoked a retreat into nations and tribes, as if to defend against the diffusion of identity that has come with having a psychic presence in many places and in many ways.

In an earlier commentary on Hillman's prescient treatment of this theme, I point out that it is not just Hestia that our online activity fails to observe. There are qualities of Hermes himself related to the comprehension of communications that are also absent. Although Hermes is singular in the Greek Pantheon for his capacity to travel between heaven and the underworld, delivering messages to all who lay between,

> the *internet* Hermes leaves passage to the underworld behind, traveling every which way but down. What we mostly do on the internet is "surf"—skim along as each new wave of information arises . . . Pause, reflection, contemplation, or depth of any kind is mostly absent.[53]

Hermetic intoxication leaves behind the god's verticality. "Hermes, we recall, is also the god of interpretation, guiding our efforts to discern meaning and significance amidst the merely informative." To acknowledge these traits is to realize that "Hermes, fully apprehended, requires a vertical imagination."[54]

The contrast between the endless horizontality of hermetic intoxication and the missing verticality of underworld familiarity, hermeneutic skill, and Hestian home-coming is effectively summed up in an observation of Flusser, who wrote, "at bottom, that is what we know today: that we can know everything except that which interests us."[55] We are thus called back to ourselves and to a kind of careful comprehension and layering of knowledge that anchors us to what is more enduring and more profound.

A critical part of what makes the one-sided hermetic presence so intoxicating is the quickness of digital life. Speed has become an unquestioned value. The roots of this go back to the mechanical marking of time and the focus on efficiency and productivity in the industrial age. Now it drives the pursuit of faster computation in the post-industrial age. Yet whereas the ability to do certain things faster would, logically, seem to free up time, ubiquitous computing has instead come to fill almost every waking moment.

Yet nowhere in this tacit agreement to continually adapt to computation is there any recognition of the limited nature of psychic energy, nor any recognition of the qualities of experience that go along with speeding things up or slowing them down. Instead, aligning with our devices leaves large chunks of our nature behind—or at least dissociates us from what these other parts of our nature are also attempting to communicate. As doing more so often adds up to being less, going faster can itself be a form of numbing. If we were more in touch with the deeply human, speed would be a constant reminder we are different from machines; instead, we are doing all we can to keep up with them, exacerbating the divisions within us and obfuscating our awareness of these divisions at the same time.

There may be certain speedy activities such as car-racing or playing sports where the body and some parts of the mind are engaged in rapid decision-making and executing actions while other parts of the mind seem to slow down. Intense focus is sometimes accompanied by the subjective perception of things decelerating. Except for these kinds of experiences, however, when the mind is tracking information or sensory input that requires it to think quickly or move between different kinds of stimuli in short spans of time, comprehension and self-awareness have been shown to decline.

In her essay "What Twitter Does to Our Sense of Time," Jenny Odell provides a vivid picture of the digital time warp most of us now experience, and how at least one form of social media sensationalizes some aspects of life while numbing others.[56] She writes that "It seemed that the more I used these platforms, the more I got psychologically adjusted to a certain social frame rate—one that happened to be ticking by with constant developments and quickly evolving outrage. It was as if opening my phone revealed a stream of time running much faster than the one in the room where I sat." This might be a sustainable practice if, in fact, we were to dip in and out of such media streams and maintain a mental contrast between the online and offline worlds. But in an era of ubiquitous computing and lives lived largely in cyberspace, the rhythms of being at home and in one's immediate community are giving way to the rhythms of social media. Occasionally, someone like Odell awakens to this:

> I found it harder to pay attention to other events and processes that took longer or played out less sensationally even if I cared about them—like the local effects of climate change, grass-roots housing campaigns or even just the details of friends' lives. I felt like my thoughts were running on shorter loops or never getting completed. Even my breaths were short, as though a full inhale couldn't fit into such tiny intervals, and my joints would ache from a state of constant anticipation. It was the feeling of a furrowed brow, applied to my entire body. Most haunting was a sense that I had no substance, and that the physical world, with all its minute fluctuations and gradual changes, was somehow losing its color and texture.[57]

As is the case with so many side effects of our adaptation to new technologies, the problem is what gets left behind. What are the long-term implications of this hypertrophic horizontality and numbed verticality?

"The Constitution of Knowledge"[58]

Tech enthusiasts often describe the internet as an extension of their minds, on hand for easy retrieval of practical know-how and rudimentary facts. Of course, even as some adopters of this view are forced to admit, to use the internet for knowledge beyond these matters entails not only shifting through a mass of information; it often means facing a mountain of competing perspectives, which has come to include the notoriously termed "fake news" and "alternate facts." For those capable of research and critical thinking, who may also have the time and disposition to carefully sift content, this situation is difficult enough. But for those lacking in such skills and unwilling to entertain a complex array of sources, the internet has become a tangled web of information, misinformation, and disinformation. The great irony of the age of information is that we may not, on the whole, be more informed.

Kelly, a leading commentator on all things technological, suggests this situation cuts both ways. He admits "every fact has its antifact," that "the Internet's extreme hyperlinking highlights those antifacts as brightly as the facts," and even that "you can't rely on experts to sort them out, because for every expert there's an equal and countervailing antiexpert." Yet he also extolls the value of "uncertainty," which he notes "is a kind of liquidity" whereby "opinions shift more" and "interests rise and fall more quickly." To punctuate this point he states, "I am less interested in Truth with a capital T and more interested in truths, plural."[59] He wraps up this mode of experience by noting, "it often feels like a *waking dream*."[60]

The postmodern character of thought, which emphasizes pluralism and elevates the subjective, clearly plays into Kelly's description. Many recognize there has been a necessary corrective to the programmed and authorized ways of knowing that dominated the early modern era, sometimes devaluing the subject and overlooking socio-cultural context. Uncertainty and complexity do have distinct epistemological and ontological value. In my own field of depth psychology, making room for unknowing helps keep the door open to the unconscious. Yet, intentional or not, there is some sleight of hand in Kelly's description: swimming in fluid forms of knowledge and drowning oceans of disinformation are vastly different. There is a similar difference between following shifting interests and being sucked into rabbit holes of distraction, between plural expressions of truth and pernicious manipulations of "post-truth," and between waking dreams and psychotic episodes. And the problem is, many are succumbing to the less conducive of these internet-derived ways of knowing.

The internet has turned out this way because it purposely flattens and fragments knowledge. It flattens knowledge by effectively giving all sources equal standing and all voices the same size megaphone; it fragments knowledge by packaging pieces of information to capture our attention in the most efficient and transactional ways possible. The flattening erodes a discernment of providence and authority; the fragmentation undermines the inclination and capacity to understand the larger contexts of knowledge. Both processes are dissociating, leading to a greater degree of numbing around the veracity of information and, eventually, to an inability to understand how the world works. When discerning what is important from what is unimportant becomes too difficult, people shut down. Jaron Lanier, the pioneer of virtual reality, noted in 2011 that, following the early years of optimistic and online creativity, "In the last decade . . . the Internet has taken on unpleasant qualities and has become gripped by reality-denying ideology."[61]

A vivid point of contrast to these prevailing online conditions is provided by what Rauch calls the "constitution of knowledge," which he describes as the epistemic foundation of liberal society. This constitution is based on

> a dense network of norms and rules, like truthfulness and fact-checking; and they depend on the expertise of professionals, like peer reviewers and editors . . . the entire system rests on a foundation of values: a shared understanding that there are right and wrong ways to make knowledge.[62]

After expounding upon this means by which we have, for the most part, collectively determined what is factual and real, Rauch turns to address "the most unpleasant epistemic surprise of the twenty-first century;" namely, the way "digital media have turned out to be better attuned to outrage and disinformation than to conversation and knowledge."[63]

As we have acquiesced to this undermining of knowledge and been driven to distraction, the epistemic foundation Rauch has described has begun to crack at its base. For those directly involved in the advent of social media and other online information technologies, the willful neglect of this phenomenon is damning. As Rauch puts it, "design flaws and perverse incentives in the digital information environment actively favored epistemic anarchy."[64] "We forgot," he writes,

> that staying in touch with reality depends on rules and institutions . . . We forgot that information technology is very different from knowledge technology. Information can be simply emitted, but knowledge, the product of a rich social interaction, must be achieved.[65]

Forgetting what supports the accumulation of knowledge is easy when power and money trump other values: "The commercial internet was born with an epistemic defect: its business model was primarily advertising-driven and therefore valued attention first and foremost . . . a no-holds-barred race to attract eyeballs."[66] Noting the way innovations such as "like" and "retweet" buttons amplified the effect, Rauch quotes one industry insider as saying "We might have just handed a four-year-old a loaded weapon."[67] One outgrowth has been the creation of a new online tribalism, wherein the click of a button can simultaneously create the feeling of belonging in one place and enmity in another. The resulting polarization is just one of the formidable barriers to genuine knowledge and understanding that has been erected.

In this context we might again consider the psychological difference between using a tool and becoming tools of that tool. It may also be time we stopped imagining the internet as a vast neutral realm and start seeing, as Rauch has, "how digital technology can fine-tune appeals to outrage and target them to the most receptive audiences." As he observes, "It learns what we click on and what other people similar to us click on, then builds a virtual avatar of us, then feeds whatever our avatar wants more of."[68] By "running liberal science in reverse," digital media "inverted the social incentives which the reality-based community depends on." Whereas, "The Constitution of Knowledge checks before transmitting," digital media "reward(s) instantaneity and impulsivity."[69] It has, in effect, created minds that are uncertain on one level and all-too-certain on another, minds that move from being zoned out to being overrun by the id, with genuine thinking disappearing into the chasm between them.

Whereas Kelly claims "the propensity of the Internet to diminish our attention is overrated," he also acknowledges that "smaller and smaller bits of information can command the full attention of my overeducated mind." He also notes, "I happily swim in this rising ocean of fragments."[70] I would suggest such attentive, happy online swimming can occur only when a person can bring some semblance of the constitution of knowledge along with them. That is, they have the epistemological style and practice by which to sort and arrange fragments of information. They can orient themselves in a sea of competing perspectives because their minds have, through education, socialization, and professional activity, been attuned to a system of checks and balances that are anchored in a consensus understanding of reality. The internet itself, however, does not provide this, and those who have never learned to swim just drown. Worst of all, many of those who drown somehow manage to think of themselves as great swimmers.

In the decade or so since Kelly penned his reflection, antifacts have been shown to act in a more viral manner than facts, appealing to political and cultural biases and aided by algorithms designed to keep people online. Authoritarian leaders and conspiracy theorists have learned to utilize this phenomenon, inflaming culture wars and politicizing everything they set their sights on. Tech corporations have also realized that the distortion of reality is more profitable than the advancement of knowledge. The one organization that Rauch and others frequently cite as circumventing this undermining of knowledge is Wikipedia. Fittingly enough, this is a non-profit outfit that relies on the checks and balances of experts in each subject area.

Kelly's interest in "truths, plural" and his penchant to "flow through this slippery Web of ideas" is conducive to creative thought only when you are capable of finding the shoreline, you are savvy about the way the world works, and you understand how genuine knowledge is generated. When these capacities are absent, however, people are just tossed about in a sea of agenda-driven worldviews, some of which are deliberately designed to maximize confusion and undermine consensus understanding in order to create more traction for falsehoods. Of course, journalists and academics can also succumb to political, social, and economic agendas. However, both the free press and institutions of higher learning use methods of layering knowledge as well as processes of peer engagement that work to generate genuine understanding, flag distortion, and highlight outliers. Amateur sleuths, independent scholars, and mavericks also play their part, and sometimes prevent consensus knowledge from calcifying. But effective contributions to the constitution of knowledge depend on a willingness to be familiar with epistemic standards, critical voices, and relevant sources. This layering of knowledge involves sorting through ideas and discerning perspectives in order to handle information in a conducive way.

What all of this decidedly human and community-based discernment of information does is add vertical bearings to the horizontal expanse of data and information, bearings that take us from information to knowledge to understanding. As Zuboff has convincingly shown, social media platforms have accelerated this absent verticality by organizing their platforms in ways that tend to erase the qualitative differences between news stories. She points to

> Facebook's decision to standardize the presentation of its News Feed content so that "all news stories looked roughly the same as each other . . . Whether they were investigations in *The Washington Post*, gossip in the *New York Post*, or flat-out lies in the *Denver Guardian*, an entirely bogus newspaper."[71]

Such efforts result in what Zuboff calls "equivalence without equality."[72] When the online world either deliberately or effectively produces such equivalence without equality, and people are no longer equipped to differentiate between sources of information, the result is an uninformed if not disinformed citizenry.

An absent verticality of understanding leaves us with an informational free for all that is expanding on all sides. Unlike Kelly, most have neither the book knowledge nor the technical chops to endure the experience of being pulled in all directions and come out the other side with a complex, nuanced, and substantive sense of the facts and their meaningfully connection. Whereas we might have once imagined the internet augmenting human knowledge in a Wikipedia-like manner, allowing something akin to a grand palace of comprehension and insight to be built in cyberspace, the medium has itself furthered habits of dissociation, splintering rather than layering what we come to know and how we come to see.

The entire post-industrial world is susceptible to the dissociative engine the online world has become. But America may be especially vulnerable to the dynamics of "doubling" and fundamentalism Lifton details. In his treatise on the nation's cultural-historical trajectory, Mações contends, "The principle of unreality is an answer—a specifically American answer—to the shallowness of life in a modern liberal society."[73] In the previous chapter we already examined the frontier as an image that both connects and propels the nation. Here we may locate another connective element, one that Mações contends shows what the West Coast innovators and Midwest churchgoers have in common:

> What contemporary America exhibits is a world of worlds where the high-tech utopia of San Francisco exists side by side with those parts of the country where many more people believe in heaven, hell and angels than in the theory of evolution. The European will say: there is no truth; the American: there is no truth, so everything is true. The difference between Europeans and Americans is that the former see the great narratives of nation, religion or money as fictions to be abandoned while the latter embrace them all the more for being fictions.[74]

As I indicated at the start of this book, we must contend with rather than reject outright the fictive leanings of the psyche. But it is just no longer workable to take these fictions literally or fail to see what Lifton has pointed out in terms of the shadow side of the protean self. Both the fundamentalist and the conspiracy theorist are cut from the same cloth: that of wanting to bypass the complexities of life through the embrace of packaged meanings and paranoid explanations. When this fictive function of the psyche manifests in such defensive ways, it stalls the culture-making process and undermines the bridge between nature and civilization. People are then prone to swing between the need to satisfy raw instinctual needs and institute repressive social structures. Mações thus seems right in asserting that "American populism . . . is in love with the idea of designing and building imaginary worlds,"[75] but he may be wrong when he also writes these worlds are "highly structured simulations that are just as complex and rich as the real world."[76] In actuality, the more reality is manipulated, the more resulting narratives tend to revert to caricatured heroes and stereotypical expressions of otherwise complex mythemes. America may be destined to become a giant theme park, but it is apt to remain unconscious of the inner workings and underground structures that make such escapist adventures possible. It will, instead, allow the fabricated world to replace the real world. Replacing actual persons with fabricated ones is then but a short step away.

Medicated Numbing

Medicine has become complicit in the syndrome of dissociation and numbing. Both psychotropic medications and opioid painkillers have been prescribed at sharply rising rates during the first two decades of the millennium. In the first decade alone, in the United States, prescriptions for psychotropics increased over 20%, with one in five adults being on at least one of these medications.[77] It was also found at the time that "almost four out of five of these prescriptions for psychotropic medications are written by physicians who aren't psychiatrists."[78] Over a decade later, on the heels of the COVID pandemic, nearly one in four Americans were taking these medications.[79] Even more telling, between 2000 and 2010, total opioid analgesic prescriptions doubled,[80] with one study finding that nearly 90% of prescription opioid users were self-medicating "anxiety or fear . . . depression or sadness . . . anger or frustration."[81] This same study found that "dispositional mindfulness"—the ability to attend to thoughts and feelings in a non-judging way—"was inversely associated with opioid self-medication."[82]

Psychotropic medications can support a person's capacity to tolerate their suffering enough to facilitate psychotherapy and preserve everyday functioning. They may prove necessary when psychotherapy is unavailable or would likely be ineffective. However, when doctors and patients reduce psychological symptoms to neurological malfunctions and then treat them accordingly they also erect barriers against the integration of feelings and emotions. These barriers—explicit in neurochemical explanations of psychological ailments and implicit in the adoption of chemical-based treatment plans, as well as actualized in the numbing effect of these medications—belong to the broader syndrome of dissociation playing out at the socio-cultural level. More frequently than not, patients are told their depression or anxiety is the result of a "chemical imbalance"—reducing a psychological condition to a physical illness in need of a drug-based response.

In Davis's compelling study of persons taking psychotropic medications and adapting to a neurochemical explanation of their psychological symptoms, he describes a change "in the terms in which ordinary people imagine self and suffering, its causes and its resolution,"[83] as well as a "growing break with an earlier way of imagining suffering in terms of mental life and interpersonal experience ... we might term 'psychological' or 'psychosocial.'"[84] It is a change to what he calls the "neurobiological imaginary."[85] In discussing his findings, he describes one of the study's participants he calls Lisa as someone who sees "getting in touch with her feelings as a guide to her truth," yet has come to accept "a *disengaged* perspective" wherein "she can *distance herself* and view interactions almost as though they were happening to someone else."[86] As Davis writes:

> Lisa has come around to the misfiring brain explanation because it quiets all of her worries at once. It affirms that she is not being inauthentic, for the strong and unruly emotions she experiences are not actually hers. The emotional register of those with a "normal" brain is the appropriate standard ... the only standard. And self-control is not the issue; it cannot touch what is in fact wrong. She is dealing with a physical disorder ... .[87]

The drugs depersonalize feelings; the explanation negates psychological context and etiology. Summarizing his understanding of the experience of Lisa and other participants in his study, Davis observes the persuasive power of "psychopharmaceuticals as interpreted through the neurobiological imaginary," moving "toward an 'essentially mechanistic' view of ourselves." In conjunction, he suggests this view is "subserv[ing] a larger cultural aspiration of liberal selfhood, a promise of freedom, yet in a way that diminishes the person." Working in and around this mindset, he sees "a move toward the de facto authorization of a solitary and empty conception of the will."[88]

By showing how the neurobiological view overlaps with what he calls "our troubled quest for self-mastery,"[89] and how these outlooks coincide with an eroding understanding of inner life and psychological complexity, Davis is also exposing the growing inclination to discount the psyche altogether, an inclination that has taken root within psychology itself. For while psychology has provided most of the counseling and psychotherapy undertaken by at least a portion of these patients being treated with psychopharmaceuticals, influential movements within the field have also spent decades replacing understandings of the inner world of the psyche with mechanistic models of the mind and cognitive-behavioral management techniques tailored to the economics of managed care and the training of psycho-technicians. Among other impacts, for many professionals in the helping professions and laypersons alike, such reductive and mechanistic approaches have turned the inner reaches of human nature into foreign territory.

When the dust settles on this biological reductionism and cognitive-behavioral mechanization in the broad field of psychological care, it will come to be seen as one of the worst examples of scientism to which we have subjected ourselves, aiding and abetting the wholesale neglect of our deeper nature. Instead of following symptoms into their psychosocial basis, "illness comes gradually to be defined in terms of that to which it 'responds.'" Andrew Lakoff refers to this as "pharmaceutical reason," which has become "the underlying rationale of drug intervention in the new biomedical psychiatry," holding "that targeted drug treatment will restore the subject to a normal condition of cognition, affect, or volition."[90] He suggests that this stance stems from psychiatry's attempt to become a "viable technical practice."[91]

Such stances have stopped us from being psycho-logical, removing a sense of the psyche from the way most of us think about life. The logic of working backwards from the effects of medication have also lead to deeply flawed assumptions. Following an extended journalistic odyssey into neuropsychology and the psychopharmaceutical industry, Johann Hari's blunt assessment is that "We have been systematically misinformed about what depression and anxiety are."[92] Among other findings, he points out that "In the biggest study of serotonin's effects on humans, it found no direct relationship with depression."[93] Hari interviewed dozens of leading researchers on the neurochemical basis of depression and anxiety. The assessments he received ranged from "There's no evidence there's a chemical imbalance" at work in depression and anxiety to "Almost everything you were told was bullshit . . . The serotonin theory 'is a lie.'"[94] When Hari pivots to search for more substantive causes of depression in particular, it is in response to a comment by one psychologist who simply noted, "we're such an utterly disconnected culture." This leads him to ask: "What if depression is, in fact, a form of grief—for our own lives not being as they should? What if it is a form of grief for the connections we have lost . . ."[95]

For nearly a century, we have used forms of psychological defense in the guise of psychological treatment and failed to connect what ails us to the greater story of our lives, both individually and collectively. When we are deprived of the understanding that psychological symptoms are tied to socio-cultural problems, our capacity to change these problems is also hobbled. Arguing that doctors are prescribing psychotropic medications at such high rates because they "have taken on the responsibility of curing everyday unhappiness,"[96] Ronald W. Dworkin also notes that "small changes in the mind may have serious social consequences." Further, "when a man silences his misery through artificial happiness, he also silences his conscience."[97] It may thus be no stretch to say that society is being robbed of a critical early warning system.

Depth psychologists such as Jung, who expanded our understanding of the human unconscious by showing the way individual problems have archetypal backgrounds, instigated a form of psychotherapy in which the patient comes to see suffering as a two-way street. By grasping the piece of the universal human story behind our sufferings, whether pertaining to love relations, childhood wounds, thwarted creativity, or any number of timeless themes, we can alleviate if not eventually heal the acute alienation that often accompanies the psychological pain. The alienation and pain occasion a kind of existential revelation—a glimpse of the way archetypal patterns, traditionally related to as gods, enter life and draw us into their dramas. One higher purpose of suffering in this manner appears in the production of art, literature, and other forms of symbolic expression. Such expressions thicken the cultural container and provide storied pathways by which subsequent generations may negotiate the most challenging facets of being.

At least some neurobiologists back the idea that psychological pain has a deeper purpose, both individually and collectively, and thus prompt the need to find integrative rather than dissociative forms of alleviation. As Antonio Damasio puts it:

> Pain and pleasure are the levers the organism requires for instinctual and acquired strategies to operate efficiently. In all probability they

were also the levers that controlled the development of social decision-making strategies. When many individuals, in social groups, experienced the painful consequences of psychological, social, and natural phenomena, it was possible to develop intellectual and cultural strategies for coping with the experience of pain and perhaps reducing it.[98]

After noting that emotion is the human organism's way of registering and consciously experiencing pain, Damasio notes, "suffering offers us the best protection for survival, since it increases the probability that individuals will heed pain signals and act to avert the sources or correct their consequences." He goes on: "If pain is a lever for the proper deployment of drives and instincts, and for the development of related decision-making strategies, it follows that alterations in pain perception should be accompanied by behavioral impairments."[99] We should not skate by this crucial point: The experience of pain is vital for the "proper deployment" of our deeper drives and instinctual impulses—including the tinkering instinct. Moreover, if we alter the way in which we experience pain and the emotions surrounding it, we risk impairing our behavioral responses. As Damasio puts it at the beginning of a subsequent text, "Feelings of pain or pleasure or some quality in between are the bedrock of our minds."[100] To dissociate from pain and habituate to contrived forms of pleasure may therefore be a formula for slowly losing our minds.

Along similar lines, the psychiatrist Julie Holland has raised the alarm over the number of women taking psychotropic medications, the ease with which they are prescribed, and the unwarranted fear of emotionality behind this phenomenon. She writes: "The pharmaceutical industry plays on that fear, targeting women in the barrage of advertising on daytime talk shows and in magazines." Holland continues: "the increase in prescriptions for psychiatric medications, often by doctors in other specialties, is creating a new normal, encouraging more women to seek chemical assistance." Her diagnosis? "This is insane."[101]

Her focus on women highlights the cultural devaluation of the kind of emotional sensitivity and feeling response that can play a more primary role in their psychology. She notes, "we are under constant pressure to restrain our emotional lives." The core concern is that this natural dimension of women's psyches, necessary in responding to events and experiences in ways that foster personal growth, is being chemically incapacitated. Men are obviously vulnerable to the same syndrome, but women are more likely to be prescribed these medications, underscoring this cultural devaluation. Holland writes, "when we are overmedicated, our emotions become synthetic. For personal growth, for a satisfying marriage and for a more peaceful world, what we need is more empathy, compassion, receptivity, emotionality and vulnerability, not less."[102]

Holland qualifies her comments by acknowledging such drugs can improve some people's lives. But they can also be unnecessary. The picture emerging from the increased use of psychotropic medications suggests not only a pattern of over-prescription, but over-prescription as a symptom of a vicious cultural circle in which the new normal may be a chemically engineered one. This will, over generations, also leave us with a different kind of human being. As Holland writes:

> more serotonin might lengthen your short fuse and quell your fears, *but it also helps to numb you, physically and emotionally* . . . Some people on SSRIs have also reported less of many other human traits: empathy, irritation, sadness, erotic dreaming, creativity, anger, expression of their feelings, mourning and worry.[103]

Piecing together what Damasio is telling us about human neurobiology, what Jung offers in terms of the integrity of the psyche, and what Holland is observing in the field, we are forced to consider the larger frame here, recognizing that the mere alleviation of pain and relief of psychological suffering without understanding its root cause or social context may be an ultimate disservice to both the individual and society. Any psychologist interested in the actual roots of individual suffering will help the patient recognize the intrapersonal, interpersonal, and cultural conditions behind such suffering—in order to bring enough consciousness to these conditions to instigate effective and lasting change. It is such consciousness and the meanings that derive from this which allow an acceptance of the human predicament. When the patient realizes that pain has a purpose, and that their psyches may be generating opportunities for deeper and wider awareness as well as for behavioral and relational change, they suffer differently and can experience genuine transformation—in themselves and perhaps even extending into the world around them.

Division and Integration

The psycho-cultural patterns and trends addressed in this chapter indicate that post-industrial dissociation has been layered on top of industrial dissociation. That is, while the fragmentation of the digital age has its own influences, it also functions as a defensive response to the modern loss of meaning and existential disorientation. To reverse this defensive response would thus involve peeling back these defensive layers and facing an underlying alienation from the needs of the deeply human. As Baudrillard noted in his early work, alienation has been built into the structure of society, and is maintained by economies that commodify all things, including ourselves.[104] Heidegger sees the same dynamic in terms of the enframing involved in modern technology. It is in such ways we remain stuck in horizontal planes of becoming, rather than recovering vertical axes of being.

Alienation is the felt experience of psychological division. It is a vertical matter. Any effective path to authentic rather than neurotic suffering requires consciousness of this divided verticality and confronting rather than numbing the depressions and anxieties that are its results. To return to one of Jung's axioms of human maturation, we must experience this inner disintegration before a new integrity can be found. Yet socially, economically, and technologically, contemporary life seems designed for avoiding this kind of inner confrontation and authentic suffering, instead opting for the hypertrophy of horizontality.

Without more vertical awareness, the consumerist way of life, which is geared to externality rather than internality, will surely keep the cognitive-enhancement and body-modification industry of tomorrow well afloat. Just as people are prone to seek the latest and greatest emblems of material success, when the time comes, they will likely buy whatever reengineered brain chip or body part promises higher status. As material pursuits are already driven mainly by the search for status and power, and as biological enhancements are likely to be tailored for improved intellectual and physical performance, these products and services are likely to be even more highly valued than regular emblems of material success. Those who take full advantage of these concrete self-improvements will also be likely to have a significant social advantages. Beyond mere association with objects that convey a certain image, image-consciousness will be directed towards a literal remaking of oneself.

Aside from the psychological implications, such dynamics are likely to create an even more stratified society. The gap between rich and poor has been widening everywhere. Yet, so far at least, the advantage of being wealthy has not penetrated the essence of being a person or the degrees of our humanity. However, wealthy parents are already pushing to give their offspring a biomedical edge through elective use of stimulants and other performance-enhancing drugs. Genetic engineering for desirable traits in children is currently banned, but is likely to emerge at some point, even if via high-end black markets. Playing out such scenarios, McKibben has written about the " 'disturbingly comfortable fit between the techno-eugenic vision' and our consumer society,"[105] and that "there's no obvious line between repair and improvement."[106] He states that *"we stand on the edge of disappearing even as individuals,"*[107] that "the chance for emotional growth, for becoming 'real' in some deep sense, would dwindle," and that we must "try to chart the emotional geography of this world we are so near to stepping into."[108] Yet how could we even begin to chart such geography without authentic contact with our emotions?

In the interest of shaking off the numbing effect of the digital world, it may therefore be time to stop talking about soul loss and start talking about *soul murder*. This is a term psychoanalysts have used to describe the way narcissistic parents destroy their offspring's sense of self.[109] Unable to properly mirror and support the individuality of their children, these parents, like vampires, suck the psychological life from them, typically in some perverse attempt to stave off their own inner emptiness and reinforce their own outer facade. Something akin to this essential dynamic may now be playing out on a cultural level.

If it "takes a village" to raise a child and whole societies begin to act like narcissistic parents, instilling superficial external ideals and creating virtual hothouses to finesse the persona rather than form character, collective soul murder will result. Villages vary. But in the digital age, the primary village is the online world, and each young adult has more or less unlimited exposure to it. Images of success and notions of well-being no longer come from local heroes and community elders but from entrepreneurs, pop stars, actors, sporting heroes, and influencers with the most online presence. This presence has itself become the measure of success, elevating the devious and notorious alongside the dignified and noble, often leaving young people with little sense of the difference. Baudrillard writes that "Somewhere, we all mourn this stripped reality, this residual existence, this total disillusion."[110] Reflecting on this postmodern philosopher's insights, his translator comments, "We must attempt to liberate ourselves from our own fascination with the lowest common denominator of existence: the banality of existence itself."[111] Baudrillard refers to these tendencies as "the equivalent of the suicide of the species."[112] Perhaps this begins with soul murder.

Tech corporations and the governmental agencies that fail to regulate these corporations could be susceptible to charges of soul murder. We may all, even if unwittingly, be accessories to such murder. When Francis Haugen, aka the "Facebook Whistleblower," testified before congress, she presented compelling evidence that this company put profits ahead of both socio-political stability and the psychological wellbeing of adolescent girls (a focus of internal studies). Rather than indulging in the fantasy of bad actors at work, which is an easy enough charge, we might consider the way corporations are themselves organized to compartmentalize and dissociate, especially when it comes to finding ways to disconnect from their infliction of harm, with their leaders taking minimal responsibility for their companies as a whole. Along such lines it has been observed that corporations frequently display psychopathic behavior, showing little capacity for empathy and little remorse for destructive actions.[113] Even when this trait is not directly attributable to individual office holders—although frequently it can be—it is the systemic lack of accountability that fosters the inhuman actions of the corporation as a whole. To the extent the technosphere is shaped by corporate psychopathy, and to the extent this environment shapes the minds of young people, is the extent to which we might reasonably charge these so-called "tech giants" with soul murder—even if such charges are brought only in the courts of social science and public opinion.

To the work of social scientists and cultural commentators who have noted the prominence of dissociation and numbing in contemporary life, I am thus attempting to add the understanding of how this style of consciousness is smoothing the way for a posthuman future, with innovation providing the vehicle and existential escapism and normative psychopathology providing the fuel. With an attention economy, driven by surveillance capitalism, enabling the commodification of ourselves, and a growing inability to feel the effect of these trends, we are taking our first steps toward such a future. We provide the attention, or allow it to be taken from us, which generates the data used to alter our thinking and behavior. Money and power flow to those with the most incentive to maintain the system. These are the ways our dissociative tendencies will provide the opening for a futurist turn toward more profound forms of human alteration.

Without embracing all we know of ourselves and the extent of the psychic fields we inhabit, we are likely to allow the prevailing current to carry us "like a river that wants to reach the end, that no longer reflects, that is afraid to reflect,"[114] as Nietzsche puts it. The ability to reflect upon the river of ideas and sensibilities carrying us is thus the task before us, a task requiring the capacity to feel the divisions of our psychic makeup, whether these are inherent to the deeply human, part of the history of ideas, or part of modern existence. From the collision of the conscious and unconscious, to the psychological implications of Cartesian philosophy, to the separation of spirit and nature, we are, as contemporary citizens of this world, rife with divisions and charged with making them more conscious.

In psychotherapy, any expansion of consciousness or restoration of integrity necessitates moving beyond dissociative tendencies. No deeper understanding, transformation of emotion, or capacity to restory life is possible without first overcoming the tendency to ignore or displace underlying psychological problems. Patients must in the end befriend their problems, not become more detached from them. This is a matter of feeling as much as thinking, and these same broad conditions apply to the collective situation. Only by first wrestling with our divided nature and awakening to what we are leaving behind may we arrive at the integrative stance able to stem the tide of self-commodification and offer a contrasting vision of becoming to the posthuman one. The next chapter will thus work toward a clearer grasp of the divisions underpinning the current push to computerize the mind and roboticize the body.

Notes

1. William James, *The Will to Believe and Other Essays in Popular Psychology* (New York: Longmans Green, 1907), 69.
2. Yuval Noah Harari, *21 Lessons for the 21st Century* (New York: Spiegel & Grau, 2018), 271.
3. Robert W. Rieber, *Manufacturing Social Distress* (New York: Plenum Press, 1997), 70.
4. Vilēm Flusser, *Post-History* (Minneapolis, MN: Univocal, 2013), 17.
5. I first described this difference in an essay entitled "Numb." In Stanton Marlon ed., *Archetypal Psychologies* (New Orleans: Spring Journal Books, 2008), 351–367.
6. CW 18, par. 149.
7. Melina R. Uncapher and Anthony D. Wagner. "Minds and Brains of Media Multitaskers: Current Findings and Future Directions." *PNAS*, vol. 115, no. 40, 9889–9896.
8. See Johann Hari, *Stolen Focus: Why You Can't Pay Attention—and How to Think Deeply Again* (New York: Crown, 2022).
9. See Nicholas Carr, *The Glass Cage: Automation and Us* (New York: W. W. Norton, 2014).
10. CW 17, par. 334.
11. CW 10, par. 361.
12. Jean Baudrillard, *The Illusion of the End* (Stanford, CA: Stanford University Press), 108.
13. Ibid., 108.
14. Ibid., 109.
15. See, for example, *Ghost in the Shell*, directed by Mamoru Oshii (1995; Santa Monica, CA: Lionsgate, 2020), Blue-ray Disk.
16. *Forbidden Planet*, directed by Fred M. Wilcox, featuring Walter Pigeon, Anne Francis, Leslie Nielsen (Metro-Goldwyn-Mayer, 1956). Film.
17. *Altered Carbon*, Richard K. Morgan (New York: Del Rey, 2003).
18. Robert Jay Lifton, *The Protean Self: Human Resilience in an Age of Fragmentation* (Chicago: University of Chicago Press, 1993).
19. Ibid., 9.
20. Ibid., 10.
21. Ibid.
22. Ibid., 11. Italics in original.
23. Ibid., 192. Italics in original.
24. Ibid., 206.
25. Ibid., 207.
26. Ibid., 208.
27. Sherry Turkle, *Alone Together: Why We Expect More from Technology and Less from Each Other* (New York: Basic Books, 2011), xi.
28. Ibid., xii.
29. Carr, *Glass Cage*, 207.
30. Jonathan Haidt, "A Guilty Verdict." *Nature*, 2020, vol. 578, 226.
31. Talbot Brewer, "The Great Malformation." *The Hedgehog Review*, vol. 25, no. 2, summer 2023, 119.

32. Ibid., 120.
33. Ibid., 109.
34. Herbert A. Simon, "Designing Organizations For An Information-Rich World." In *Computers, Communications, and the Public Interest*, edited by M. Greenberger, (Baltimore, MD: The Johns Hopkins Press, 1971).
35. Brewer, "Malformation," 113.
36. Ibid., 114.
37. Max Frisch quoted in Rollo May, *The Cry for Myth* (New York: W. W. Norton, 1991), 57.
38. Hari, *Stolen*, 6.
39. Ibid.
40. Ibid., 8.
41. Jean Baudrillard, "Telemorphosis." In Drew S. Burk trans. *Telemophosis Preceded by Dust Breeding* (Minneapolis, MN: Univocal, 2011).
42. Ibid., 30.
43. Ibid., 31.
44. Ibid., 51.
45. Ibid., 45.
46. James Hillman, *Mythic Figures*, Uniform Edition, vol. 6.1 (Putnam, CT: Spring Publications), 259–275.
47. Ibid., 263.
48. Ibid.
49. Ibid., 264.
50. Ibid., 269.
51. Ibid., 270.
52. Ibid., 272.
53. Glen Slater, "Hermetic Intoxication and Dataism." In Joanne H. Stroud and Robert Sardello eds., *Mythic Figures: Conversing with James Hillman* (Dallas: Dallas Institute of Humanities, 2018), 71.
54. Ibid.
55. Flusser, *Post-History*, 42.
56. Jenny Odell, "What Twitter Does to Our Sense of Time." *The New York Times*, December 11, 2023.
57. Ibid.
58. I take this phrase from Jonathan Rauch, *The Constitution of Knowledge* (Washington, DC: The Brookings Institution Press, 2021).
59. Kevin Kelly, "The Waking Dream." In John Brochman ed., *Is the Internet Changing the Way You Think?* (New York: Harper Perennial, 2011), 19.
60. Ibid., 20. Italics added.
61. Jaron Lanier, "The Maximization of Neoteny." In John Brochman ed., *Is the Internet Changing the Way You Think?* (New York: Harper Perennial, 2011), 277.
62. Rauch, 5.
63. Ibid., 17.
64. Ibid., 119.
65. Ibid., 125.
66. Ibid.
67. Ibid., 128.
68. Ibid., 132.

69 Ibid., 133.
70 Kelly, 21.
71 Shoshana Zuboff, *Surveillance Capitalism: The Fight for the Human Future at the New Frontier of Power* (New York: Public Affairs, 2019), 507. Quote within quote, Nicholas Thompson and Fred Vogelstein from *Wired*, February, 2018.
72 Ibid., 509.
73 Bruno Maçães, *History Has Begun: The Birth of a New America* (New York: Oxford University Press, 2020), 138.
74 Ibid., 139.
75 Ibid., 140.
76 Ibid., 141.
77 Brendan L. Smith, "Inappropriate Prescribing." *Monitor on Psychology*, June, 2012, vol. 43, no. 6. https://www.apa.org/monitor/2012/06/prescribing
78 Ibid.
79 Center for Disease Control "Household Pulse Survey." April, 2020–April, 2022. https://www.cdc.gov/nchs/covid19/pulse/mental-health-care.htm
80 Brian D. Sites, Michael L. Beach, and Matthew Davis, "Increases in the Use of Opioid Analgesics and the Lack of Improvement in Disability Metrics Among Users." *Regional Anesthesia & Pain Medicine*, Jan.–Feb. 2014, vol. 39, no. 1. https://www.ncbi.nlm.nih.gov/pmc/articles/PMC3955827/#:~:text=After%20 accounting%20for%20changes%20in,1.06)%20(Table%202)
81 Eric L. Garland et al., "Low Dispositional Mindfulness Predicts Self-Medication of Negative Emotion with Prescription Opioids." *Journal of Addiction Medicine*, January–February 2015, vol. 9, no. 1. https://www.ncbi.nlm.nih.gov/pmc/articles/PMC4310788/#:~:text=The%20use%20of%20prescription%20opioids,to%20self%2Dmedicate%20anger%20or
82 Ibid.
83 Joseph E. Davis, *Chemically Imbalanced: Everyday Suffering, Medication, and Our Troubled Quest for Self-mastery* (Chicago: University of Chicago Press, 2020), 7.
84 Ibid., 14.
85 Ibid., 18ff.
86 Ibid., 21. Italics added.
87 Ibid., 21–22.
88 Ibid., 177.
89 Ibid. Subtitle.
90 Andrew Lakoff, *Pharmaceutical Reason* (Cambridge, UK: Cambridge University Press, 2005), 7.
91 Ibid., 6.
92 Johann Hari, *Lost Connections: Uncovering the Real Causes of Depression—and the Unexpected Solutions* (New York: Bloomsbury, 2018), 14.
93 Ibid., 29.
94 Ibid., 30.
95 Ibid., 43-44
96 Richard W. Dworkin, *Artificial Happiness: The Dark Side of the New Happy Class* (New York: Carroll & Graf Publishers), 7.
97 Ibid., 14.

98 Antonio Damasio, *Descartes' Error: Emotion, Reason, and the Human Brain* (New York: Avon Books, 1994), 262.
99 Ibid., 264.
100 Antonio Damasio, *Looking for Spinoza: Joy, Sorrow, and the Feeling Brain* (New York: Harvest, 2003), 3.
101 Julie Holland, "Medicating Women's Feelings." *The New York Times*, March 1, 2015.
102 Ibid.
103 Ibid. Italics added.
104 Jean Baudrillard, *The Consumer Society* (Thousand Oaks, CA: Sage Publications, 1998). Originally published in 1970.
105 Bill McKibben, *Enough: Staying Human in an Engineered Age* (New York: Times Books, 2003), 21. Marcy Darnovsky quoted within.
106 Ibid., 29.
107 Ibid., 46. Italics in original.
108 Ibid., 50.
109 See Leonard Shengold, *Soul Murder: Thoughts About Therapy, Hate, Love, and Memory* (New Haven, CT: Yale University Press, 1999).
110 Baudrillard, "Telemorphosis," 50.
111 Ibid., xiv.
112 Ibid., 46.
113 As I will take up further on, psychopathy, even more destructive than narcissism, involves an extreme form of dissociation. See Chapter 10.
114 Frederick Nietzsche, cited in David Michael Levin, "Psychopathology in the Epoch of Nihilism." In David Michael Levin ed., *Pathologies of the Modern Self* (New York: New York University Press, 1987), 41.

Chapter 6
BORGED

> The body is a unity of actions, and if a part of the body is split off from the action, it becomes "alien" and not felt as part of the body.
> A. R. Luria[1]

Borg is a noun that has morphed into use as a verb. A contracted form of "cyborg," which is itself a contracted form of "cybernetic organism," the term has entered the vernacular to describe a certain action: to *borg* something means to enhance its technology or make it cyborg-like.

Although some who have electronic or mechanical implants already identify themselves as cyborgs, this is not yet a widespread phenomenon, particularly as most of these implants restore rather than enhance function. However, when it comes to our general mindset, it is a different matter. Both the way we look at ourselves and relate to the world suggest we are already effectively fusing with our devices, imagining a form of self-enhancement, and the technocracy enveloping us ensures this fusion continues. The idea we are already "functional cyborg[s]" or "fyborgs" has been floated.[2] How we live has already been radically modified by technology. Thus, in some sense, as Turkle wrote in 2011, "We are all cyborgs now."[3] To the extent our cyborg becoming has already been built into our mindset, *we have been, and are being, borged*. How we have arrived at this point is the focus of this chapter.

Perceiving minds as computers and bodies as machines has been the main instigator of this borged state of mind. In the industrialized world, brains are now regularly thought of as hardware and thoughts as software. Mental states are frequently described with terms such as "wired," "plugged-in," "hacked," and "programmed." Events and knowledge are channeled through the digital devices that are always on or around our person. "Memories" are being stored in the cloud and decisions are being outsourced to algorithms. From birth to death we maintain our health and are frequently kept alive by being attached to various machines. We strap ourselves into an apparatus to exercise and recreate. Tech companies are finding more ways for us to seamlessly overlap virtual and actual experience.

At the same time these changes are occurring, advances in AI and robotics are instilling in us the idea that computers may have minds and machines could have bodies. The advances are real enough, but our perception of what they amount to also reflects a largely unconscious feedback loop in human–computer interactions. As Lanier puts it:

> When developers of digital technologies design a program that requires you to interact with a computer as if it were a person, they ask you to accept in some corner of your brain that you might also be conceived of as a program.[4]

The taking of technology into the human is thus coinciding with the humanizing of technology.

It might be said that the immediate future will be determined by whether computer minds and mechanical bodies are held literally or metaphorically. Each is of course a metaphor. However, this has not prevented medicine, cognitive science, and influential strands of neuropsychology from taking these metaphors literally. To this day vast amounts of money, time, and energy are poured into demonstrating this literal understanding, and vast amounts of innovation seem to bear out these efforts.

Ties between the way metaphors capture our imagination and the way we relate to the world are aptly summed up by Lent:

> Metaphors matter . . . The root metaphors cultures use to make sense of the cosmos encourage patterns of thought that permeate daily life. They hide in plain sight within our cognition, becoming so entrenched in our thinking that we forget they are metaphors and begin to believe

them as fact, along with the logical entailments that arise from them. As George Lakoff puts it, "Metaphor plays a very significant role in determining what is real for us . . . Metaphorical concepts . . . structure our present reality. New metaphors have the power to create a new reality."[5]

Our capacity to discern artificial and human entities will also be heavily influenced by how we perceive fabricated minds and bodies. As androids designed by roboticists appear more human and display emotional responses, including sexual desire, people will start overlooking the distinction between flesh bodies and silicon bodies.[6] In *Love + Sex With Robots: The Evolution of Human–Robot Relationships*, David Levy argues that "robots of the mid-twenty-first century will possess human-like or superhuman-like consciousness and emotions."[7] Yet tellingly, he notes, these "robots of the future will learn by watching what makes us happy and grateful and will sense our desires and satisfy them,"[8] and that "the appearance of the androids will be almost as important as, if not more important than, their technical capabilities."[9] What is more certain than androids developing consciousness and having emotions, however, will be the human inclination to attribute these qualitative states to them.

Following Levy's line of thought, if there is one thing we are always eager to believe it is the perception we are desired. Even if such desire does not actually exist, the perfect compliance and made-to-order appearance of these artificial humans will hasten the effect. Being lonelier and more socially inept—states induced by our initial turn to digital devices—will also create susceptibility to this kind of programmed interest. And if we are already inclined to think of ourselves in programable terms, when we start relating to entities that affirm our wishful thinking and feeling, we will be more likely to accept that we too are merely mechanized beings. Such seems to be the case for the famous roboticist Hiroshi Ishiguro. After decades of interacting with his android creations, he is inclined to think that human emotions "are nothing more than responses to stimuli and thus subject to manipulation."[10] Emotions have always been "subject to manipulation," but the "nothing more than" is where our borged mindset is on full display.

As has been the case with most technology, it is the promise of more control that beckons: control of our own minds and bodies, control of the minds and bodies we grant the artificial entities with which we will seek more intimate interaction, or control of reality itself. Imagine virtually every aspect of life becoming a matter of programming and design, guided by willful desire. While ultimate agency makes this prospect most enticing, it is also a recipe for existential solipsism, in which psychological maturation and emotional intelligence will be undermined. By focusing on what we can control—rational intelligence, physical form, virtual personality—we will be more prone to dissociate from what we cannot control—raw emotions, the bonds of psyche and soma, the complexities of real relationships, and the use and misuse of personal information. This will make us more insecure, driving a search for even more control and more powerful technological solutions. Along the way we will also be deprived of the serendipity and spontaneity that less fabricated interactions with the world and with other beings provide.

The psychodynamics of this pattern are clear: control appeals to the ego, which has a habit of turning its back on other parts of the psyche and pretending our animal nature has no claim on us. As this posture grows, we become alienated from the pattern of inclusion that governs the psyche as a whole. To the extent this is so, the psyche and our posthuman pursuits will be on a collision course. Between now and any final departure from the deeply human, the resulting split within us will become ground zero for the collapse of our ecology of mind and what promises to be a normative madness.

The task of this chapter is therefore twofold. First, we need to spell out the incomplete and illusory character of our borged minds and to understand the kind of reductive thinking that has captured them. Second, we must critically examine the associated focus on the pursuit of perfection and immortality, which springs from a thinly veiled religious search for spiritual transcendence. In such ways we may see through the posthuman thinking already shaping us.

―

Computing is the way of the world. Short of some cataclysm sending us back to the Stone Age, this is not about to change. Yet despite our almost total reliance on computers, they have one glaring flaw, which is also not about to change: these so-called "thinking machines" always mistake the map for the territory. And in our aligning our thinking with theirs, we have begun making the same mistake. If we want to live in the actual world rather than the map-world, we need to understand this.

To be fair, for computers there are only maps—only grids of abstract data that represent actual dimensions of life. Using programmed instructions, computers convert the speech, text, and imagery of human communication into bytes, which they then process, before converting these bytes back into speech, text, and image. What we are prone to forget, however, is that the state of being informed that begins and ends this process rests entirely with us. Computers may also execute actions in other machines, but the significance of these actions is also only known by us. It is, technically, even wrong to say that computers "process information" because at no point are they actually informed.

To be informed is to tie information to experience. If a computer applies the brakes on a self-driving car, it is not in order to save a life or even to protect the car; it is because an object activates a visual sensor, connected to a chip, programmed to do one thing with one kind of input and something else with different input. When a computer calculates a sudden movement in the stock market, it is indifferent to the direction and degree of that movement, the real-world meaning of which is only apparent to those who receive this information and know the value of goods and services that the graphs and numbers signify. Such computer actions have no significance beyond the human context. Thus, despite all appearances, computers do not have the slightest idea of what they are doing or why. Actions, words, and pictures mean something to us, but they do not mean anything to computers. Lanier states it plainly: "A computer isn't even there unless a person experiences it. There will be a warm mass of patterned silicon with electricity coursing through it, but the bits don't mean anything without a cultured person to interpret them."[11]

In short, computers have no skin in the game. Without programming and inputting and collecting data, represented by layer upon layer of ones and zeros, there is nothing going on inside a computer. There are only circuits and wires and the housings that contain them. The computer has no being, only doing, and that doing is instigated and directed by us. Even with the neural networks and self-programming capacity of AI, this computerized intelligence is oriented to the tasks we elect. Whether AI is working to exploit loop-holes in the banking system or to design nanobots to cure cancer is of no real concern to it. As long as computers lack the ability to tie data to life, they will only ever be dealing with maps, not the territory those maps represent. So if we are to become more like computers, we must realize it is the territory we will be leaving behind—not only the mere sense perception of the earth, but the whole qualitative terrain of existence, filled as it is with the complexity and wonder of encounters with other living beings, grounded in what we call common sense—which happens to be a great hurdle when it comes to the holy grail of computing: the creation of artificial general intelligence.

Nonetheless, the technocracy enveloping us is also rigging the game. We need only consider the screen time that fills the day to appreciate the way computers occupy the space between us and most everything that happens. Most everything we see, hear, and know is now shaped more by algorithms than natural rhythms. Efforts to further adapt body and mind to this map-world thus begin to seem like logical steps forward.

In a recent social science experiment, a group of teenagers were asked if they would rather break a bone in their hand or see their smart phone shatter. Roughly half the participants chose to break a bone. They rationalized the choice in terms of being able to stay connected while their hand was being repaired.[12] For many a digital native, the instrumentality and utility of staying online trumps that of having a healthy hand. This may seem like a quirky snapshot of contemporary culture, but the shift in values captured by it have vast evolutionary implications. Human faculties are being sacrificed to the map-world. This downgrading of common sense is also evident in the phenomenon of drivers deferring to their GPS navigation devices even when they are obviously wrong. Refusing to take in the lake or unfinished bridge before their very eyes, drivers have been known to follow their devices right off the road. If most of us put the maps before the territory, we may all be headed for something of the same fate—literally and metaphorically.

Most analysis of this dawning posthuman age focuses on the enticement of the cybernetic. While this focus is vital, we must also consider our declining relations with the organic. Becoming cyborgs will not just stem from adopting the map-world; it will also stem from departing the phenomenal world. It is in this sense the cyborg is the apotheosis of the dissociative syndrome detailed in the last chapter—a remaking of the human enabled by a disconnection from the deeply human. From the cyborg's perspective and thus the posthuman one, this departure is a necessary if cloaked part of the operation.

Even if this human–machine hybridization takes decades, many of us will start seeing the results in our lifetimes. However, on the timescale of evolution, it will be the equivalent of a quick elective surgery, and anesthetization will be a key part of the procedure: to feel the cut more directly, to actually register the loss of faculties or comprehend the side effects of the new attachment, it may be a different story. From the posthuman perspective, the focus is on enhancement and augmentation— new parts, added capacities, expanded perceptions. But from the psychological standpoint, particularly one that attends to the psyche as a whole, the focus must be on numbing and amputation. The sensual body? Titanium will last longer. Difficult emotions? Who needs them. The capacity for empathy? Recommended but optional. For this surgery to proceed, the fully and deeply human must first be subjected to anesthetization.

Yet how can we stand outside the belly of this techno-beast that has already begun to swallow us? We must get to know the beast.

Reduced

The belief system that sustains the rise of Dataism and the "map-world" that sustains this it is a reductionistic one, ultimately leaning on mechanistic and atomistic modes of understanding. Among other things, these modes allow us to hyphenate bio-tech and info-tech, as well as imagine their ultimate overlap. It is this overlap that permits many techno-scientists to assume they have finally placed their finger on what nature has been doing all along and what we will now take over. In this vein, Steven Rose, surveying his own field of biology, sees "reductionism as ideology," describing it in this context as

> the tendency ... to insist on the primacy of reductionist over any other type of explanation, and to seek to account for very complex matters of animal—and above all of human—behavior and social organization in terms of a reductionist precipice which begins with a social question and terminates with a molecule—often a gene.[13]

This "fashionable gene's-eye view of biology," Rose notes, "conceives living organisms as nothing but 'lumbering robots.'"[14] Given this robotic image of nature, what could or should prevent our efforts to reengineer these living organisms?

As Rose's description conveys, few things have shaped biological reductionism more than scientific materialism, which Barrett contends is "the pervasive current that flows all around modern philosophizing." As he points out, "materialism need not be explicitly professed as a creed; it becomes the de facto philosophy of an era reaping great triumphs in the physical sciences and in technology and pushing more and more of its energies into those fields." Barrett adds that "the amazing proliferation of machines and apparatus seem somehow solid and real; and in comparison with these, any meditation about matters like the human soul are bound to seem ghostly, insubstantial."[15]

In the scientific mind's eye at least, matter is all that matters, and matter is reducible to components such as molecules and atoms, the mechanistic dynamics of which are in turn comprehendible by chemistry and mathematics, which are these days thoroughly ensconced in computation. When these components are broken down even further, the dynamics can be mysterious and speculative, defying even the most sophisticated calculation. However, this does not prevent many scientists from clinging to reductive models, the latest versions of which reflect a computational race to the most manipulatable bottom. By the late nineteenth century science was contending that anything that could not be observed was not real; by the later twentieth century anything that could not be calculated was not real.

Richard Dawkins provides a strident expression of where this belief system or ideology has taken us: "Life is just bytes and bytes and bytes and bytes of digital information. Genes are pure information—information that can be encoded, recorded and decoded, without any degradation or change of meaning."[16] For Dawkins, this reduction "is not a metaphor, it is the plain truth. It couldn't be any plainer if it were raining floppy disks."[17] Existence is just data, of the kind that computers process; we just need to adjust our perception in order to see all the ones and zeroes that may be arranged as orchids and zebras or anything else we find ourselves contemplating. People, as well as everything they think about and create, must ultimately come down to the same thing.

Lent's commentary on these statements by Dawkins offers the opposing view. He points to a number of biologists who hold there are "intrinsic principles to life that fundamentally differentiate it from even the most complicated machine." And yet, Lentz acknowledges, the convergence of these views "has not prevented Dawkins's mechanistic view of life form being broadly accepted as the truth in mainstream thought."[18] To cast aside the critical role that metaphor plays in science,[19] and to thus take literally metaphors that reveal the zeitgeist more than the subject at hand, lies at the very heart of our borged view of reality. As the philosopher of consciousness John R. Searle points out, one "difficulty is peculiar to our intellectual climate right now, and that is the urge to take the computer metaphor of the mind too literally."[20]

Dawkins exemplifies the mad scientist syndrome described earlier: looking at reality in one ultra-reductive way and taking no account of the vast number of perceptual and imaginative modes of the human psyche. Yet the flip side of this reductionist coin is the reanimating and personifying fantasies that inevitably reenter the picture, demonstrating how these rejected modes can unconsciously shape what is supposed to be pure science. This is particularly evident in Dawkins's early and most impactful book *The Selfish Gene*.[21] Here, as Hayles observes of this book, we are met with "a work ... deeply informed by anthropomorphic constructions" wherein Dawkins approaches his topic by "overlay[ing] onto the genes the strategies, emotions, and outcomes that properly belong to the human domain."[22] An animated, poetical mode of thinking, reflecting the soul's natural means of comprehending reality, ultimately finds its way back in, albeit at the entirely wrong level.

Reductive thinking is not always this blatant. But such examples can train our eye to catch it at work in more measured techno-scientific outlooks. The views of Nick Bostrom, one of the leading posthuman philosophers and a founder of the World Transhumanist Association, exemplify a more surreptitious reductionism. In his essay "A History of Transhumanist Thought," Bostrom provides a comprehensive account of the currents of thinking behind the push to remake ourselves.[23] While cognizant of the potential pitfalls of human–computer merger, he advocates for this development and is among those who think "human level artificial intelligence ... could happen within the first half of this century."[24] He also provides an illustrative example of the attempt to marginalize the psyche and mind–body–world connection in transhumanist/posthumanist thought.

For example, although Bostrom relates the rise of transhuman ideas to enduring images in myth, religion, mysticism, and romantic philosophy, aptly highlighting the timeless reach for transcendence, he peels away the cautions and admonitions that have always been built into these efforts. He notes the Greek concern with hubris evident in the exploits of figures such as Prometheus and Daedalus, the medieval Christian critique of alchemical attempts to create homunculi and panaceas, the warnings imbedded in the Jewish legend of the Golem, as well as the range of works from twentieth-century futurists and novelists responding to the Frankenstein factor of the nineteenth century. But he dismisses as unscientific and immature the uneasy feelings and perceived transgression associated with these attempts to transcend ourselves, as if such themes have no business conversing with the efforts of technoscience. The result is a selective history of ideas and a compartmentalization of human ideation that dismiss the fully and deeply human.

Bostrom couches both early and ongoing responses to Enlightenment scientific philosophy as "reactions against the rule of instrumental reason and the attempt to rationally control nature."[25] However, he conveniently overlooks the fact these reactions appear not only in the form of religious, philosophical, and literary protest; they also stem from the way nature has herself responded to these outlooks and efforts. Whether coming from without or within, instrumental overreach and rational control have given rise to objective, observable counter-effects and valid counter-narratives. Ecologies of both earth and mind have suffered significant damage as science has gone from being a valuable, even primary, perspective to a summative account of all that is real. This not because applications of reason and judicious attempts to control certain natural processes are inherently flawed, but because our turn to these capacities at the expense of other approaches has become exclusionary and one-sided. In the end, this bias must also count as *scientifically* problematic. Bostrom contends that "we need to learn to think about 'big-picture questions' without resorting to wishful thinking or mysticism" and that "these questions should be addressed in a sober, disinterested way, using critical reason and our best available scientific evidence."[26] But here "wishful thinking" and "mysticism" are euphemistic labels for what lies beyond the paradigm of a narrowly-construed technoscience. Matters such as integrity, dignity, consciousness, beauty, justice, and empathic regard must also enter into "big-picture questions," if not lead such questioning. "Critical reason" and "scientific evidence" are not always the best means to grasp these matters, which can shrivel up and disappear if "addressed in a sober, disinterested way." In plainer terms, we are well beyond the point of needing to take an interdisciplinary approach to all matters that concern the deliberate alteration of human nature.

In Bostrom's accounting, the reductive, scientific stance of transhumanist/posthumanist philosophy is pronounced. The subtext of his stance is that we are here to serve the development of science, not the other way around. Whereas we may be here to serve the evolution of consciousness, this constitutes a broader, deeper, and ultimately more mysterious matter. It is also one that goes to the core meaning of human existence. And to the extent being more conscious pertains to the task at hand, there are a number of other habits of mind comprising the borged approach to reality that must be addressed.

Automata to Androids

Humanoid robots, appearing first in mythology, then in science fiction, and now in life, displaying relational capacities and personalities, occupy a critical place in the normalization of our cyborg prospects. The more like us robots appear to be and the more AI appears capable of humanlike interaction, the less resistance we will have to becoming more like them.

Our perception of sentience in animated machines, which can be activated by any indication of agency, is amplified by the personifying propensity of the psyche. Google recently had to fire an engineer who seemed to jump the gun by declaring the company's "Language Model for Dialogue Applications," or "LaMDA," to be sentient, presumably because of its rather advanced conversational capacity. A spokesperson for the company was forced to describe this viewpoint as "anthropomorphizing."[27]

Prior to the objectifying gaze and material reductionism of modern science, crafted objects such as weapons and tools were regularly seen to have personalities. To this day we revert to this mode of perception when we have lived with a machine for a long time or become dependent on its smooth functioning. Conversing with (or cursing at) gadgets that thwart our intentions is not unusual: we name old cars, which seems to help when coaxing them to life. Because we spend so much time with them, computers frequently end up on the receiving end of our pleas and admonitions. This innate tendency, which can bring a soulful sense to what we might otherwise regard as lifeless, evidences a deep proclivity to want to live in an animated world. But the same tendency can be a conduit for illusion.

The possibility of intelligent machines actually becoming sentient is something I will explore in a later chapter. For present purposes, it is the tendency to imagine sentience where is does not actually exist that we must consider, because this phenomenon is smoothing the way to human–machine hybridization, allowing the masses to converse with non-human persons and tech insiders to think of neural networks as minds.

For many addressing this trend, digital assistants, chatbots, social robots, and other artificial presences are simply realizing an enduring motif of the imagination, evident all the way back to the ancient Greeks. These artificial antecedents are frequently invoked to offset contemporary unease and wariness and instill a sense of inevitability about this realization. As the argument goes, this long prefiguring of artificial beings challenges those who see them as aberrant outgrowths of technological overreach, because innovation is just catching up to a vision of human creativity long in the making. However, a close reading of this prefiguring suggests a different meaning.

In general, myths, legends, fairytales, and other traditional stories do not just portray the heroic quests, spiritual journeys, and other acts of forging ahead against the odds; these tales also describe the dangers and pitfalls of such pursuits. Heroes are frequently confronted with gods and monsters undoing their naivety and putting a dent in their hubris. Even a passing familiarity with this record of the cultural imagination reveals the way these personifications of colliding forces in the psyche highlight the detriment and destruction that comes with the failure to observe existential and cosmic limits. The mere prefiguration of artificial beings reveals very little. However, the narrative surrounding these prefigurations reveals a great deal.

It thus becomes important to recognize that nearly all traditional stories of artificial beings are cautionary tales. This is not to say they offer clear moral and ethical guidance, but they do show that certain propensities constellate, countering forces in the psyche, the dismissal of which is typically a recipe for a fateful reckoning. As Adrienne Mayor points out in her study of myths and robots:

> One cannot help noticing that all of the automata used to inflict pain and death in ancient mythology belonged to tyrannical rulers, from King Minos of Crete and King Aeetes of Colchis to Zeus, the Father of Gods and Men, who chuckles in anticipation of his cruel "trap" for humans.[28]

Mayor's detailed account of the myth of Talos, an android-robot created by Hephaestus, emphasizes the way "artificial beings are seen as the products of divine artisanship, not just divine will." Talos was one of three gifts made by the smith-god at the behest of Zeus and given to his son, Minos. Functioning as a guardian of Crete, Talos comes to an end through the cunning of Medea, after being engaged by Jason and the Argonauts. Introducing her study, Mayor notes that "in the rich trove of tales from the ancient mythic imagination, one can discern the earliest traces of the awareness that manipulating nature and replicating life might unleash a swarm of ethical and practical dilemmas."[29]

"Dilemmas" may understate the matter. Especially in the approach to the modern world, tales from the Golem of Prague to Frankenstein underscore the upending of values and the ultimate suffering of these artificial beings. We should thus tread carefully when lining up the automata of history as if they have just been waiting to be released into life in some kind of predestined manner. Their preexistence in the mythic imagination is in no way a license for their realization; in fact, it is more often accompanied by a rather large warning label.

The Quest for Immortality and Perfection

At the start of his meditation on the nature of the universe *Until the End of Time*, physicist Brian Greene writes, "In the fullness of time all that lives will die."[30] Placed in such a cosmological frame, it is hard to conceive of a more pointed statement about the intrinsic relationship between life and death, and the way it pervades all existence. This principle of the universe as we know it finds its extension in the reality of the psyche too, giving rise to a dynamic that appears just as unassailable: the ability to fully live depends on the capacity to face death; the capacity to face death also depends on the ability to fully live. The experience of life and death are deeply intertwined. Maturation means worn-out versions of ourselves must die. A purposeful life, in tune with the rhythms of cosmos and psyche, involves an ability to contemplate endings, endure losses, and keep possibility and limitation in creative tension. By contrast, a suspended or provisional life grows mostly from the denial of death and avoidance of loss.

Yet the urge to beat death is a primary motivator for many strands of technoscience and is central to the posthuman philosophy of transcendence. In part an extrapolation of contemporary attitudes that have made prolonging human life a priority, this philosophy is also fueled by the declining belief in an afterlife. The more death is regarded as a final end, the more we do to try to conquer it. As many commentators have pointed out, and I have addressed in previous chapters, the promise of a posthuman overcoming of bodily limitation and the possibility of digital forms of consciousness offer an alternate religious fantasy—another entry into a spiritual existence, with science as the god that could grant us eternal life.

Immortality is a very big carrot. Religion in general and Western religion in particular have shown just how big, with the promise of eternal life offered to offset the death of the body. This promise has been at the core of great historical and cultural change and been used to motivate and manipulate great chunks of humanity. In the decades around the turn of the millennium, the biblical prophecy of apocalypse and the apocalyptic possibilities of a politically unstable and disaster-prone world have occasioned the sense of living in the End Times and an expectation around what is known as the rapture. Now the religion of technology is promising to help us transcend the destructive prospects that technology has itself set in motion.

Both religious fundamentalism and techno-religion offer pathways to some form of ultimate being or state of perfection; both involve abandoning the mortal body, the flawed and limited nature of which is regarded as an obstacle to a more perfect form of existence; both grow out of the decline of more traditional religious orientations and the erosion of the cultural imagination; both literalize the archetypal fantasy of a realm wherein existence may continue on a spiritual plane; both give rise to a zealotry that suppresses critical questions; and both require massive leaps of faith.

Religious fundamentalism may adhere to religious dogma to the point of caricature, but it tends to reject the intricate symbology of traditional religion, the sophistication of mature spiritual philosophies, and the kind of psychological grounding found in deeper mystical outlooks that focus on discovering the sacred in earthly life. In so doing, fundamentalism displaces the integrative power of the psyche's religious function, which is concerned with reconciling the conscious mind and the larger forces of existence. Fundamentalism instead adopts a defensive and compartmentalizing posture, its hyper-religiosity forming a rigid shell to protect again the soul loss and conundrums of modern "secular" life. Inflated claims to possess the truth and perceived mandates to orchestrate divinely authorized socio-political transformations complete the picture.

As I argued in the previous chapter, pursuing a literal departure from the deeply human is fed by a dissociative reaction to the emotions and feelings rooted in the body, as well as to the psychic realities that grow from these things. Akin to religious fundamentalism, posthumanism does not want to deal with the messy complexities of human experience. Instead, posthumanism wishes to rise above this level and exchange the human condition for the pristine heights of intellect and spirit, denying that the formation of identity, personality, mind, and character are intrinsically tied to living in flawed, imperfect, and mortal bodies.

When Baudrillard contemplates "the lethal illusion of perfection," he points to the ubiquitous presence of "objects from which wear and tear, death or aging have been eradicated by technology." Reflecting on an artifact from the dawn of the digital age, he refers to "the compact disc," noting "it doesn't wear out, even if you use it. Terrifying, this. It's as though you never used it. It's as though you didn't exist. If objects no longer grow old when you touch them, you must be dead."[31] Of course, to merge with such objects would mean we too could overcome such wear and aging. Things can be perfect if they are abstract and ephemeral enough, but not when organic matter is involved.

Having the ability to exchange artificial bodies or other forms of physical "shells" *ad infinitum* or becoming some kind of digitalized entity in cyberspace separates sentient being and personhood from the body, granting a kind of immortality. On one hand, we will transcend the limits and vulnerabilities of physical life; on the other, we may choose the way we do manifest, according to what we desire—assuming desire even makes sense at that point. Perhaps the sleight of hand in this line of thought is the very use of the term "we," for even suspending the question of whether such a transcendent form of life is possible, the kind of entity or life form that would likely result from such a development would barely resemble "us." What would simple pleasures or anything beyond logical thought even mean? By peeling away the physical limitations and psychic imperatives that come from an embodied life in a phenomenal world, little that is us may remain.

Robot Love

Rollo May writes, "it is both significant and surprising in Greek myths that many times significant persons are offered immortality but choose mortality instead."[32] He goes on to share the story of Zeus falling in love with a mortal, the wife of Amphytrian, a Greek general. Hermes eventually helps Zeus fulfill his passion by disguising the ruler of Olympus as Amphytrian. After the god and the mortal make love and engage in conversation, Zeus returns to Olympus troubled, telling Hermes:

> she will say, "When I was young, or when I am old, or when I die." This stabs me . . . We miss something . . . We miss the poignancy of the transient—that sweet sadness of grasping for something we know we cannot hold.[33]

Immortals pining for "the poignancy of the transient" is a recurring motif in myth and story. Those possessing the supposedly ultimate gift are drawn to what we possess—temporality, limits, and even loss. They envy the longing sensation that creates soulful being, wherein the weathering of things and vulnerability of bodies gives rise to bittersweet imaginings and feelings of affinity with the living. In this archetypal realm, the personified forms we call gods are forever wanting to enter and incarnate in the mortal realm. Their spirit wants realization in this life. Gods and angels, imagined shapers of thought and action, want our recognition—our consciousness—one of the main ways they enter life. One of the highest values, viewed from the soul's perspective, is tracing the universal and timeless engaging with the here and now. Lusting after some delicious mortal or lamenting about what is missing on the immortal plane is a metaphorical reflection of this value. God's "only son" entering into ordinary existence while retaining his divinity, uniting the mortal and immortal, is probably the most recognizable exemplar of the motif, one that changed the course of civilization.

While we imagine becoming something like gods, living on in our digital minds and replaceable bodies, the figures that intimate this possible future keep sending us the same signals as the immortals eyeing entry into this world. In the film *Bicentennial Man*, Robin Williams plays a robot on a quest to become human.[34] The quest begins with a desire to be creative and empathic, two qualities that were not part of his original programming. After some years of developing a capacity for social interaction, his owners eventually grant him freedom to pursue his own interests. With a titanium body and endless "upgrades" at his disposal, the robot-man first embraces his virtually unlimited life. Then melancholy sets in. The robot-turning-man tires of watching generations of those around him age and die. Moreover, he realizes that without a flesh and blood existence he cannot seek or experience the deep attachments and love that come with being fully human. This path eventually leads him to give up his virtually immortal existence and exchange it for a mortal one. The paradox he must live into is the realization that the fullness of life comes with the limits of life.

Bicentennial Man depicts a romanticized variation on a prominent theme in stories of androids, cyborgs, and other humanoid figures: they are rarely content with their existence. A stark, impersonal core to their being is a frequent source of cognitive dissonance. Though their silicon features and robotic limbs take the human form to new levels, they are prone to be empty shells, unable to imagine, emote, or play in the way of those they serve. A palpable tension exists between what they are and who or what they are driven to become. Unlike the Bicentennial Man, the soul-searching is frequently to no avail and their existence remains a kind of manufactured limbo.

As humans imagine into a future of manufactured beings, artificial intelligences, and robot companions, the imagination has kicked back a series of figures whose goal is to become more human, and experience humanness through the search for love. The Golem of Prague and Frankenstein's creature are early examples of this. Mary Shelley's tale of scientific hubris, which coincided with the dawn of the industrial world, tapped into a collective intuition; whether it lingers beneath the surface or takes the form of overt destructiveness, fabricating beings in our own likeness invites monstrosity. Untethered from careful gestation and containment in cultural and social structures, life and consciousness turn out to be powerful forces. Humanity is born at the crossroads of evolution and culture, and the inhuman appears when these paths fail to intersect. Sometimes this denial of the human leads these fabricated beings to turn on their creators, with humans becoming slaves and tools in the service of some form of mechanized hierarchy.

It is hard to escape the irony that as the imagination leaps ahead it sees the fabricated beings posthumanists are aspiring to become wanting to be more like us. Their existence, as we have come to imagine it, lacks love and meaning, both of which are reliant on the shared limitations and vulnerabilities of embodied, mortal life. As Dreyfus notes, "all meaningful discourse must take place in a shared context of concerns."[35] An absence of those concerns is also an absence of meaning.

We would be wise not to cordon off love and place it in some psychological compartment that has nothing to do with general intelligence and therefore nothing to do with what kind of humanoid robot we are likely to construct, or what kind of cyborgs we may become. Eros is implicated in almost everything that concerns having a psyche or being psychological because it is what draws us toward both persons and things: we relate to what attracts us and, by implication, dwell on what repels us; our understanding of the world is mediated by the presence or absence of eros. Dreyfus observes it is "human needs and propensities which give the facts meaning, make the facts what they are,"[36] and eros is at the heart of these needs and propensities. Another way to describe this process is in terms of relevance. Dreyfus writes:

> We can and do zero in on significant content in the field of experience because this field is not neutral to us but is structured in terms of our interests and our capacity for getting at what is in it. Any object which we experience must appear in this field . . . Since we create the field in terms of our interests, only possibly relevant facts can appear. Relevance is thus already built in.[37]

Insofar as they pertain to universal human modes of perception, this "field of experience" and the "interests" that structure it are not primarily personal and subjective but archetypal and objective. "Relevance," as Dreyfus calls it, is in this context a synonym for value or meaning, which is finally rooted in the archetypal basis of mind. Importantly, though, once we get beyond the Cartesian divisions and other philosophical reductions that deny the existence of such fields of experience, we face the problem of where, exactly, these fields and their structures are to be located. As Dreyfus offers, "When we are at home in the world, the meaningful objects embedded in their context of references among which we live are not a model of the world stored in our mind our brain; *they are the world itself.*"[38] What we think of as the world is actually the way the psyche is activated by our being in the world. This is what Jung called psychic reality.

One final point on this theme: the dichotomy of character we find among the cyborgs, robots, and androids of possible futures tends to follow that of love and power. In contrast to those figures who seek to be more related or even devote themselves to an attempt to understand and experience love, an array of artificial humans and humanoid beings either find and exploit their power or act as instruments of some totalistic system. These second types are inevitably destructive.

Belonging to some distant, intergalactic past while also hinting of a possible human future, Darth Vader of *Star Wars*[39] is a fitting example. Perhaps the first cyborg to enter the popular imagination, this figure has become so well known that contemplating his significance risks cliché. But Vader's storied role as the central character of "The Empire"—a technocracy of the darkest extremes—would also be a glaring omission. In the present context, Darth Vader's final pivot from power to love also represents a critical turn, occurring in unison with the removal of his black mask and mechanical life-support. After being described by the wisdom figure of the films Obi-Wan Kenobi as more machine than man, Darth Vader thus achieves a climactic return to "humanness," simultaneously rescinding his fusion with technology. This is not only an outstanding depiction of the psychological dichotomy of mechanized beings; it is one of the most iconic moments in cinema, which has shaped the cultural imagination in a remarkable way. The *Star Wars* "universe," as it has come to be called, resonates not only because of its portrayal of the battle between good and evil and planet-hopping heroes; it is peppered with a cast of characters that are both unique and archetypal, timely and timeless; it is also a late-modern portrait of the battle between creative and destructive technologies we are all now in.

The presence of this dichotomy of mechanized beings carries forth a critical focus on love and power in the history of depth psychological thought. As Jung writes:

> It is interesting to see how this compensation by opposites also plays its part in the historical theories of neurosis: Freud's theory espoused Eros, Adler's the will to power. Logically, the opposite of love is hate, and of Eros, Phobos (fear); but psychologically it is the will to power. Where love reigns, there is no will to power; and where the will to power is paramount, love is lacking. The one is the shadow of the other . . .[40]

If poetry and science are ever to come together, it will be in this realm of understanding. Any liaison between the poet and the scientist will necessarily involve exploring what lies between love and power. At stake is whether or not the world we know and utilize will remain the world we belong to—the world we love.

The Spiritualization of Cyberspace

The etymology of "desire" is traceable to *de sidere* and means "from the stars." This connection stems from the ancient Roman practice of *considerare*, which was to study the stars with care, as in looking for some kind of astrological guidance. "It soon came to be used more generally with the sense 'to observe carefully, to examine,' before developing into the figurative use 'to reflect upon.'"[41] And, curiously enough, as noted earlier on, to carefully consider or look again is the meaning of the term *religare*, which is one of the roots of "religion."

What these etymological connections suggest is, first, desire comes from the archetypal patterns of the psyche, which the ancients projected onto stellar constellations. And, second, that desire calls one into reflecting upon these universal origins. Such a startling web of meanings points to a critical psychological truth: desires may be acted out, but only desire reflected upon is ultimately fulfilled, joining action to a larger, impersonal frame of understanding. In the widest sense of the term, a study of one's desires may thus be considered as a religious activity. Indeed, for the more pagan imagination, wherein the strongest of human appetites and desires find correlates in gods, spirits, and daimons, returning impulses to their divine source was the main activity of religion.

An individual who experiences desire and fails to engage with such a process may therefore be considered sacrilegious. He or she is likely to be buffeted about by impulsiveness, encountering frustration and the indifference of others. Behavioral norms and moral principles may limit the fulfillment of unmitigated desire and, at best, force an individual into the mode of reflection. However, sometimes the response to such restrictions on thought and action simply leads to repression, which sets up psychic divisions and vicious cycles of raw desires colliding with restrictions, being repressed and resurfacing in the same unmetabolized form. This vicious cycle is the one most evident in fundamentalist religions and other moralistic communities.

What desires need is reflection in larger pools of existential meaning, which are ultimately rooted in the archetypal patterns of the psyche. Desire detached from this larger realm of meanings and seen only in an egoic light is not only a set-up for frustration and ultimate failure; it robs the individual of contact with the deeper reaches of the soul. These considerations, articulated in relation to personal experience so we can more easily grasp the dynamics in question, apply to posthuman desires, which are not so much personal as communal.

In posthumanism, the desire for perfection, immortality, and physical and intellectual augmentation are articulated and pursued as if they are logical extensions of advancing technology. In reality they exemplify the unconscious return of spiritual aims. And as both collective manifestations and supposed extrapolations of purely techno-scientific understandings, they are even less subjected to any deep reflection. When everyone wants something, the impulse tends to be normalized and legitimized. The hive-mind kicks in and resistance is more easily circumvented. This is a recipe for unconscious thought and unconscious action. These collectivized posthuman spiritual desires may be the least reflected upon elements of the movement, but have been well documented by previous commentators.

In *The Religion of Technology*, David F. Noble gathers a number of comments from technologists that capture the thinly veiled spiritualism of their efforts: "Cyberspace will feel like Paradise ... a space for collective restoration [of the] habit of perfection"; "human beings could enjoy a god-like instant access"; virtual reality "fulfills the need 'dwell empowered or enlightened on other, mythic, planes.'"[42] When our minds are able to fully enter cyberspace and take up residence in virtual worlds, it will be like "a Heavenly City, the New Jerusalem ... a place where we might re-enter God's graces ... laid out like a beautiful equation."[43] And as one programmer quips, "I am God to the universes I create."[44] Such inflation points not only to the potential power, but also to the lack of spiritual awareness in the form of absent wariness or restraint. This absence is on full display with the fantasy "that the accelerated advance of Artificial Intelligence [is] the only salvation for mankind,"[45] a view that leads some Silicon Valley influencers to argue for the removal of any regulation of AI, lest the utopian future it promises be delayed.

As Erik Davis astutely writes in his book *Techgnosis*, "mystical impulses sometimes body-snatched the very technologies that supposedly helped yank them from the stage in the first place."[46] He argues:

> Such thinkers and tinkerers are loosed in a world of possibility whose profound metaphysical and religious dimensions they are often incapable of handling, let alone recognizing; as such, they find themselves unconsciously drawn to the soul's most adolescent fantasies of transcendence and immortality.[47]

In his Jungian take on this theme, *The Enchantments of Technology*, Lee Worth Bailey provides plenty of backup for the notion that "Myth has not been dissolved by reason. Myths are the most powerful moving forces behind technology."[48] At the same time, of course, "techno-society still neglects the mythos hidden in the logos."[49] Part of the problem then becomes the way "Technology's desire-filled mythic heavens are haunted by its hells, with only the rickety promise of faith in progress to pull it out of its despair."[50] In the form of both hubris and totalism, there is a large price to pay for the displacement of the religious function.

Two characteristics of the posthuman mindset clearly stand out at this point in the discussion of our borged state of mind. The first is the radical materialism that allows for such a smooth equating of computer and brain, brain and mind, therefore mind and computer. It is this equivalency that moves AI from being a valuable tool to a viable contender for human replacement. The second is the unconscious, yet pervasive, spiritualization of artificial intelligence and digital technology just noted—the attribution of sublime, even divine, knowledge to the exponential growth of data-processing and the related view of cyberspace as a heavenly realm.

We need to see the necessary link between these two characteristics, wherein the de-mystifying and de-spiriting of matter gives rise to a reanimated, re-enchanted fixation on our own godlike creations. The belief that consciousness itself can either be generated by or transferred into digital technology appears to be the immediate stepping stone to these lofty aspirations, involving hard-nosed computer scientists in the most philosophical and religious of questions. But even aside from the possibility of creating conscious machines lies the idea that AI will be a kind of salvation, capable of solving all present and future problems.[51]

The Cartesian quality of these perspectives is difficult to overlook. On the one hand, there is "nothing but" atoms, molecules, genes, and bytes; on the other, a spiritual transcendence and power to perfect existence. This division immediately prompts the question of the absent third—the vacuum between these poles where the reality of the psyche and a sense of soul may appear. As spirit and soul are often used interchangeably, confusion abounds when we start to discuss these notions and the distinction between them. Here I suggest we follow the lead of James Hillman in his contrasting of spirit's quality of ascent and soul's affinity for descent, associating spirit with what is rarefied and soul with what is embedded.[52] The spirit seeks clarity and precision; it is attached to the directed and penetrating light of consciousness.

Posthumanism's focus on consciousness rooted in rational intelligence, and the development of this intelligence unencumbered by animal instinct and sensual experience, reflects spiritual goals. Soul, as Hillman and Thomas Moore have pointed out, involves more mystery and complication. It is associated with qualitative states in which imagination, emotion, and significance are simultaneously stirred. Far from an arbitrary distinction, this contrast between spirit and soul begins with the pre-Socratics and may be traced all the way through the Neoplatonic thinkers of the Renaissance, to the romantic philosophical tradition and the rise of depth psychology in the late nineteenth century. And the history of this contrast tracks the history of Western thought. Soul, as the qualitative state of contact with the psyche's depth, has been occluded by both the mainstream Christian view and the Enlightenment philosophies of science, which have in different ways polarized spirit and matter, leaving little room for the psychic field that lies above the merely material and beneath the otherworldly. Moreover, Christianity has coopted the soul into its spiritualized resistance to world and body, packaging it as an individualized, ghostlike entity that leaves the body at death and is then subjected to the competing claims of God and the devil.

Even in the absence of overt religious concern and the embrace of secular streams of thought, many writers and thinkers remain under the influence of this Christianized view. Popular visions of animated robots and serious scientific views often contain an unquestioned tendency to view the essence of existence and consciousness itself as things that are potentially separable from the gestalt of mind–body–world relations and transportable to various other mediums. Technophilia and posthumanism require this notion to breathe life into the ones and zeroes of the digital realm and make conscious machines.

This soul–spirit distinction allows us to look back over the terrain we have traversed thus far and add another layer of understanding. Whereas it may be difficult to describe a causal relationship between these themes, it is clear that the confusion and dissociation of the digital age as a response to late-modern disorientation and anxiety is a continuation and reiteration of modern soul loss. This loss appears to stem from as well as perpetuate the polarization of spirit and matter. In spite of numerous attempts to form epistemologies and psychologies that challenge this polarization—and here I want to underscore phenomenology and depth psychology in particular—an unbroken line of thought extends from the Cartesian philosophical and Protestant religious roots of the scientific revolution to today's posthuman expectation. In mainstream, secular outlooks as well as more ardent religious ones, the soul has been neglected and a vicious, antagonistic circle is the result.

What interrupts this vicious circle is the ensouling of the spiritual impulse on the one hand, and the ensouling of dead, objectified matter on the other. In other words, to recover the sense of what the human condition and its psychic inheritance so desperately needs, which is to realize that the spiritual-religious impulse is rooted in our psychological nature, and that our encounter with the world of things is permeated by patterns of the imagination and impassioned bonds. The psyche's religious function appears in our thinking whether we accede to it or not. And the mind must be trained in a particularly narrow and privileged way to dispense with its innate and acculturated ties to nature and the things of the world. Once again this takes us back to the overarching point of becoming conscious. For it is not that we must forgo a primarily secular or scientific outlook; it is that we need to understand how such stances can forget what it is that makes us most deeply human, and then proceed to construct a future in that state of forgetting.

Scientific Presentism

Posthumanists such as Kurzweil reinforce the idea of becoming cyborgs by joining the evolutionary process to the history of technology as if these things belonged to one long process of biological data-crunching and adaptation to mechanical devices. If we have evolved on this basis, as the argument goes, why should merger with our tools not be a natural step, as if our own redesign is a minor variation on an age-old theme. This perspective gives rise to notions such as "cyborg anthropology," a term its originators admit to being an oxymoron, yet one that makes the shift from *anthropos* to cyborg appear seamless. Such a field engages

> the argument human subjects and subjectivity are crucially as much a function of machines, machine relations, and information transfers as they are machine producers and operators. From this perspective, science and technology impact society through the fashioning of selves rather than as external forces.[53]

Here, becoming the tools of our tools is just the way of things and has been since our distant ancestors picked up the first stone ax. The whole history of our species is read through mechanistic eyes. The authors are surely correct when they state that "cyborg anthropology poses a serious challenge to the human-centered foundations of anthropological discourse," but fail to see the irony that a cyborg-centered mentality, one in which the human subject is best comprehended as "a function" of "machine relations," effectively perceives a million years of evolution and tens of thousands of years of pre-mechanized culture through the lens of a few decades of computation and robotics.

Along the same lines, at the conclusion of his book *Natural-Born Cyborgs*, Andy Clark argues that becoming cyborgs would be a natural extension of our turn to technology—*natural* because, as he puts it, "in a very deep sense, we were always hybrid beings, joint products of our biological nature and multilayered linguistic, cultural, and technological webs."[54] While he admits the need for caution, noting that "to recognize the deeply transformative nature of our biotechnological unions is to at once see that not all such unions will be for the better," he nonetheless punctuates his thesis by stating, "if it is our basic human nature to annex, exploit, and incorporate nonbiological stuff deep into our mental profiles—then the question is not whether we go that route, but in what ways we actively sculpt and shape it." He then ends the book with this line: "By seeing ourselves as we truly are, we increase the chances that our future biotechnological unions will be good ones."[55] At least on this last point, I think he is right: to whatever extent we merge with machines, the direction this process takes will have everything to do with "seeing ourselves as we truly are." The problem is how Clark and other posthumanists define what we truly are, which is largely an introjection of contemporary and reductive techno-scientific conceptions.

Here we see the degree to which scientism becomes a mindset without any external vantage point, to the extent that even the evolutionary process has to be seen through the eyes of data-processing, and the most salient aspects of human development and intelligence have to do with the adaptation to various tools, culminating in the modern way of life, which is itself defined by the relationship with technology. The term "presentism," typically used in history and social studies, wherein a judgment of past persons or eras is rendered on the basis of present-day mores or sensibilities, also pertains to these posthuman apologists:

episodes in the history of human development are perceived as early iterations of modern technological processes, lines between the biological and the mechanical and the cognitive and computational are blurred if not erased, and the merger of human and machine is held to be the seamless extension of a million-year-old process. This is nothing short of a cyborg creation myth.

Kelly provides us with a perfect example of this style of thinking. In considering the evolutionary advantages provided by cooking food as opposed to eating it raw, he describes cooking as "predigestion," and then says it "acts as a supplemental stomach—an artificial organ that permits smaller teeth and smaller jaw muscles and provides more kinds of stuff to eat."[56] At first blush this is an imaginative way to describe this momentous development. However, it is also an imperious application of the view we are "natural-born cyborgs,"[57] in this case turning something such as "an artificial organ" into old news, effectively giving a sort of evolutionary permission slip for the kind of reengineered body now on the drawing board. It privileges the notion we are destined to merge with machines and reads the evolutionary process back from this standpoint. As a mode of thinking, it is no different from the kind of reading of history we might find in a particular religion, wherein certain pivotal events are seen as instrumental in the destined rise of a people or unfolding prophecy. It may seem scientific, but it is essentially rhetorical, designed to evoke a sense of seamless transition between biological evolution and posthuman redesign.

Pointing to the way certain animals have developed rudimentary tools or created elaborate nests or shelters, Kelly anchors his argument along these lines with the statement "technology predated our humanness." He then offers, "the strategy of bending the environment to use *as if* it were part of one's own body is a half-billion-year-old trick at least."[58] But we need to be clear about this: first, having a "strategy," which seems evident in all kinds of fauna and may be a rudimentary expression of tinkering and techne, is not the same as possessing technology; second, an awful lot is carried by those two seemingly inconsequential words "as if." It may well be that animals bend their environment *as if* it were part of their bodies, just as it is *as if* cooking is a form of predigestion via a supplemental stomach, or it is *as if* our ancestors developed second skins when they began wearing animal fur. The choice of simile elevates the idea these evolutionary moves were prototypical forms of using technology to change the body. Yet there is little to suggest the kind of cognition involved here was technological in any contemporary sense of that term.

I do not think Kelly is being deliberately manipulative in these formulations: he is doing his work. It is just that this work involves swimming in an ocean of largely unquestioned metaphors that make it seem as though biology is more mechanical than it is, cognition is more rational than it is, and deliberate design permeates each chapter of human development. The overall effect is to draw a very narrow line between civilization and nature, which passes only through that part of culture that corresponds to scientific understanding, leaving all the other cultural forces and imperatives to fade into the background.

Another way to read evolution—and there are surely several—is to emphasize the role of human sociability. That our species has evolved, gathered knowledge, and built things is premised at least as much on our capacity to cooperate and form intricate social structures. Lent refers to this as the "socio-cognitive or cultural niche," from which "human culture emerged as a set of shared symbols and practices that ties a group together." He goes on to describe "a new feedback loop" being contemplated in evolutionary biology: "in a process known as gene-culture coevolution, culture has shaped the human niche so profoundly that it's caused changes within the human genome, affecting the very direction of human evolution."[59] Lent also cites the use of fire and cooking as an example.

Adaptability, sociability, and strategies employed over extremely long periods of time, with endless trial and error, in semi-conscious if not unconscious ways, resulting in the use of fire to cook food is one thing; deliberate vision, conscious innovation, and understanding of cause and effect that constitutes technological invention is something else. Blurring the line between the two makes the radical alterations to body and mind now on the horizon something they are not. No human involved in early cooking had any notion of the evolutionary changes to the body that would result, and cooking was, in any case, evidently in service to far more immediate instinctual and cultural dynamics. The changes to our physiology that resulted were unforeseeable and secondary—as many evolutionary events have been.

Ideologically, the cyborg as image of human becoming has already begun to assimilate other viewpoints, producing a borged state of mind.

For those writing about the pending merger with machines, a totalizing vision underscoring the "ism" in posthumanism is starkly evident. An example may be found in Haraway's foreword to *The Cyborg Handbook*, where she writes, "Lovelock's earth—itself a cyborg, a complex, autopoietic system that terminally blurred the boundaries among the geological, the organic and the technological—was the natural habitat, and the launching pad, of other cyborgs."[60] The notion of the Gaia hypothesis as an ideational platform for a cyborg future will come as news to most commentators, including Lovelock himself, who recently wrote:

> Unfortunately, we are a species with schizoid tendencies, and like an old lady who has to share her house with a growing and destructive group of teenagers, Gaia grows angry, and if they do not mend their ways she will evict them.[61]

Revenge of the Nerds

In the introduction to the exposé of life in the tech industry *Close to the Machine: Technophilia and its Discontents* by Ellen Ullman, Lanier writes:

> There is a purpose of life in the nerd world, which is treating reality as a code, and optimizing it. Life becomes a problem-solving activity, and the problem is some sort of lack of optimization. Of course this imperative breaks down on close examination . . . Back when *Close to the Machine* came out, one of my reactions was that it might portend the beginning of a consilience many of us expected. We expected the real world, the world of relationships, experience, and mortality, to overtake the abstract little nerd bubble of Silicon Valley . . . Instead what happened was the opposite. The whole world bought in to our nerdy way of life . . . We define ourselves as being in relationships, or not, on social-networking sites as if we are setting bits in a program. The nerds took over.[62]

Lanier speaks as an insider. He is known as one of the pioneers of virtual reality. But he has more lately taken to exposing some of the assumptions and dynamics at work in the technologist's grooming of the rest of us. In his starkly titled book *You Are Not a Gadget,* he states, "It is impossible to work with information technology without also engaging in social engineering."[63] He goes on:

> Ideals are important in the world of technology, but the mechanism by which ideals influence events is different than in other spheres of life. Technologists don't use persuasion to influence you—or, at least, we don't do it very well.

Instead, arguably as a way of compensating for the lack of interpersonal skill:

> We make up extensions of your being, like remote eyes and ears (webcams and mobile phones) and expanded memory (the world of details you can search online). These become the structures by which you connect to the world and other people. These structures in turn can change how you conceive of yourself and the world.

He concludes the point by stating, "We tinker with your philosophy by direct manipulation of your cognitive experience, not indirectly through argument."[64]

It may seem irreverent but it is certainly not flippant to think that one factor driving these efforts is a "revenge of the nerd" syndrome, rooted in an idealization of the intellect as a means of compensating for social awkwardness and isolation. Whereas early childhood is most often scanned for the roots of later life problems, adolescence can be an equally or sometimes more significant phase of psychological development, and the feeling of not being fully recognized or accepted at this stage of development can linger, even when one achieves great success and widespread recognition later in life.

We cannot dismiss the idea that the psychologically wounded brainiac-outsider carries into life an unconscious or semi-conscious desire to remake the world according to his or her disposition—a world where a narrowly defined, calculative intelligence and a relative devaluation of aesthetic and social matters carry the day. An attractive spouse, a garage of sports cars, and a private jet may not be enough: the world itself has to be converted into a place in which technologists feel a certain mastery, so that others can finally be subjected to their worldview. It is "Rosebud" redux. While I am not suggesting everyone in the tech industry neatly fits this psychological profile, I suspect a critical mass of those rising to places of prominence are affected by this syndrome in one way or another. Insiders such as Lanier and Ullman would seem to concur. Others might have successfully imbedded their scientific and technological pursuits in an array of other traits and human values. But what is needed more than ever are technologists who are also humanists.

Silicon Valley and Seattle tech folk have been successful in generating an impressive counter-narrative, one hinging on their economic success and worldly recognition: being a nerd is actually pretty hip. But this subculture may be doing more to reinforce a comfortable one-sidedness rather than opening up other aspects of human development. Combined with trends toward micro specialization and the decline of the humanities in education, the success and influence of the tech industry may be normalizing an outlook at odds with being fully and deeply human and is prone to "treating reality as a code," as Lanier puts it.

Today's tech nerds, effectively taking unconscious revenge on their adolescent peers, contain something both of the mad scientist described in a previous chapter and something of the culture hero. As isolated geniuses, rejected in love, they may retreat to a lair and contrive grandiose schemes to show the world their greatness; as culture heroes they emerge with problems to highlight and technological solutions to develop, earning respect, acclaim, and megaphones that encompass the globe.

As I noted earlier, it is a common theme in sci-fi that those interested in the fabrication of humanoid robots (typically male) invariably turn their interests in the direction of creating some perfect female form, such as we see with the character of Rotwang in Lang's *Metropolis*.[65] It is similarly, if more comedically, on display in the 1980s teen flick *Weird Science*.[66] More recently, the film *Lars and the Real Girl*[67] explored the transferring of a man's search for an accepting companion onto a mail-order sex doll—though this quirky and endearing story moved its protagonist through this psychological phase and into the possibility of real intimacy.

Lanier just "rolls his eyes at the digerati in Silicon Valley obsessed with the 'science-fiction fantasy' of A.I." As this "father of virtual reality" states:

It can sometimes become a giant, false god ... You've got these nerdy guys who have an awful reputation for how they treat women, who get to be the life creators. "You women with your petty little biological wombs can't stand up to us. We're making the big life here. We're the supergods of the future."[68]

As biting as these descriptions are, we need to see through the tech-world sarcasm to the archetypal undertone. To the extent Lanier's caricature of the nerd-creator points to a psycho-social reality may also be the extent to which socially awkward men, seeking to control women, are giving acute and consequential expression to a broader inclination to dominate the feminine dimension of life. That is to say, a problem with accepting the organic side of creativity and the wisdom of nature—aspects of life traditionally personified by the goddess—seems to be shaping our future.

Algorithmic Halos

There are many helpful, even life-saving, innovations that involve devices being placed in our bodies and connected to the internet. At this point, the purpose of these devices is to restore physical function. But such innovations are also shifting the way we think of ourselves, inviting the mind to follow the body down the same path. More widely, online life is grooming us for a fuller immersion into virtual reality and inviting the prospect of direct neural connections to the internet and a more fluid view of who and what we are. Our sense of self is thus being altered from the incorporation of both hardware and software into our physical and mental being.

In his book *Intervolution*, Taylor shares his experience of having diabetes and acquiring an artificial pancreas, the effect of which is not only life-saving but also revelatory, causing him to realize how imbedded we are in the world beyond our own bodies and minds. He writes, such experiences

> transform assumptions about individuals by subverting many of the binary distinctions that have long informed our understanding of ourselves and thereby call into question what it means to be human.

Self/Other, Subject/Object, Identity/Difference, Animate/Inanimate, Human/Machine, Natural/Artificial, Body/Mind, Private/Public, Autonomy/Heteronomy. I no longer know what is mine and what is not, and I am no longer sure where "my" body begins and where it ends.[69]

For Taylor, having such a medical device places him right in the middle of the way the "Internet of Things (IoT)" is leading to the "Internet of Bodies (IoB)." He says:

> the IoT and IoB are inextricably integrated—each requires the other. When joined together and linked to the Intranet of the Body, they constitute an intervolutionary network in which the form of life following what has been known as humanity is emerging.[70]

In other words, he is describing an early phase of posthuman existence. And given how this development offers him a smoother, more reliable, way to receive the life-saving insulin that his diabetes requires, Taylor is understandably enthusiastic. He holds this experience as emblematic of the way developments in big tech are not only restoring physical health and extending lives but ushering in a new paradigm. Herein nature, culture, society, and technology form an interactive, interdependent whole—a "network of networks" that even possesses a kind of self-organizing ecology.[71]

The question that must be asked is whether such arguments, focused on technology's restorative power and extension of the evolutionary process, also function as Trojan horses. As in ages past, the gifts of technology are plentiful, but the full implications of these gifts are also mostly hidden, especially in their advent, allowing less conducive if not destructive factors to infiltrate other aspects of life. This general stream of technology has an interest in implicating itself into the functions of body and mind in a far more pervasive way and at a far earlier juncture than the present restoration of function and preservation of life necessitate. Bodies and minds are thereby prone to becoming another resource, prone to commodification, subject to being regulated rather than merely attended and supported by outside entities, to the point that what happens to these bodies and minds no longer reflects an integral sense of self or invites individual responsibility.

In a brief nod to such outcomes, Taylor acknowledges Zuboff's work on the corrosive nature of surveillance capitalism. He even quotes Larry Page describing how the collection of personal data is central to Google's mission. However, even in the face of the possible misuse of medical data, Taylor's bottom line is that "High-speed networked computers, mobile devices, miniature sensors, Big Data, and artificial intelligence are converging to create breakthroughs that offer hope for the treatment and perhaps even the cure of diseases that have plagued human beings for centuries."[72] In other words, the sacrifice of personal data and its potential misuse is trumped by the possibility of medical breakthroughs. If those breakthroughs happen to be life-saving, this argument seems convincing. But we cannot know for certain whether the physical health benefits will outweigh the psycho-social costs of these developments. We know from prior experience and from the nature of the human psyche that our focus on what is apparently conducive often deflects awareness of what is surreptitiously corrosive; our track record with technology actually epitomizes this pattern. The extension of life is a trump card in this context, making the step from the IoT to the IoB not only difficult to challenge but almost utopian in its prospects.

Few would question that chronic illness, often accompanied by pain and disability, is a worthy focus of medical research and technological innovation. Yet in the larger scheme of things, we cannot deny that illness and disease frequently call attention to larger problems; they also function as reminders of mortality and our participation with all living creatures in the great round of life and death. What ails us can sometimes be the only indication of larger social and environmental syndromes—something that is missed when we focus exclusively on technological solutions. There can be no real ecology of mind or matter without paying attention to what breaks down, acknowledging unconscious as well as conscious motivations, and realizing that being belongs to an ancient patterning of psyche and soma, containing deep intelligences that compensate and ultimately complement calculative reason and logic. If what ails us cannot be held as one of the key means by which such deeper intelligences communicate, a true and real ecology of being will not be possible.

Not so long ago Amazon released a health monitoring device called Halo. Like those of other manufacturers, this device monitors vital signs as well as sleep and exercise patterns. It is associated with applications that process 360 degree body images sent to a cloud-based AI doctor. Most innovative of all, it has a built-in microphone that tracks mood via voice tone and provides feedback on how much of the day you are feeling frustrated or depressed, suggesting the algorithm may know your mood better than you do. The upshot is that data on your physical and mental health is now being tracked by the world's largest ecommerce company. Halo may well be the next level of Zuboff's surveillance capitalism.

Let us then imagine the Halo project forward. A few years from the present, Amazon could easily team up with a pharmaceutical company that makes psychotropic medications. Wanting to go beyond current search engine marketing (SEM) techniques, where individually targeted ads for drugs appear in relation to your health-related online searches, this new corporate team decides to make and distribute a psychotropic drug-administering device. Similar to Taylor's artificial pancreas, this device monitors your mood instead of your glucose levels and Amazon's algorithms decide when you may need a neuro-chemical adjustment. Later still, pursuing an even more direct solution, another medical device is created; this time involving direct wiring to your brain—a kind of pacemaker for the mind. If our psychological problems really come down to chemical imbalances and faulty neural wiring, why not? With either kind of device, the result is basically the same: your feeling life is now monitored and controlled by Amazon and its medical partners. But who or what will be there to monitor the line between corporate exploitation and individual interest? Who or what is going to limit the expansion of this Halo-effect? How far will the insatiable profiteering of American medicine in particular end up going?

We are standing between the advent of machine-learning systems indirectly altering our neural pathways and the prospect of the "Amazon Mood Enhancer" or something similar making direct changes to our brain chemistry. Decisions are being made inside tech corporations and government agencies that are going to determine the degree to which we alter the entire landscape of existence. And most of these decisions stem from a state of mind that displays an alarming absence of understanding that existence—whether self-knowledge, how cultures shape perceptions of reality, or the interconnected character of being. Mesmerized by the glow of progress and largely indifferent to technology's rather long shadow, we move forward at an unrelenting pace, but our vision is narrowing to the point of not knowing where we are putting our feet or which step may be our last. Attending to our blind spots and instigating some collective introspection would thus appear to be a vital undertaking.

Flesh, History, World

The world is a matrix of significances our minds are structured to perceive and evaluate. Brain structure and chemistry have been derived from layer upon layer of repeated sensations and perceptions laid down over millions of years of embodied experience. Sensations and perceptions occurring in the life of an individual then activate this brain structure and chemistry. Ontogeny may not recapitulate phylogeny in a precise way, but evolution is at least echoed in the development of each human mind. Grasping the nature of thought in light of such things has been called "the embodied cognition approach."[73] Understanding the way thought always occurs in relation to the body and the world has also been emphasized by the field of phenomenology. Such approaches inform us that the world is not merely data—a calculable sum of things and events—but a field of experience that begins with how our evolved brain structure and chemistry respond to a similarly evolved and adapted body. This is a very different reality to the one processed by computers.

Such a contextualized sense of knowledge and understanding is forgotten when AI specialists imply that computers function like minds and that the time for joining computers and minds is just around the corner. Even programmers capable of taking in the bigger picture acknowledge this. Ullman states, "The world as humans understand it and the world as it must be explained to computers come together in the programmer in a strange state of disjunction."[74] Looking back on his work, Zarkadakis notes, "there was something about programming that did not quite capture the way that minds seemed to function. Something was amiss."[75] But somehow this state of disconnect between mind and machine is brushed aside in the minds of those smoothing the path to human–machine hybridization. We look at the poetry or art of the chatbot and conclude its capabilities are about to meet or surpass our own, forgetting the role that embodied, worldly experience plays in actually understanding a string of words or arrangement of colors. The difference between the layered nature of meaningful understanding (mind) and the flattened character of algorithmic organization (programming) is canceled.

On the mind side of this endeavor also lies thirty thousand years or so of civilization—the development of language, symbology, religion, mathematics, science, philosophy and the practice of relating all these things to life. The brain both functions and evolves by means of these ways of representing the world. Yet, while we have neural structures and pathways ready to process and implement these ways of seeing, interacting, and understanding, none of them exist wholly in the brain; none of them are activated without the stimulus provided by families, societies, and cultures; they exist only in a relation to a given world. What we think of as mind is therefore more like a field of thought and perception that arises only in the phenomenal encounter, via this particular kind of embodied experience, with a world oriented to accumulated knowledge and understanding that exist in traditions, stories, books, and other cultural artifacts. This is the world as it exists in human understanding.

In spite of such understanding, there is still a widespread tendency to think of the mind as a computer. This is one reason I want to preserve and inject the reality of the psyche, which exists between mind and body, and between cognition processes and the world of ideas, symbols and values that belong to the cultural field. Whereas we are apt to think of a mind as something that sits atop the instinctual drives and emotional responses that define the embodied, animal level of being, between this mind and its physical carrier there is a psychic field that translates raw emotion into differentiated feeling and image. And this process is mediated by a collective pattern of experience, in which certain values and symbols hold sway, guiding and shaping human response. The mind is connected to the body via this intermediate reality of the psyche, which is not altogether located within the individual and cannot be replicated within any bounded entity. The most we can say is that the mind is an instrument of awareness and knowledge that is attuned and responsive to this larger psychic sphere.

For all these reasons, an artificial mind, created or recreated outside such embodied contexts and unable to be attuned and responsive to the psychic sphere (because the means by which such attuning occurs is also a matter of feeling) is almost certainly not going to be comparable to a human mind. Although it may well be programmed to emulate this capacity, in an actual sense it will not even remotely navigate the cultural or social world in a similar way. The upshot of this understanding is that constructing an AI that is not just able to process vast amounts of data but is emotionally responsive, has a feeling life, or is capable of generating psychic images is most unlikely. Hardware and software are not enough. A brain in a vat is not a human being and an artificial brain in a robotic body will be no different. No amount of data or ability to process data is going to alter these fundamental considerations.

We may be tempted to confine those who believe in the creation of artificial human minds to a small group of technophilic extremists. However, commentators such as Lanier readily perceive this belief flowing through the "mainstream dominant culture of the Internet," wherein he finds

> a denial of the biological nature of personhood. The new true believers attempt to conceive of themselves as becoming ever more like abstract, immortal, information machines instead of messy, mortal, embodied creatures. This is just another approach to an ancient folly—the psychological denial of aging and dying. To be a biological realist today is to hold a minority opinion in an age of profound, overbearing, technologically enriched groupthink.[76]

Stepping Off Point

Technologists admit a cordial relationship between technologically enhanced or augmented humans and any remaining *Homo sapiens* cannot be guaranteed. Think tanks are currently hard at work on getting around this possible hiccup. Yet this has not prevented the spread of posthumanisn as a dominant view of the future. This is telling.

Once a fringe movement, posthumanism was a faith in human salvation through human ingenuity, promulgated by postmodern thinkers and hardcore techies with avid followings among the movers and shakers of Silicon Valley. But this picture has changed. Now the posthuman outlook is mainstream, extrapolating today's trends to a time when devices will be able to build even better versions of themselves than we can, looking to an era when we are fully reliant upon, if not partly defined by, artificial forms of intelligence.

Perhaps it has always been this way: we go about our ordinary lives while the great wheels of history turn. We focus on what is right before us, adapt to the latest innovations, and leave the overall direction of the world to the movers and shakers. But this time around the stakes are much higher. This time we are on the fringe of detaching ourselves from patterns of life that have for millennia defined who and what we are. This time, the change is not an organic cultural transformation, but a sudden, radical leap. We are poised to move from a largely natural reality, carved out by over the last million years of natural evolution and topped by a civilization developed over several thousand years, toward a reengineered, corporate-driven existence figured out over a couple of decades, mainly by those who poorly comprehend human nature. Beyond simple denial, we may recognize a feeling response to this prospect, one that combines a primal survival instinct and a devotion to the continuity of creative human expression and spiritual precepts, one that is repelled by the notion of bringing such a quick end to a million years of human life by rolling the dice and letting computers take the reins. As we do not know whether the experiment of conscious life is going on elsewhere in the universe, it would be a shame to mess things up here on earth.

Human beings may need rebuilding, just not in such a literal way; especially not in a way that isolates and enhances particular parts of us and therefore neglects the significance and function of being fully and deeply human. To the degree we are falling apart, however—psychically, socially, and culturally—this rebuilding project of another kind does seem entirely necessary. We might begin such an effort with a better understanding of the brain. For in our current thinking at least, this part has become more disconnected from the rest of us and more subject to reductive ways of knowing than any other.

Notes

1 Quoted in Oliver Saks, *A Leg to Stand On* (New York: Touchstone, 1984), 166.
2 Cited in Michael Chorost, *Rebuilt: How Becoming Part Computer Made Me More Human* (New York: Houghton Mifflin, 2005), 42.

3 Sherry Turkle, *Alone Together: Why We Expect More from Technology and Less from Each Other* (New York: Basic Books, 2011), 152.
4 Jaron Lanier, *You Are Not a Gadget* (New York: Alfred A. Knopf, 2010), 4.
5 Jeremy Lent, *The Patterning Instinct: A Cultural History of Humanity's Search for Meaning* (New York: Prometheus Books, 2017), 278.
6 See Alex Mar, "Love in a Time of Robots." *WIRED* November, 2017.
7 David Levy, *Love + Sex with Robots: The Evolution of Human–Robot Relationships* (New York: HarperCollins, 2007), 10.
8 Ibid., 17.
9 Ibid., 297.
10 Mar, "Love."
11 Lanier, *Gadget*, 26.
12 Delisa Shannon and Noah Friedman, "Teens Would Rather Break Their Bones Than Lose Their Phones." *Business Insider* online, May 6, 2021.
13 Steven Rose, *Lifelines: Biology Beyond Determinism* (New York: Oxford University Press, 1997), 273.
14 Ibid., 272.
15 William Barrett, *Death of the Soul: From Descartes to the Computer* (New York: Anchor Books, 1987), 57.
16 Richard Dawkins, *River Out of Eden: A Darwinian View of Life* (New York: Basic Books, 1996), 19.
17 Quoted in Lent, *Patterning*, 285.
18 Ibid.
19 See Mary Midgley, *Science and Poetry* (London: Routledge, 2001).
20 John Searle, *The Mystery of Consciousness* (New York: The New York Review of Books, 1997), 9.
21 Richard Dawkins, *The Selfish Gene* (New York: Oxford University Press, 1976).
22 N. Katherine Hayles, *How We Became Posthuman: Virtual Bodies in Cybernetics, Literature, and Informatics* (Chicago: University of Chicago Press, 1999), 228.
23 Nick Bostrom, "A History of Transhumanist Thought." https://nickbostrom.com/papers/history.pdf (Originally published in *Journal of Evolution and Technology*, vol. 14, no. 1, 2005, 8.)
24 Ibid., 8.
25 Ibid., 4.
26 Ibid., 10.
27 Nitasha Tiku, "The Google Engineer Who Thinks the Company's AI Has Come to Life." *The Washington Post*, June 21, 2022.
28 Adrienne Mayor, *Gods and Robots: Myths, Machines, and Ancient Dreams of Technology* (Princeton, NJ: Princeton University Press, 2018), 178.
29 Ibid., 4.
30 Brian Greene, *Until the End of Time: Mind, Meaning, and Our Search for Meaning in an Evolving Universe* (New York: Vintage, 2023), 3.
31 Jean Baudrillard, *The Illusion of the End* (Stanford, CA: Stanford University Press), 101.
32 Rollo May, *The Cry for Myth* (New York: W. W. Norton, 1991), 293.
33 Ibid., 294.
34 *Bicentennial Man*, directed by Chris Columbus, featuring Robin Williams, Sam Neill, Embeth Davidtz (Columbia Pictures, 1999). Film.

35 Hubert L. Dreyfus, *What Computers Still Can't Do: A Critique of Artificial Reason* (Cambridge, MA: MIT Press, 1992), 64.
36 Ibid., 262.
37 Ibid., 262–263.
38 Ibid., 265–266. Italics in original.
39 *Star Wars: A New Hope*, written and directed by George Lucas, featuring Mark Hamill, Carrie Fisher, Harrison Ford (1977, 20th Century Fox). Film.
40 CW 7, par. 78
41 From "Word Origins," by Linda and Roger Flavell. http://falsemachine.blogspot.com/2015/09/the-etymology-of-desire.html
42 David F. Noble, *The Religion of Technology* (New York: Penguin, 1997), 159.
43 Ibid., 160.
44 Ibid., 171.
45 Ibid., 154.
46 Erik Davis, *Techgnosis: Myth, Magic and Mysticism in the Age of Information* (New York: Harmony Books, 1998), 3.
47 Ibid., 124.
48 Lee Worth Bailey, *The Enchantments of Technology* (Chicago, IL: University of Illinois Press, 2005), 32.
49 Ibid., 33.
50 Ibid., 35.
51 See, for example, Mark Andreesen, "Why AI Will Save the World." Posted June 6, 2023. https://a16z.com/ai-will-save-the-world
52 See James Hillman, "Peaks and Vales." In *Senex and Puer*, Uniform Edition, vol. 3. Glen Slater ed. (Putnam, CT: Spring Publications, 2005), 71ff.
53 Gary Lee Downey, Joseph Dumit, and Sarah Williams, "Cyborg Anthropology." In Chris Hables Gray ed., *The Cyborg Handbook* (New York: Routledge, 1995), 343.
54 Andy Clark, *Natural-Born Cyborgs: Minds, Technologies and the Future of Human Intelligence* (New York: Oxford University Press, 2003), 195.
55 Ibid., 198.
56 Kevin Kelly, *What Technology Wants* (New York: Viking, 2010), 21.
57 Clark, *Natural*.
58 Kelly, *What*, 21. Italics added.
59 Lent, *Patterning*, 21.
60 Donna Haraway, foreword. In Chris Hables Gray ed., *The Cyborg Handbook* (New York: Routledge, 1995), xiii.
61 James Lovelock, *The Revenge of Gaia* (New York: Basic Books, 2007), 47.
62 Jaron Lanier in Ellen Ullman, *Close to the Machine: Technophilia and Its Discontents* (New York: Picador, 2012), x–xi.
63 Lanier, *Gadget*, 4.
64 Ibid., 5–6.
65 *Metropolis*, directed by Fritz Lang, featuring Alfred Abel, Brigitte Holm, Rudolf Klein-Rogge (Kino Lorber Films, 2003). DVD. Originally released 1927.
66 *Weird Science*, directed by John Hughes, featuring Anthony Michael Hall, Ilan Mitchell-Smith, Kelly LeBrock (Universal Pictures, 1985). Film.
67 *Lars and the Real Girl*, directed by Craig Gillespie, featuring Ryan Gosling, Emily Mortimer, Paul Schneider (Metro-Goldwyn-Mayer, 2007). Film.

68 Quoted in Maureed Dowd, "A.I. is Not A-OK." *The New York Times*, October 30, 2021.
69 Mark C. Taylor, *Intervolution: Smart Bodies, Smart Things* (New York: Columbia University Press, 2021), 32.
70 Ibid., 18–19.
71 Ibid., 154. Taylor goes so far as to suggest that such a network creates a Batesonian "ecology of mind."
72 Ibid., 27.
73 Francisco J. Varela, Evan Thompson & Eleanor Rosch, *The Embodied Mind* (Cambridge, MA: MIT Press, 2016), xix.
74 Ullman, *Close*, 21.
75 George Zarkadakis, *In Our Own Image* (New York: Pegasus Books, 2015), x.
76 Jaron Lanier, "The Maximization of Neoteny." In J. Brockman ed., *Is the Internet Changing the Way You Think* (New York: Harper Perennial, 2011), 277.

Chapter 7
HALF-BRAINED

> Neuropsychology is admirable, but it excludes the *psyche*—it excludes the experiencing, active, living "I".
>
> Oliver Sacks[1]

To this point I have adhered mainly to the psychological perspective, examining the prospective merger with machines in terms of the field of psychical experience surrounding the human subject, typically referred to as the *psyche*. Whereas the bounds of the psyche may be impossible to fathom, its key characteristics are discernable in behavioral, imagistic, and ideational patterns. To the degree these patterns are universal, timeless, and carry a certain magnitude of value, they are archetypal. Rooted in instinctual impulses, yet manifested in artistic, spiritual, and philosophical expressions, such patterns are the means by which we relate to the world and to each other. The psyche is especially defined by the way these so-called "lower" and "higher" aspects of being are dynamically bridged. We spend our lives negotiating the tension of these poles, sublimating instincts into creative actions and transforming raw emotions into meaningful images. It is in this process we realize *we are psyches*: we are the unique and impermanent manifestations of these universal and timeless patterns.

Reflecting on technological pursuits in terms of old and new, part and whole, ascending and descending, gods and demons, crafting, hunting, tinkering, fragmentation, integration, and so on, I have attempted to make psychological sense of these pursuits and to understand the shape of related thoughts and actions. Absent this psychological vision, we are flying blind—failing to notice what propels us or where the ground of existence is located. In the posthuman context, where a disregard for the psyche is a prerequisite for adopting a computation model of mind, this is acutely problematic. Unlike the psyche, computer minds are not rooted in the movement from the lower to higher realms of experience and will therefore neither generate nor be oriented towards the archetypal images that convey and cradle the deeply human. The result is likely to be a stream of thought that not only places utility and efficiency far ahead of the meanings and values but creates an alternative, fully technocratic, way of being, placing us in the service of machines and their map-world.

Posthumanists may well ask, why remain beholden to the nature of the psyche? Why not design minds that are not rooted in human psychology and therefore are not limited by the nature of the psyche? The answer to this question involves the generation of meaning. In the last chapter I set out the way the psyche stems from a mind–body–world configuration that is more unconscious than conscious. As the capacity to generate meaning is bound to this configuration, meaning, by definition, lies beyond computation, because meaning relies on existential limitation and symbolic expressions stemming from this. As Kauffman puts it, "A central failure of the 'mind as a computational system' theory is that computers, per se, are devoid of meaning. They are purely syntactic."[2] Any path to a posthuman future built on this theory of mind is one wherein meaning is made optional—if not redundant—rather than intrinsic. This alone should give us pause to think.

If this contrast between the computer-mind and the human psyche does not seem significant or concrete enough for the hard-nosed futurist, it turns out that neuroscience also backs the psyche-centered perspective. Whereas its early discoveries appeared to favor a computational model of cognition, at least in some quarters the field has matured to provide plenty of support for the importance of meaning, the unconscious, imagination, and symbolization—not just in general but specifically in understanding how human intelligence and consciousness actually work. It is this confluence of depth psychological and neuroscientific understanding of what makes a mind and occasions a psyche to which we now turn.

Those hoping to merge minds and machines are obviously interested in the brain, especially in replicating its function using digital technology. Doing so would not only be a major advance in AI; it would offer a clearer path to bridging human and artificial intelligence. According to some, it could even open the door to a sentient AI. These aims are frequently anchored by the "neural substitution argument," which holds that the complete substitution of neurons by electronic equivalents would replicate human intelligence and biological consciousness. This argument has been advanced to the point of imagining digital versions of ourselves, or "mind children," as Moravec has called them,[3] existing as disembodied beings in cyberspace. It is assumed that brains alone make minds and brains are really just biological computers. As posthumanists such as Minsky put it, "If you look at the brain you find 400-odd different pieces of computer architecture with their own busses" as well as "a lot of independent gadgets which do separate things and send signals that are informational rather than chemical."[4] Even "the Oedipus Complex" is imagined as "some piece of machinery."[5] If it turns out a computer can fully replicate brain function, there would seem to be little stopping either computers or us from becoming disembodied digital minds. But can it?

Investment in the view that brains are biological computers also derives from the assumption that minds and brains are equivalent. However, this equivalency has never been established. Davis observes that

> many of the claims about the relation of mind and mental states to brain are not really scientific at all and cannot themselves be tested in any empirical way. They rest not so much on a theory as on changed assumptions about human being.[6]

Even further:

> While the promise of neuroscience responds to a widespread yearning for concreteness and a promise of unambiguous solutions to intractable problems, the explanatory force of its insights and their actual productivity is not nearly enough to explain the public appeal.

He contends, "Something else, something in our common culture, is afoot."[7] Davis says this "something else" is that we have been "informed that our understanding of ourselves as complex persons is fundamentally in error," and that "the heralds of this pronouncement are remarkably diverse and come from regions of the intellectual world that engage in no direct conversation." Each of these influential perspectives, Davis concludes,

> calls us to reject an internal, meaning-making, qualitative, first person view of ourselves for some version of an external, disengaged, and mechanistic perspective in which we think of ourselves or crucial aspects of ourselves independently of our intentions, inner experience, our evaluative outlook, and our place in the social and cultural world.[8]

This point goes to the heart of my main argument: it is not only advancing technology but a diminishing sense of human depth and complexity that is clearing the posthuman path. While a mind is clearly dependent upon a functional brain, just as a car is dependent on a functional engine, reducing minds to brains is another matter, one that philosophically and psychologically astute neuroscientists also question. An engine does not make a car, nor does it alone generate the experience of driving. Any conception of mind that accounts for the holistic character of thought and feeling will negate this reduction.

Yet there is more to this problem of mind–brain equivalency, which may be where the rubber really meets the road, because it turns out the mind-as-biological-computer theory is not entirely true to its own terms: nearly all iterations of this theory ignore half the brain. By focusing on the left hemisphere—the side most associated with logic and reason—those seeking to create artificial minds not only tend to ignore the neuropsychology of the right hemisphere; they overlook the primary role it plays in the life of the mind, especially the way emotional intelligence, pattern recognition, imagination, and the capacity to generate meaning underpin mental activity. Right hemisphere function appears to be posthumanism's inconvenient truth, marking a fork in the road between the deeply human and the posthuman.

Evasion of right-hemisphere brain function makes the posthuman view of the mind literally half-brained. This approach goes back to the nascent cybernetic research and theorizing of the 1950s, when Pitts and McCulloch tried to demonstrate "the equivalence between a biological neuron and a logical function." This work led to the building of "the first electronic neural net" and eventually to the idea "a logical machine could imitate a brain and, ultimately, attain sentience."[9] The tree of cognitive science spread out from this point, with branches in computation, neuroscience, and psychology. However, in summarizing this early history, Zarkadakis describes the tilt:

> The claims of pioneering AI research were founded on the premise that logic was the active ingredient of intelligence. There was a strong cultural element at play in this claim; the essence of "humanness" was assumed to be about the ability to reason and do clever things such as solve complex logical problems. Emotions were ignored as part of a lower, and rather uninteresting, "animal" or "primitive" aspect of being human.

Zarkadakis then draws a direct line from this oversight to the posthuman project: "Focusing on logic, the AI pioneers suggested that the advent of artificial intelligence was a matter of scale. As computers became better at performing so their intelligence would increase until it reached, and surpassed, that of humans."[10] The fork in the road between the deeply human and the posthuman was thus established at the very start of AI research.

It is difficult to imagine how we as a species could end up supporting such a one-sided evolutionary path. However, thinking of bodies as machines and minds as computers we have, over time, traveled quite far in the direction of left-hemisphere dominance, grooming ourselves for posthumanism in the process. At least some neuroscientists are offering important insight into how this grooming process has played out, describing the devaluation and underutilization of right-hemisphere faculties in the modern era. If this trend endures and becomes entrenched, they argue, it may alter the way the brain itself evolves and change the character of hemispheric relations.[11] In other words the brain may already be rewiring itself for a cyborg existence. When posthumanists such as Minsky critique philosophers for using "ordinary, naive ideas like *understanding* and *consciousness*,"[12] it is evident in at least some circles this rewiring is well underway.

By brushing aside the discrepancy between having psyches and obtaining a type of intelligence that can keep up with AI, posthumanists are effectively suggesting we develop different minds. With these minds, meaning and purpose will not matter as much as expanding knowledge, utility, and control; understanding and wisdom will matter less than the speed of information retrieval; humans will have fulfilled their evolutionary function by creating machines that will go beyond us. But if meaning and purpose are absent, if the capacity for cooperation, friendship, intimacy, and other satisfactions of the soul are all left behind, what would be the point of existence?

The pioneering and imaginative neurologist Oliver Sacks encountered the neglect of existential and psychological integrity early in his career, noting that "classical neurology in general . . . striving to establish a rigorous science of function . . . exclude[d] any observations beyond the realm of function." Overarchingly, "neurology had no room for anything existential."[13] Saks goes on to describe how he helped advance neuropsychology, armed with a new credo: "the organism is a unitary system."[14] The question he kept in mind was, "what is a system to a living self?" Summing up his experience of these fields at the time, he states:

> neuropsychology, like classical neurology, aims to be entirely objective, and its great power, its advances, come from just this. But a living creature, and especially a human being, is first and last active— a subject, not an object . . . Neuropsychology is admirable, but it excludes the *psyche*.[15]

I have already set out the inadequacies of this "entirely objective" stance in relation to the mind–body–world relationship (phenomenology) and in relation to neurochemical reduction in our culture of dissociation and numbing (psychotropic medication). Here the point is that even when posthumanists turn to the brain in the hopes of mastering and then rewiring this biological machine, they do so with a technophilic one-sidedness that follows the mechanistic inclinations of early neuroscience but ignores many of this field's recent understandings— especially those related to emotion and consciousness.

Several projects are now underway to replicate the brain in artificial form, hoping to produce a computer intelligence that functions like a human mind. Although those involved in these efforts do not deny the complexity of neural structures and processes, the idea endures that replicating these structures and processes in the form of circuitry and software will eventually lead to an artificial mind. As neural networks based on models of cognitive function are already a prominent feature of AI, this recreation of cortical function is regarded as something of a natural extension of innovative efforts.

The recently successful large language model (LLM) constitutes another approach to AI. By combing through vast amounts of text-based data and making predictions about which words best follow other words, chatbots utilizing LLM are capable of producing impressive creative output, including coherent college-level essays, moderately engaging stories, computer coding, and distillations of complex data. To judge by this output, something like an intelligence ready to exceed human capability may seem just around the corner. Despite appearances, however, these chatbots are still examples of the way AI mimics rather than replicates human intelligence. As the AI specialist and neuropsychologist Gary Marcus has noted, a chatbot "doesn't know the connections between the things that it's putting together"; its assembly of cohesive text "doesn't mean it really understands what it's talking about." Unlike humans, Marcus notes, chatbots do not make the kind of "internal models of the world" necessary for genuine thought and understanding.[16]

The synthesized and rearranged material that chatbots produce has been described as "plagiarism" by Noam Chomsky.[17] Together with another linguist and an AI researcher, Chomsky asserts that ChatGPT and other learning programs "are stuck in a prehuman or nonhuman phase of cognitive evolution," what these programs predict "will always be superficial and dubious," they exhibit a "moral indifference born of unintelligence," and are "unable to balance creativity with constraint."[18] The high-level fakery of the chatbot has even been called "bullshit" by Ezra Klein.[19] And Marcus agrees with Klein's blunt assessment, noting "these systems have no conception of the truth . . . they're all fundamentally bullshitting in the sense they're just saying stuff that other people have said . . . It's all just auto complete, and auto complete just gives you bullshit."[20]

In contemplating the AI revolution, this kind of skepticism is most warranted. While LLM-based technologies have great promise when it comes to distilling vast amounts of information and transforming scientific research, we do not know what our indifferent use of artificial creativity may augur. The potential to erode cultural processes and disrupt social processes is most apparent. One immediate challenge is to understand and then overcome the spell that AI easily casts. Bearing in mind the kind of thinking that derives from right-hemisphere brain processes will likely help us do this.

Hemispheric Relations

Some years ago a colleague involved in the early development of AI shared an example of the kind of hurdle he encountered—the ability for a computer program to recognize a chair. If all chairs were designed with a back, a seat and four legs, the matter might have been straightforward. But how do you program a machine, or how does a machine learn to recognize modern chair designs—a plastic chair shaped like an "S," for example? Whereas AI might have by now mastered this challenge, this type of problem still dogs developers. Being on the verge of allowing fully self-driving cars on the roads has meant AI systems have had to learn to differentiate between countless kinds of obstacles and driving conditions. But they still face the same essential incapacity evident in the problem of chair recognition—one borne out by the persistence of automated driving accidents and mishaps.

Humans recognize various kinds of chairs because we look for places to sit. We look for these places because we need to sit. Identifying a chair is, for us at least, bound up with this need, which is not just related to having a body in the sense of being an object in space, but to having a particular kind of body and living in particular cultures. No doubt context also matters: there are places we expect to find chairs. As Dreyfus notes, "chairs would not be equipment for sitting if our knees bent backwards like those of flamingos, or if we had no tables as in traditional Japan or the Australian bush."[21] Whereas a computer or a robot may become quite good at identifying chairs by scanning thousands of examples and having that data on hand, not knowing what it means to sit also means never comprehending what a chair is, exactly. That is, to identify an object as a chair is one thing—and a remarkable feat for a computer—but understanding "chairness" is something else. Such understanding requires familiarity with the flesh, tradition, and culture.

This may seem like a trivial example. But multiplied in relation to the myriad of understandings that constitute being in the world, one can grasp what AI is unlikely to grasp any time soon. Absent these kinds of embodied and contextual understandings, how do we expect AI to achieve anything like a comprehensive grasp of the values and concepts that pervade our lives?

Thoughts and feelings, perceptions and sensations are dependent on neuronal activity, just as a functional mind is dependent on a functional brain. Having or being a psyche is, by extension, equally dependent on the function of this organ in our heads. Yet the mind and psyche are not just neuronal. Progressive neuroscientists know this, in part by realizing the preference for atomistic and analytical versus integrative and synthetic approaches to human mentation is tied to the divided structure of the brain itself. They step back and consider the field's findings in terms of the overall character of human experience: they do not just employ the reductive solipsism associated with isolating and elevating left hemisphere processes. One such researcher is Iain McGilchrist, who pursues this bigger picture in terms of increasingly asymmetrical relations between the cerebral hemispheres.

Pushing past the popular and sometimes misleading caricatures of left- and right-brain psychology, and acknowledging advances in understanding that focus on the cooperation of the hemispheres, McGilchrist presents a compelling case for a differentiated view of each region's function: the right hemisphere is largely responsible for our capacity to form images, generate contextual understanding, and have empathy, while the left hemisphere supports abstract thinking, language ability, and calculative intelligence. He states: "There is a plethora of well-substantiated findings that indicate that there are consistent differences—neuropsychological, anatomical, physiological and chemical, amongst others—between the hemispheres." He adds, by way of underscoring the significance of these findings, "such a coherent pattern of differences helps to explain certain aspects of human experience, and therefore *means* something in terms of our lives, and even helps explain the trajectory of our common lives in the Western world."[22]

Neuroscientist Daniel Siegel also discusses the respective role of the hemispheres in relation to both general states of mind and human maturation. He notes the way the right hemisphere is apt "to process things quickly and more directly in relation to the body and the external world," that it is "able to perceive patterns within a holistic framework" and create "the overall meaning of events." Whereas the right hemisphere appears to generate "a bottom up perspective,"[23] the left hemisphere has a predilection for "top-down processing."[24] This differentiation also helps us see how the left hemisphere ends up deceptively prioritizing its own modalities.

Here we begin to discover the way even neuroscientific findings present an unlikely challenge to the posthuman outlook, the one-sidedness of which seems to be evident in terms of the most concrete and recent scientific research on cognition. In this regard McGilchrist lays out two crucial insights: First, because hemispheric differences and interactions are behind thinking in general, they are also behind thinking about brain function itself. He says from the start, "we are obsessed, because of what I argue is our affiliation to left-hemisphere modes of thought, with 'what' the brain does." This obsession is behind the "machine model" of the brain, which he notes "gets us only some of the way; and like a train of thought that gets one only some of the way is a liability." What would it mean to step off this train of thought? It would mean being concerned with "the manner in which" the brain does what it does, "something no one ever asked of a machine." He thus states, "I am not interested purely in 'functions' but in ways of being, something only living things can have."[25] We may recall Saks's similar focus.

For McGilchrist, "the most fundamental difference between the hemispheres lies in the type of attention they give to the world." He suggests that the "disposition towards the world and one another" is "fundamental in grounding what it is that we come to have a relationship with." Here we find neuroscientific backup for Jung's focus on "psychic reality," which, together with our modern ability to literally rearrange molecules, split atoms, and alter the atmosphere, means the way we think about existence can and will end up transforming that existence. The implications are profound when it comes to remaking ourselves by joining our brains to computer chips, for this will inevitably transform consciousness itself. Conveyed in the title of his book *The Master and His Emissary*, this potential for radical change brings us to the crux of McGilchrist's argument:

> Each [hemisphere] needs the other. Nonetheless the relationship between the hemispheres does not appear to be symmetrical, in that

the left hemisphere is ultimately dependent on, one might almost say parasitic on, the right, though it seems to have no awareness of this fact. Indeed it is filled with an alarming self-confidence . . . it is as if the left hemisphere, which creates a sort of self-reflexive virtual world, has blocked off the available exits, the ways out of the hall of mirrors, into a reality which the right hemisphere could enable us to understand. In the past, this tendency was counterbalanced by forces from outside the enclosed system of the self-conscious mind; apart from the history incarnated in our culture, and the natural world itself, from both of which we are increasingly alienated, these were principally the embodied nature of our existence, the arts and religion.[26]

The mechanistic approach, which reduces human beings to complex machines and provides the foundation for the notion of upgrading and enhancing ourselves through technological tinkering, exemplifies the "alarming self-confidence" of the left hemisphere. In this regard, it may be the failure to think deeply enough about our thinking that is literally creating a half-brained future. This left hemisphere solipsism, a "self-reflexive virtual world . . . a hall of mirrors," is precisely the stream of consciousness in which the posthuman outlook thrives. While pothumanism appears to think very hard about the world, about human life and its biological basis, it thinks less about its own thinking and it does not comprehend its aims in context—neither the broad psycho-historical context nor the body–mind–world one.

Posthumanism is a quintessential expression of the left hemisphere emissary abandoning its right hemisphere master. The right hemisphere may still operate in the background; it just is not valued or properly considered. This oversight roughly corresponds with the tendency to disregard the role of the unconscious also. Hidden motivations and cloaked ideas thus tend to run amok in the posthuman outlook. The emissary denies the master's power, remaining unconscious of what is operating in the background of its thinking. This state of mind is in stark contrast to what Siegel has described as the neurological correlate of wellbeing:

What research findings can be synthesized to suggest and what we can propose, in fact, is that an emergent quality of living and a vital and flexible life may come from an openness to bilateral functioning involving many ways of knowing. The brain is designed to integrate its functioning.[27]

This brings us to McGilchrist's second line of argument: the modern world has slowly been diminishing the role of the right hemisphere, a process that may be altering neural pathways. While early on noting that "the structure of the brain reflects its history; as an evolving dynamic system, in which one part evolves out of, and in response to, another,"[28] McGilchrist eventually comes to argue the change in modern modes of thought and perception has created "a world increasingly dominated by the left hemisphere, and increasingly antagonistic to what the right hemisphere might afford."[29] In the long term, even without direct manipulation, this could mean a significant shift in cortical function, one that obviously smooths the way for a merger with computational intelligence. In the short term, what occurs is a displacement of vital and enduring sensibilities and primary modes of perceiving the world and understanding ourselves. As McGilchrist puts it:

> In our time each of these [modes] has been subverted and the routes of escape from the virtual world have been closed off. An increasingly mechanistic, fragmented, decontextualized world, marked by unwarranted optimism mixed with paranoia and a feeling of emptiness, has come about, reflecting, I believe, the unopposed action of a dysfunctional left hemisphere.[30]

This neuropsychological snapshot of modernity corresponds with several of the precursors of posthumanism I have already named. A "decontextualized world" is one disconnected from the past and without a symbolic life. "Unwarranted optimism" pertains to the belief that reason and logic will solve every problem, supporting technological solutionism and Dataism as the bases of future existence. At the same time, "paranoia" and "emptiness" result from being haunted and undone by what we keep leaving behind.

According to Giddens, "the nature of modern institutions is deeply bound up with the mechanisms of trust in abstract systems."[31] Such trust has become more than a habit of mind: it has become the institutional ideal, pervading our understanding of both what we perceive and the way we perceive. Neuroscience itself has, historically, not only pursued its subject matter via abstract mechanistic modeling and the metaphors of computation but has also focused its research on the left hemisphere. As Saks puts it, "There are a thousand papers on the left hemisphere for every one on the right, and yet disturbances and disorders occur in both."[32] When the effect of these one-sided understandings are added, we end up not only with a distorted model of human cognition, but our unconsciousness of the values and processes of mind associated with the right hemisphere become a neglected wilderness of the mind.

This brings us back to the vicious circle of technophilic response to psycho-social fragmentation, which can be restated in neuroscientific terms. Siegel puts it thus: "The blockage of right-hemisphere processes from consciousness and from engaging with others may be an adaptive 'defense' against feeling anxious and out of control."[33] He is describing the neurology of the individual who retreats into linear thinking and rationalism in the face of interpersonal difficulty and volatile emotions. Yet this is also the default character of a very broad swath of individuals who comprise today's collective, especially many responsible for technological trends. It also describes those who offer formulaic solutions to complex human problems and set in motion various kinds of fundamentalist and totalitarian solutions to human existence. Without adequate means of approaching and metabolizing raw emotion, free-floating anxieties, and complex cultural-historical problems, people cling to rationalistic explanations and reductive thinking.

Neurotic Implications

The posthuman vision of a cyborg future and what lies beyond may be based on extrapolated trends in infotech and biotech, and such a vision may seem like an inevitable development given scientific paradigms and computational advances. But, as we have seen in this and previous chapters, posthumanism also diminishes if not neglects different kinds of intelligence and other characteristics that define our existence, putting the mind and intelligence into boxes that are easier to equate to computational processes. This operation adds up to something far more unnerving than a different style of consciousness or a shift in the evolutionary process. Psychologically, it shows every indication of a defensive splitting of self-awareness. It is, in a word, neurotic—moving us more in the direction of pathology than delivering on the promise of greater liberation and creativity.

We have already seen how technophilia goes hand in hand with dissociation, generating a normative psychology disconnected from critical modes of thought and experience. It should now be evident that posthumanism is the ultimate expression of this technophilic dissociation: posthumanism wants to take an obsessive love affair with technology and turn this into a flawed marriage, with the aim of producing offspring altogether dissociated from their human ancestors.

It is neither rash nor reactionary to consider these developments a form of madness, especially if we follow G. K. Chesterton's view that "The madman is not the man who has lost his reason. The madman is the man who has lost everything except his reason."[34] McGilchrist similarly uses the term "madness" to describe the mindset that neglects the masterly role of the right hemisphere. Referring to the work of psychologist Louis Sass, McGilchrist writes, "Sass explores the idea that 'madness . . . is the end-point trajectory [that] consciousness follows when it separates from the body and the passions, and from the social and practical world, and turns in upon itself.'"[35] In depth psychological terms, such a split between consciousness and passion is at the root of all psychological problems; it is a disturbance of the psyche's core dynamic, which is to regulate relations between these dimensions of being.

McGilchrist goes on:

For Sass, as for Wittgenstein, there is a close relation between philosophy and madness. The philosopher's "predilection for abstraction and alienation—for detachment from body, world and community", can produce a type of seeing and experiencing which is, in a literal sense, pathological.[36]

Further,

there is a loss . . . of the pre-reflective, grounding sense of the self . . . a loss of the pre-reflective sense of the body as something living and lived, a loss of the immediate physical and emotional experience which grounds us in the world.[37]

Along similar lines, Siegel suggests that "the field of mental health can reframe its compendium of disorders as revealing the chaos and rigidity of nonintegrated conditions."[38] In the course of defining psychological dissociation, Andrew Samuels, Bani Shorter, and Fred Plaut write about an approach "which fragments in order to 'analyse' when a holistic, all-embracing attitude would be more productive." They continue, "Western society's dependence on science and technology and on a certain 'rational' style of thinking illustrate this point of view."[39] All are making the same essential point: a mind detached from embodied knowledge and emotional responsiveness is a mind prone to mental disorder.

Posthumanism is not just a philosophy that engages in abstract thought; it ultimately promotes a separation from the conditions of life that generate coherent thought in the first place, including the effective abandonment of whole-brain function, which ultimately invites an abandonment of the body. To understand its trajectory of thought in this way, we must consider the scenario frequently imagined as the culminating point of the reduction of mind to brain and brain to computer, laid out by Moravec, who writes about "a postbiological world dominated by self-improving, thinking machines [which] would be as different from our own world of living things as this world is different from the lifeless chemistry that preceded it."[40]

The scenario, described by Moravec elsewhere, begins with "a 'brain in a vat,' sustained by life-support machinery, and connected by wonderful electronic links, at will, to a series of 'rented' artificial bodies at remote locations or to simulated bodies in artificial realities."[41] However, given "the brain is a biological machine not designed to function forever ... Bit by bit our brain is replaced by electronic equivalents ... The downloading of a human mind into a machine."[42] Yet, even Moravec admits, "A human totally deprived of bodily senses does not do well," so everything is arranged so that "a person may sometimes exist without a physical body, but never without the illusion of having one."[43] In thinking the scenario through, Moravec is further forced to see that "A human would likely fare poorly in such a cyberspace." In contrast to

> the streamlined artificial intelligences that zip about, making discoveries and deals ... a human mind would lumber about in a massively inappropriate body simulation, analogous to someone in a deep diving suit plodding along among a troupe of acrobatic dolphins.[44]

The next step would thus be "to replace some of our innermost mental processes with more cyberspace-appropriate programs," resulting in a "bodiless mind" that "could in no sense be considered any longer human."[45]

We must note the breezy way Moravec assumes the need to adapt to cyberspace and consider the degree to which such adaptation and functional disembodiment have already begun. It is equally telling that in imagining many of the dynamics involved in this digital existence, he offers, "The closest parallel is the growth, evolution, fragmentation, and consolidation of corporations, whose options are shaped primarily by their economic performance."[46] Corporations! A more fitting analogue for a world in which all gives way to expansion, growth, efficiency, and lack of concern with the quality of life would be hard to find. This image of a fully digital mind-world beyond the human is the ultimate abstraction. It is the final step in the project of separating intelligence, consciousness, and other qualities of mind from the embodied life that has always defined being human.

In light of these problems, the matter posthumanists do not consider but we must is this: At what point in this prospective transition to becoming disembodied minds would we in fact lose our minds? Once again, neuroscientists prepared to study the brain as it actually exists and functions—that is, in the body–mind–world context—have much to tell us.

One scientist who has most effectively articulated the inextricable relationship between mind and body is Damasio. In a book dedicated to what he calls "the feeling brain," Damasio focuses on the way the mind depends on feeling, feeling depends on emotion, and emotion has physiological origins. Essentially, minds are built on feelings, and feelings stem from bodily responses. No body, no mind. As he puts it, "Emotions and feelings are so intimately related along a continuous process that we tend to think of them understandably, as one single thing." And yet, as he puts it, "Emotions play out in the theater of the body. Feelings play out in the theater of the mind."[47]

One stark and contemporary illustration of the way altering the flesh can affect the mind has become evident in the use of Botox, which changes the musculature of the face in such a way as to disrupt the capacity to perceive the emotional states of others. We might think the problem here would involve the expression of emotion—and the macro alteration of facial presentation might well make this so. However, in another example of neuroscience accounting for embodied life, it has been revealed we empathize with others in part by registering micro-changes in our own facial musculature. That is, in unconscious ways, our faces mirror the emotions of those we encounter and communicates this to our brains. When Botox immobilizes these micro-muscular events, it numbs the capacity to empathize.[48]

It seems easy to imagine how the accumulated use of such physical alterations will likely impact emotional responsiveness and feeling life, especially if we expand the effect across successive generations, eventually disrupting pivotal developmental process in the mother–infant mirroring process, among other psychodynamics.

Including emotion in its researches is one of the most important steps neuroscience has taken. It puts to rest once and for all the notion that a healthy or functional human being could result from some kind of program of operant conditioning—a Skinner-boxed environment—or have an intellectual life that is not accompanied by interpersonal attachments and feeling responses. As Siegel declares early on in his summary of recent neuropsychological findings, "emotion serves as a central organizing process within the brain."[49] And later on, "Emotion directly influences the functions of the entire brain and body, from physiological regulation to abstract reasoning."[50] In the present context, the questions these findings raise are of great significance: If emotion has such a vital role in abstract reasoning, how do we imagine artificial intelligence will be able to reason in a human manner? Would disembodied "mind children" (Moravec) think in a way that promotes and sustains a continuity of culture and civilization? To the extent the answers to these questions are negative ones is the extent to which we are also obliged to reconsider the trajectory we are on. Otherwise the question becomes, why allow the world to move in a direction that will likely undo everything humanity has come to create and understand?

Perhaps a careful and thorough consideration of emerging conceptions of the mind will help us with these questions. Mark Solms's work on thinking and consciousness,[51] which approaches these matters from the perspective he calls a "neuropsychoanalysis,"[52] does not bode well for either humanoid robots or mind children. Without emotion, which is what an embodied existence generates, Solms convincingly argues there is unlikely to be much consciousness. Solms actually challenges the whole view that sees the cortex as the part of the brain that generates consciousness. This is, of course, the part of the brain most associated with intelligence, and the one that AI specialists seek to reproduce. However, even as Solms puts forth this emotion-based theory of consciousness, he cannot help but wonder— surprisingly in my view—whether emotions and thus consciousness could also be generated in a lab.[53] In other words, what he assumes to be the relocation of the consciousness-generating parts of the brain may one day lead to an emotionally equipped and therefore (according to his theory) conscious AI. Here again we seem to hit up against the great phenomenological error made by almost everyone in the field of AI, which is underestimating the degree to which emotion and consciousness are generated via embodied responses to people and things and the memory thereof. Emotional life is based on experiences of proximity and distance, pain and its absence, attachment and loss, being physically comforted or deprived, and the kinds of psychic templates or internal models of life these experiences generate. It may be the case that consciousness can be generated only via brain activity, but it does not necessarily follow that brain activity is the exclusive basis of consciousness.

A developed feeling life challenges the notion that brains alone generate consciousness. As Damasio's work shows, while feelings appear to involve the metabolization and reflection upon raw emotional responses, they also involve cultural perceptions and understandings. Sadness is qualitatively altered by the capacity to legitimate or find meaning in such feeling. Here we run into the qualitative nature of consciousness, which, as I will suggest in a later chapter, holds clues as to what consciousness is. Feeling is inseparable from the contextual understanding provided by right hemisphere function, and that function is mediated by a person's immersion in the complexities of culture. A sophisticated feeling life is cultivated and mediated by words and images; it involves the metabolization of emotional responses in the context of interpersonal relations and cultural meanings. Consciousness may begin with emotion and evolve with feeling, but it requires more than flicking a few neurological switches to develop into anything like the kind of self-understanding and empathic regard we generally associate with its higher levels.

The psychological and neurological imbalance evident in the currents of thought that feed into posthumanism raises the question of whether civilization will be capable of assessing the totality of its situation let alone progress with a viable sense of humanity. Even if we are headed into a posthuman future, how will we get there? What happens to social discourse and human behavior further along the path of left-hemisphere isolation? At least while we remain flesh and blood humans with emotional lives, feeling worlds, and psyches that produce dreams and fantasies that go beyond the rational mind, we must find ways to metabolize and understand these things. We cannot simply put them aside or on hold.

It will not work in the end to extract our future designs from the history of thought and cultural expectations either. As much as some may want to step outside the human psyche to redesign our species, there is no way to do this. We may attempt to ignore the right hemisphere, emotional intelligence, the instinctual and archetypal basis of mind, and the history of ideas, but these things are constantly shaping our understanding in unconscious ways. The tail that grows when our back is turned will still end up wagging the dog. Such a disconnection from primary processes of mind will only insure the return of the id monsters the one-sided process creates, and it is unlikely civilization would survive such an unconscious repossession by unwieldy emotions and untethered fantasies.

Turn-ons and Turn-offs

Part of the confidence in reproducing something akin to the complexity of the human cortex lies in the seemingly simple nature of neurons. These fundamental components of the brain appear to work like switches, firing or not firing. They therefore seem replicable in computer circuits because the transistors that comprise such circuits function as switches—on and off modes correspond to the binary ones and zeros of computer bits—which digitize the elemental instructions that make computers run. This neuron–transistor correspondence is for many the very basis on which computer and brain function may be equated, as this comment, slipped into a breezy description of computer processing, conveys:

That's likewise how humans work. The brain is the processor. The nervous system is the set of all the bus in a computer (they're basically cables that connect components together). The volatile memory is our short-lived memory, whereas the storage is our long-lived one. We use our short-lived memory to remember the state we're in while doing some tasks, just like the RAM is used by a processor to store instructions, which are sequentially fetched and which are similar to our tasks.[54]

Decades of computer and cognitive science have encouraged this reductive parallelism, producing very influential models of artificial and human intelligence, reinforcing a vision of human–machine likeness to the point of dogma. However, a deepening appreciation for the complexity of human neurology, roadblocks in the advance of AI, and renewed questions about the reducibility of mind to brain have also undermined this dogma.

One hurdle for the neuron–transistor equivalency would seem to be sheer numbers. Employing computational terms, as Damasio puts it, "There are several billion neurons in the circuits of one human brain." As well, "The number of synapses formed along those neurons is at least 10 trillion, and the length of the axon cables forming neuron circuits totals something on the order of several hundred thousand miles."[55] However, following Moore's Law, as the number of transistors in the average computer processor has grown exponentially over several decades, passing through tens of billions in recent high-end microprocessors and into the trillions for the most advanced computer chips, the math of technological advance has bolstered the confidence in building an artificial intelligence that finally functions like a human brain. So, one might thus say that massive numbers alone do not deter the attempted equivalency.

A more significant hurdle, however, is the biological nature of the neuron itself. As Tallis notes, "a nerve impulse is a wave of physical and chemical excitation passing along a neuron, analogous to (although quite different from) an electrical current going along a wire."[56] The synaptic gap that transmits signals between neurons is also more complex than a straight link in an electrical circuit. These may be thought of as "complex way stations where activity may be added up, subtracted or modulated before it passes on to the next neuron."[57] And an even greater hurdle may be the now oft-cited flexibility of the brain known as neural-plasticity. Combined with the simultaneous and interactive character of neural activity involved in even the most rudimentary activity, we end up with an understanding of cortical activity that is quite unlike what we find in computer circuits. Gathering enough electronic neurons on computer circuits is thus one thing, but organizing them to create the symphonic arrangement behind the most rudimentary thought processes and then having them function like human biology using non-biological material is another. It would probably be easier to make a Ferrari out of what you might find at your local hardware store—at least the materials would be roughly the same.

Even if the fundamental architecture of the brain could be reproduced in some other form, based on an accurate map of neuronal structures, reverse engineering the complexity of neural connections and the factors involved in their firing is a phenomenon of altogether different magnitude. Unlike a complex machine, the brain does not adhere to the same kind of linear cause and effect relationships. While certain "mechanisms" may be observable—likely because it is a mechanism being looked for in the first place—there are other processes that determine the way such discernible mechanisms operate. Once some of these processes are acknowledged, we have moved into similar biological territory as the discovery of the multiple factors thought to be involved in gene regulation—something that recently robbed researchers of the heady feeling of comprehending and thereby altering these so-called blueprints of life. The factors involved in activating millions of neurons in different parts of the brain over the course of minuscule periods of time thereby appear to lie beyond the paradigms of either mapping or reverse engineering. Continuing his meditation on this complexity, building on a point about the interaction of micro and macro "circuits" in the brain, Damasio writes:

> The product of activity in these [micro] circuits is a pattern of firing that is transmitted to another [macro] circuit. This circuit may or may not fire, depending on a host of influences, some local, provided by other neurons terminating in the vicinity, and some global, brought by chemical compounds such as hormones, arriving in the blood. The time scale for the firing is extremely small, on the order of tens of milliseconds—which means that within one second in the life of our

minds, the brain produces millions of firing patterns over a large variety of circuits distributed over various brain regions.[58]

This interactive and simultaneous activity of human neurology barely resembles the linear cause and effect operation of even the most complex computers. The AI specialist may argue that the computerized brain of the future will not be working in a vacuum but in conjunction with detailed input from the outside world—conveyed via sensors and perception devices. A degree of functionality if not intelligence will likely be attainable this way—enough to drive a car or replace human beings in a myriad of other practical situations. However, these sensual and perceptual inputs will lack the organic character of human responsiveness in the same way the electronic brain will lack the biology of the neuron. This qualitative difference may not be a problem when it comes to building an intelligence for delivering the mail, but when it comes to delivering babies it will be a different story.

Half-Brained Ideas

In Kurzweil's monograph on the brain–mind–intelligence relationship, dubiously titled *How to Create a Mind*, he downplays if not erases the differentiated function of the hemispheres, writing of this meta-structure as "another level of redundancy in the brain." He avers that the hemispheres, "while not identical, are largely the same" and, while referring to the way they "specialize to some extent," in mechanistic fashion he suggests "these assignments can also be rerouted, to the point that we can survive and function somewhat normally with only one half."[59] Although he implies it may not really matter what half we might be able to function with "somewhat normally," we know by now that in the posthuman outlook it is the left half that counts, and it is the only half Kurzweil seems to be using when we assesses the brain–mind–intelligence relationship along such lines.

Kurzweil's ensuing discussion of patients whose hemispheres have been cut at the corpus callosum, focuses almost exclusively on sensate perception and body function—on the rudimentary if not robotic ability to have functionality in the world. In the index of his book, there is only one rather telling entry pertaining to the hemispheres: "hemispherectomy." This is indicative of his whole attitude toward the meta-structure of the brain, citing its redundancy and treating the neural cortex as one giant web of computational powers, ready for enhancement and rearrangement. In this way Kurzweil epitomizes the colonizing and controlling mentality of the left hemisphere and becomes exhibit A for the syndrome McGilchrist has set out.

What posthuman thinkers fail to see is that understanding the brain, the mind, and the psyche, requires three very different levels of discourse, and that the connections between each are very far from being clearly definable and even further from being reducible to a discreet biological process analogous to computation. Whereas we may agree that having a brain is a necessary anatomical organ for thought and consciousness, it does not follow that whatever goes on in this organ, especially in this organ alone, is equivalent to these things. A brain removed from its embodied state, which involves sensate contact with the phenomenal world, instinctive and emotional responses to that world, and neural patterning that results from cultural immersion and social interaction is not a mind. And a being without a mind cannot be a soul.

Whereas there are now many ways of understanding this necessarily interconnected anatomical aspect of mind, including neuroscientific research, posthumanists and other technologists still tend to think the brain may be plugged into a body like a central processing unit may be plugged into a robot. This not only shrinks everything of psychological significance to this one organ; it also effectively imagines that organ working in isolation. It is such compartmentalized thinking that gives rise to notions such as reverse engineering the brain by replicating its neural connections by digital means.

In a way, posthumanists are in a bind when it comes to the role of the right hemisphere, the unconscious, emotion, and so on, because much of their futuristic expectation involves a notion of mind and intelligence that has no necessary ties to our animal ancestry, the flesh that defines us, or the phenomenal world. As the sensations and perceptions that derive from these realms form our current basis of mind, the idea of radically changing and then departing from both body and planet requires an equally radical departure from this basis. To get to the point where both the body and the world as we know it can be jettisoned for an existence that defies mortality and allows viable adaptation to other worlds—whether in cyberspace or outer space—the dimensions of the psyche that are intrinsically bound to an instinctive, embodied, earthly existence have to be devalued or dismissed. This describes a soulless future.

Posthumanists may try to have their cake and eat it too, by coming up with elaborate ways to fabricate the pleasures and creativity that come from the non-rational quarter. As Kurzweil has suggested, we will be able to engineer in helpful and pleasurable emotions while cutting the rest from the picture. But anyone with even a passing familiarity with the nature of psyche understands this kind of selective relationship to the deeper reaches of mind is not feasible, let alone conducive to any kind of psychological integrity or equanimity. It is also an unlikely path to happiness, which has been shown to be associated with cultures that are satisfied with ordinary, everyday challenges and are able to persevere with difficulties. Perhaps we have already begun to test this kind of fabricated experience in our entry into virtual worlds; but there, too, a great deal of choice is offered rather than exposure to the way things in life just "happen." The question is whether the virtual will alone suffice, and whether the departure from the real will catch up with us. So far, as we have seen, a psychological hangover from regular bouts of artificial relationships and other kinds of substitute stimulation is discernible.

Even more overarching, the question of preserving human dignity and freedom in the face of the commodification of people themselves is now a major problem. Posthumanists have yet to show just how humans will merge with machines without first being encoded and shaped by corporate interests and market forces. This is yet another dimension of how we get "from here to there." As Schneider puts it, "even if microchips could replace parts of the brain responsible for consciousness without zombifying you, radical enhancement is still a major risk. After too many changes, the person who remains may not even be you."[60] The idealized framing of technological advancement focused on human enhancement fills the pages of many a posthuman treatise. But how this enhancement is supposed to take place without contending with experiments going awry and economic disparities vying with whatever vision of progressed civilization happens to prevail is another matter.

Reading the Brain

Much of the confidence invested in neuroscientific explanations of brain function is the result of MRI-based research. Typically, the experimenter will ask a research subject to perform a simple task. The subject's brain activity is recorded as the task is performed, so that a correlation can be made between the recorded activity and the performed task. The logical assumption is the area of the brain demonstrating the most activity must play either a special or an executive role in the execution of the task. Tallis points out, however, that what this brain imaging process actually measures is "additional activity associated with the stimulus,"[61] leaving minor and more diffuse brain changes unobserved. Moreover, even for simple tasks such as finger-tapping, results can vary enormously both across subjects and in terms of test–retest correlations. In a somewhat sarcastic swipe at neuroscientific attempts to identify complex dimensions of lived experience with particular neural pathways and regions, Tallis writes, "For most of us, finger-tapping is less, rather than more, complex than being in love."[62]

Although some more recent MRI-based experiments have advanced to the point that researchers are eyeing the possibility of infallible lie-detectors and even "thought-reading" capabilities, Tallis concludes his discussion of the fallacies involved in MRI experiments by stating, "ordinary consciousness and ordinary life lie beyond the reach of imaging technologies, except in the imagination of neuromaniacs."[63] When the results of much imaging research is scrutinized, the widespread use of computer-derived metaphors for understanding brain and mind function—circuits, hard-wiring, information processing—begins to look more like a predisposed mode of comprehension than an accurate way to model what is taking place. Researchers of this bent seem to be wanting to see computational processes where these do not exist.

Despite the above, strides have been made connecting robotic limbs to patterns of thought, returning otherwise lost mobility and dexterity. Now, as Moises Velasquez-Manoff has documented, this research is moving in the direction of computers reading specific thoughts. Research subjects look at a series of images and the neural activity is recorded. This work has become so precise that these researchers have begun to "reverse engineer the algorithm they'd developed so they could visualize, solely from brain activity, what a person was seeing."[64] But the moral, ethical, and legal implications of such technology are enormous, particularly if there comes a time when brain scanners of some description are connected to our computers or mobile devices, or our so-called "neural circuits" are somehow directly wired to the internet.

To extrapolate from present-day trends, we can imagine what kind of feedback loop would be created if tech companies could directly track our thoughts rather than just our online behavior. As one pioneer in this area is quoted as saying, "humanity was at the dawn of a new era, one in which our thoughts could theoretically be snatched from our heads."[65] This kind of direct brain–computer connection is imagined working the other way too, with people being able to pull information from the internet or stream content straight into their heads. Among others, Elon Musk has already set up an outfit called Neuralink, aiming to do just that.

On the way to this kind of seamless mind–computer interface lie a number of more immediately feasible yet problematic applications. One neurosurgeon, Casey Halpern, is hoping to use this technology for impulse control; that is, to treat various kinds of addiction or compulsive behavior by implanting electrodes in parts of the brain involved in these behaviors. An algorithm then learns how to intervene between patterns of thought and behavioral response. "The goal is to help restore control."[66] As is often the case with the medicalization of psychological matters, we are entreated to think of all compulsive or addictive experience as a brain disease rather than a psychological problem. Whereas this stance is often designed to destigmatize such problems, it also reduces them to seemingly random biological maladies and deprives affected persons of insight into their overall life situation and psychological history.

In some extreme cases, where addictions and compulsions become life-threatening, and other means have failed, radical medical treatment seems appropriate. But as we have already seen with psychotropic medication, the slippery slope of such approaches is they can short-circuit the process of becoming conscious of underlying complexes, the awareness of which is key to a person's psychological maturation. Reflecting on a proposed brain implant to restore impulse control, the neurobiologist Rafael Yuste says, "this technology could blur the boundaries of what we consider to be our personalities."[67] Velasquez-Manoff asks, "What happens if people are no longer sure if their emotions are theirs, or the effects of the machines they're connected to?"[68] He goes on to raise the larger psycho-social implications in terms of "what could happen when brain-writing technology jumps from the medical to the consumer realm,"[69] as well as the potential for disruptive widening of the socio-economic gap. Again, we come to the question of how well the enhanced will be able to sustain relationships and build communities with the unenhanced. *Homo sapiens* may eventually need to make room for some kind of humanoid species, advantaged by their modifications and enhancements, dedicated to consolidating an even more entrenched technocracy.

Imagining a time when corporations may gain direct access to our brains, Velasquez-Manoff asks, "How will you know if your impulses are your own, or if an algorithm has stimulated that sudden craving for Ben & Jerry's ice cream or Gucci handbags?"[70] On this point he cites Yuste as noting, "there's a line that you cross once the manipulation goes directly to the brain, because you will not be able to tell you are being manipulated."[71] This is the key point. What begins with a focused medical technology, designed to correct a debilitating problem, whether physical or psychological, slides easily into the arena of elective enhancement, in the same way plastic surgery or the wider personality-tweaking use of psychotropic medications have already done. And whether it happens because of algorithms and information gathering, or because of the pressure to enhance oneself to succeed in the surrounding techno-culture, what is compromised in the end is the sense of self. And what will society become with so many compromised individuals walking about. Who will be capable of stepping outside the realm of modified thought to get a grasp of the actual state of things?

Technologists and researchers who focus only on solving psychological problems by biological means and end up reducing the sense of self and most other aspects of life to neural circuitry often show little interest in many of these larger psycho-social questions. So, they also tend either to downplay or ignore ethical concerns. One researcher interviewed by Velasquez-Manoff quipped, "those [ethical] discussions should not at any point derail the imperative to provide restorative neurotechnologies to people who could benefit from them." Another noted, "the only way out of the technology-driven hole we're in is more technology and science ... That's just a cool fact of life."[72] Cool for whom? Tinkering away in the lab while society collapses around you seems dissociative, as does concluding that every specialized innovation is going to seamlessly meld with some kind of overarching technological progress.

But what if "more technology and science" in this context means simply making bigger and better hammers because psychological problems keep being seen as nails? As the neuroscientist and developmental psychologist Marc Lewis puts it, "the current trend of labelling psychiatric problems brain problems has not panned out, and it has obstructed other channels of investigation that could be hugely valuable." Quoting a colleague in the same article, he says "despite many decades of considerable research efforts into uncovering underlying biological mechanisms, we have not identified specific and reliable markers for many of the most prevalent mental disorders."[73] Nonetheless, in very influential and well-funded circles the desire to find technological solutions drives the need to identify problems defined by neuromania (Tallis): in these circles, our lives are our brains, and scientists are going to use electrodes and computers to rectify and alter whatever problems and shortcomings we try to attribute to them, irrespective of the wider implications of this way of thinking or the exploitation these advances will likely enable.

Why Intelligence is Not Enough

As recent history has shown, the mere presence of smart people and the wide use of intelligent machines is not enough to counter ideology, tribalism, and greed. Ironically enough, it is the innovations of these smart people and the algorithmic machine learning they have developed that have accelerated the grip of all three tendencies. Whereas these so-called "cultural creatives" have given us smart phones and self-driving cars, which have to do with information processing, they have been tone-deaf when it comes to sustaining the social institutions and the foundations of democracy that will be necessary to cradle the pace of innovation and turn the information tide into streams of knowledge and understanding. Abstract, calculative intelligence, assumed to be the pinnacle of the evolutionary process, will not be able make up for the erosion of emotional and moral intelligence necessary to advance human and planetary wellbeing. Without other faculties working alongside artificial forms of intelligence, including empathic regard and ethical responsibility, we will lose the ability to perceive and regulate the life-process. In the end, we will not be capable of situating how we think and what we do within contexts of ultimate concern.

The capacity to look back as we look forward appears critical. Episodes of narrowed intelligence devoid of these other faculties fill history books. Apocalyptic stories about machine uprisings turn around a similar theme; namely, the willful blindness of obsessive scientists and megalomaniacal geniuses who are content to support the tyrannical rule of regressive overlords—as long as they can continue their research.

Devaluing and disregarding the fullness of human intelligence and the processes behind this are the largely unrecognized building blocks of a cyborg future. Yet, as I have argued from the beginning, such a fragmented basis for a prospective way of life is only half the story. The other half has to do with the kind of obsessive-compulsive behaviors and paranoid thinking likely to result from the neglected and festering state of that to which we fail to attend. As this psychic instability grows, it will prompt many to cling all the more to reductive, mechanistic explanations and techno-solutions. This is the thoroughly neurotic nature of what is now construed by many as the expected direction of civilization. This is the psychopathology of posthumanism: rejection of the deeper underpinnings of thought and feeling in order to merge with the computer-mind inviting a more entrenched selective vision of life. In the course of this second part of the book I first considered the normative embrace of dissociation and numbing, I then turned to the denial of the mortal and embodied character of existence, and now we have examined the role played by the one-sided understanding of human-brain function.

In the final chapter of this second part, I will explore a fourth aspect of this selective vision, one that has become ingrained in many of our contemporary institutions; namely, the way posthumanism has been enabled by postmodern philosophy and the reductive theories of mainstream psychology. As we turn to these factors, it is worth remembering that awareness of where our conceptions of being have become narrow and selective is, at the same time, an awakening to what is being overlooked—which is the first step in their recovery and revitalization. This notion will be more explicitly engaged in Part 3, when we explore the compensatory power of the unconscious, its push to recollect what we have forgotten and its prodding to renew our contact with the archetypal patterns of evolutionary and cultural development. Invoking this larger arc of thought may serve as a much-needed reminder that in diagnosing these problems we are also moving toward the piecing together of what has been torn apart.

Notes

1 Oliver Sacks, *A Leg to Stand On* (New York: Touchstone Books, 1984), 177.
2 Stuart A. Kauffman, *Reinventing the Sacred: A New View of Science, Reason, and Religion* (New York: Basic Books, 2008), 192.
3 Hans Moravec, *Mind Children: The Future of Robot and Human Intelligence* (Cambridge, MA: Harvard University Press, 1988).
4 Marvin Minsky, "Why Freud Was the First Good AI Theorist." In Max More and Natasha Vita-More eds., *The Transhuman Reader* (Oxford: Wiley-Blackwell, 2013), 69.
5 Ibid.
6 Joseph E. Davis, *Chemically Imbalanced: Everyday Suffering, Medication, and Our Troubled Quest for Self-mastery* (Chicago: University of Chicago Press, 2020), xi.
7 Ibid., xi–xii.
8 Ibid., 177.
9 George Zarkadakis, *In Our Own Image: Savior or Destroyer? The History and Future of Artificial Intelligence* (New York: Pegasus Books, 2015), 256.
10 Ibid., 257.
11 See Iain McGilchrist, *The Master and His Emissary: The Divided Brain and the Making of the Western World* (New Haven, CT: Yale University Press, 2009).
12 Minsky, "Freud," 175. Italics added.
13 Saks, *Leg*, 174.
14 Ibid., 175.
15 Ibid., 177.
16 Ezra Klein and Gary Marcus, "Transcript: Ezra Klein Interviews Gary Marcus." *The New York Times*, January 6, 2023. https://www.nytimes.com/2023/01/06/podcasts/transcript-ezra-klein-interviews-gary-marcus.html
17 Noam Chomsky, Ian Roberts, and Jeffrey Watumull, "Noam Chomsky: The False Promise of ChatGPT." *The New York Times*, March 8, 2023.
18 Ibid.
19 Klein and Marcus, "Transcript."
20 Ibid.
21 Hubert L. Dreyfus, *What Computers Still Can't Do: A Critique of Artificial Reason* (Cambridge, MA: MIT Press, 1992), 37.
22 McGilchrist, *Master*, 3. Italics in original.

23 Daniel J. Siegel, *The Developing Mind: How Relationships and the Brain Shape Who We Are*, 2nd ed. (New York: The Guilford Press, 2015), 256.
24 Ibid., 255.
25 Ibid., 3–4.
26 McGilchrist, *Master*, 6.
27 Siegel, 258.
28 McGilchrist, *Master*, 8.
29 Ibid., 391.
30 Ibid., 6.
31 Anthony Giddens, *The Consequences of Modernity* (Stanford, CA: Stanford University Press, 1990), 83.
32 Saks, *Leg*, 178.
33 Siegel, *Developing*, 255.
34 Quoted in Raymond Tallis, *Aping Mankind: Neuromania, Darwinitis and the Misrepresentation of Humanity* (Bristol, CT: Acumen, 2011), 207.
35 McGilchrist, *Master*, 393.
36 Ibid.
37 Ibid., 395.
38 Siegel, *Developing*, 336.
39 Andrew Samuels, Bani Shorter, and Fred Plaut, *A Critical Dictionary of Jungian Analysis* (New York: Routledge, 1986), 47.
40 Moravec, *Mind*, 5.
41 Hans Moravec, "Pigs in Cyberspace." In Max More and Natasha Vita-More eds., *The Transhuman Reader* (Oxford: Wiley-Blackwell, 2013), 179.
42 Ibid.
43 Ibid.
44 Ibid., 180.
45 Ibid., 181.
46 Ibid., 180.
47 Antonio Damasio, *Looking for Spinoza: Joy, Sorrow, and the Feeling Brain* (New York: Harvest, 2003), 28.
48 L. C. Bulnes et al., "The Effects of Botulinum Toxin on the Detection of Gradual Changes in Facial Emotion." *Scientific Reports*, vol. 9, 2019. https://www.nature.com/articles/s41598-019-48275-1
49 Siegel, *Developing*, 9.
50 Ibid., 158.
51 Mark Solms, *The Hidden Spring: A Journey to the Source of Consciousness* (New York: W. W. Norton, 2021).
52 Ibid., 36ff.
53 Ibid., 270ff.
54 Comment on Quora in response to the question "What do we mean when we say computers use zeroes and ones?" https://www.quora.com/What-do-we-mean-when-we-say-computers-use-zeros-and-ones
55 Antonio Damasio, *Descartes' Error: Emotion, Reason, and the Human Brain* (New York: Avon Books, 1994), 259.
56 Tallis, *Aping*, 16.
57 Ibid. 20.
58 Damasio, *Descartes'*, 259.

59 Ray Kurzweil, *How to Create a Mind: The Secret of Human Thought Revealed* (New York: Penguin, 2012), 225.
60 Susan Schneider, *Artificially You: AI and the Future of Your Mind* (Princeton, NJ: Princeton University Press, 2019), 7.
61 Tallis, *Aping*, 77.
62 Ibid.
63 Ibid., 82.
64 Moises Velasquez-Manoff, "The Brain Implants That Could Change Humanity." *The New York Times*, August 30, 2020.
65 Ibid.
66 Ibid.
67 Ibid.
68 Ibid.
69 Ibid.
70 Ibid.
71 Ibid.
72 Ibid.
73 Marc Lewis, "Why the Disease Definition of Addiction Does Far More Harm Than Good." *Scientific American* online blog post, February 9, 2018. https://blogs.scientificamerican.com/observations/why-the-disease-definition-of-addiction-does-far-more-harm-than-good/

Chapter 8
PATERNITY

In the poststructualist, postmodern, and cultural studies wings of the humanities, you will find . . . [an] injunction to abandon the languages we have traditionally used to make sense of ourselves . . . an image of the self in pieces, a decentered and discontinuous assemblage of experiences . . . The self is externally determined, merely an effect of roles and discourses that are instrumental, performative, and shaped by systems of power.

Joseph E. Davis[1]

Psychology, in the first half of the twentieth century, was dominated by a positivistic-mechanistic-reductionistic approach which can be epitomized as the robot model of man.

Ludwig von Bertalanffy[2]

The last three chapters have shown how dissociative, reductive, and one-sided modes of thinking background the drive to meld humans and machines. My thesis, evident throughout these chapters, is that a partitioned understanding of human nature and intelligence stands behind much of the effort to augment and enhance ourselves by technological means, an effort moving us in the direction of becoming cyborgs. Still, other factors have also wittingly and unwittingly seeded this posthuman trajectory.

A more fluid view of reality itself, originating in *postmodernism*, has provided the intellectual spirit necessary for remaking human nature according to conscious desires and prevailing social conditions. More precisely, a posthuman existence has found traction in influential arenas of cultural discourse because our understanding of humanness has been disrupted if not undone by postmodernism. Furthermore, not only does the term "posthuman" originate in postmodern writing, but *postmodern writers have provided posthumanism* with a philosophical frame, putting forth the cyborg as the ultimate expression of the socially constructed individual. Given postmodernism originally sought to challenge the dominance of scientific objectivity as well as the colonizing and anthropocentric impact of industrialization, there is not a small irony to this turn of events. However, an awkward alliance has emerged between science's aim to remake human nature and postmodernism's aim to deconstruct the idea we have an innate, determinative nature. This alliance has found a great deal of traction in a time of simulation and virtual reality, wherein technoscience has provided a vehicle for constructing and reconstructing a sense of self.

An altogether different stream of thought, which began with behaviorism and eventually combined forces with cognitive science, has been equally influential. A focus of American psychology over the past century, this is often referred to as cognitive-behavioral psychology. Offering what are touted to be effective psychotherapeutic methods, behaviorism was founded on the basis of discounting the importance of psychological interiority or an inner world as the basis of psychology. Cognitive science arose sometime later, directly related to computation models of the mind. The neural networks of AI, which have become dominant in technology, are rooted in cognitive models of human thought, which have been dominant in psychology. This coincidence has provided further impetus for the prospect that AI and human neuronal activity may one day be unified. If minds and computers work more or less the same way, why not?

Cognitive-behavioral psychology produced an impressively mechanistic conception of human nature, better suited to the observation and measurement of experimental method than other models of the mind. Indeed, it is hard to avoid the impression that this is a theory that has deliberately reduced behavior and thought to terms that scientific positivism and computational design could work with. Together with postmodernism, cognitive-behaviorism has made the human condition seem highly malleable and therefore amenable to scientific mastery and technological control. And this has, in turn, spawned the idea of reprograming minds and reconstructing bodies. This chapter will thus examine the role these influences play in the posthuman outlook.

Postmodern Base Notes

Postmodernism has called into question both foundational ideas and overarching narratives, a process that has cut two ways. On one side, it has shown how the privileging of certain perspectives over others can distort knowledge, successfully demonstrating how subjective, social, and historical elements shape ways of knowing and conceptions of truth. Once these elements are given proper weight, objectivity becomes a problematic notion. We might say, along such lines, postmodernism has tweaked our understanding of how we understand, especially in terms of the role of power structures and other sociopolitical realities. On the other side, there is some emerging consensus that postmodernism has taken its concern with the conditions of interpretation and understanding too far, in the process detaching knowledge from organic, common-sense meanings and from the way social and cultural perspectives congregate around certain time-tested ideas. As David Harvey reflects, postmodernism is discernible by "its total acceptance of the ephemerality, fragmentation, discontinuity, and the chaotic." Even further, "Postmodernism swims, even wallows, in the fragmentary and the chaotic currents of change, as if that is all there is."[3] This has resulted in an ethos in which all forms of knowledge become open to deconstruction. Considered against the backdrop of the human psyche and self-knowledge, however, this deconstructive attitude seems deeply one-sided and in denial of the way cultures and individuals are all but defined by ultimate values, timeless meanings, and enduring truths.

Whereas postmodernism might have awoken us to the role of subjectivity in ways of knowing, which is part of what I have been referring to as the vertical axis of reality, it has also truncated this axis, failing to effectively consider either instinctual patterns or spiritual principles in its grasp of the human condition. In doing so it has delegitimized two of the most crucial points of orientation needed for making sense of the world and our place in it. Postmodernism therefore leaves little room for the location of connective lines between these poles of human existence, such as we find in Jung's archetypal perspective and in the symbolic function of art and literature. A skepticism of coherent accounts of the timeless and universal dimensions of human nature has been the result. Derrida's efforts to deconstruct foundations of meaning took particular aim at Platonic conceptions of universal forms. According to the hard-nosed postmodernist, the slipperiness of meanings begins with the very terms "human" and "nature," which are subject to a myriad of differing interpretations, all of which ultimately betray some kind of historical bias and social agenda. And postmodernists have certainly made their point, exerting an unmistakable influence on the humanities and cultural studies over the past fifty years.

By mirroring if not amplifying twentieth century existential disorientation, postmodernism has attempted to convince us reality is more a willful contrivance and less an expression of enduring patterns of being, which has instilled the sense that human existence may be more malleable than it actually is. A pandemonium of fabricated imagery and virtual worlds appears to reinforce this orientation. Meaning, we are led to believe, is almost entirely conditioned by socialization, and the world can be changed by exposing these conditions and remedying their flaws. Lacking both deep roots and a discernable telos, such a conceptualization of the human condition becomes solipsistic, even incestuous. The associated philosophy can sometimes appear to exist for its own sake, feeding on its own terms and ideas, without service to higher or broader aims. Yet it simultaneously manages to capture a cultural-historical mood, which has become a springboard to the questioning of reality and radical self-invention of the digital age.

By questioning traditional forms of knowledge and understanding, postmodernism has rolled out the welcome mat for the idea that redesigning ourselves by following our desires will free us from the shackles of biology and acculturation. But posthuman images also tend to support the premises of postmodernism. As Graham puts it, "New relationships between 'the human, the natural, or the constructed' (Haraway), therefore reveal the very categories 'humanity,' 'nature' and 'culture' as themselves highly malleable." The upshot is that "technologies call into question ontological purity." In this vein, films such as *Blade Runner* reflect the way "humans have come more and more to resemble machines in their high-tech, alienated, urban wasteland surroundings."[4] Postmodern descriptions of the human condition therefore seem to be validated by depictions of where society is headed.

Almost in spite of these darker depictions of transcending human and nature constraints, an ethos has resulted in which digital reality and the prospect of radical change to embodied life is seen as fruitful if not salvational. An unlikely nexus of excitement and possibility has thus formed between an influential wave of postmodern philosophical discourse, a broad embrace of cultural pluralism, a loosening of tradition wisdoms, and the prospect of technological innovation disrupting traditional conceptions of what it is to be human.

In picturing the post-industrial collective mindset within which postmodernism plays a large role Baudrillard proposes an image of a "referential sphere," and the conditions required for orbiting around this:

> A degree of slowness (that is a certain speed, but not too much), a degree of distance, but not too much, and a degree of liberation (an energy for rupture and change), but not too much, are needed to bring about the kind of condensation or significant crystallization of events we call history, the kind of coherent unfolding of causes and effects we call reality.[5]

Baudrillard is telling us that the quality of consciousness necessary to generate a sense of reality and historical import require an optimal mix of proximity and distance. He then makes clear the ontological and epistemological problem that has unfolded in the form of an orbital breakout, a state in which "we have flown free of the referential sphere of the real and of history." He writes:

> Once beyond this gravitational effect, which keeps bodies in orbit, all the atoms of meaning get lost in space. Each atom pursues its own trajectory to infinity and is lost in space. This is precisely what we are seeing in our present-day societies, intent as they are on accelerating all bodies, messages and processes in all directions . . . Every political, historical and cultural fact possesses a kinetic energy which wrenches it from its own space and propels it into hyperspace where, since it will never return, it loses all meaning . . . Every set of phenomena, whether cultural totality or sequence of events, has to be fragmented, disjointed, so that it can be sent down the circuits; every kind

of language has to be resolved into a binary formulation so that it can circulate not, any longer, in our memories, but in the luminous, electronic memory of the computers.[6]

Baudrillard is capturing the general untethering of being and knowing. He is also describing the more specific condition of the postmodern state of mind, filled with data no longer anchored in a firm sense of reality, the presence of historical continuity, or the patterns of the deeply human. Instead, he depicts a succumbing to an acceleration of information technology and ways of knowing no longer subject to differentiate values. An aggregation of epistemologies has broken free of an ontological gravity, sending us spiraling off into a universe of constantly updating, competing, and evolving ideas with little to no convincing points of orientation. The orbit is broken in part because by challenging the grand narratives of modernism, digital postmodernism has facilitated the view that stories and meanings come easily, without ontological ground, and may be twisted any which way and dissolved in the acid of hypercontextualism and deconstruction.

One important upshot of this lost orbit and orientation is a lost ability to discern the difference between reality and facsimile, authenticity and fabrication. As Baudrillard put it, "we shall never again know what anything was before disappearing into the fulfillment of its model . . . We are leaving history to move into the realm of simulation . . ."[7] This would become the theme of his best-known work, *Simulacra and Simulation*,[8] wherein he takes this embrace of simulated reality and suggests it has become more real to us than actual reality; it has become hyperreal. And if our ongoing investment in hyperreality keeps exceeding our ability to know what is real, we be apt to embrace ubiquitous fabrication right alongside ubiquitous computation, pushing all the way into fabricated versions of ourselves.

Conditions and Philosophies

Although they are often difficult to untangle, especially because they play off and reinforce each other, it helps to at least attempt to separate the postmodern *condition*, seemingly initiated by the breakdown of traditional meanings and the existential disorientation of late modernity, from the postmodern *philosophies* that have created epistemologies to meet this condition and to some degree validate it. It is these philosophies that have embraced posthuman possibilities as means of furthering their aims, seeing these possibilities as ways in which humanism, gendered identity, and other perceived strictures of mind and body can be undone. As postmodern philosophical discourse is enamored with the modes of play and performance, it is not always clear whether this embrace of posthumanism is merely opportunistic or whether these treatments are invested in historical outcomes. What is clear, however, is that posthuman aspirations to the desired reconfiguring of body and mind align with a postmodern philosophical investment in making both conception of reality and ways of understanding more fluid.

Pointing to the essence of postmodern philosophies, Iris Murdoch simply stated, "There may be no deep structure."[9] Posthumanists have soaked in this sensibility and used it as an invitation to engage in human redesign, especially by way of thinking that intelligence need not be rooted in the patterns of a naturally evolved human nature or in the expressions of culture that mirror and metabolize these patterns. By attempting to upend the past and its ways of thinking, postmodernism lost its ties to modernism and thus to its legitimate moment of cultural correction, becoming instead overly indulgent in its capacity to break things. Like protesters who turn into anarchists, these philosophies have succumbed to the notion that undoing meaning is itself a meaningful undertaking, and by dethroning the Enlightenment-inspired conception of the modern human, they have greased the wheels of posthumanism.

Postmodernism has prepared us for posthumanism, passing on a style of knowing that has placed what it means to be human in a state of flux, thereby throwing open the door to the prospect of remaking ourselves. The intellectual mood of postmodernism both reflects and inculcates digital culture. Both are characterized by a mercurial groundlessness that captures the mood of the digital age, and both are sustained by fluid meanings based on a multiplicity of competing sources and contexts. Postmodernism has also collapsed the lines between pop culture and high-culture, between the voice of authority and that of the street-poet. While this has placed a not unneeded check on certain voices and sources of authority, it has also expanded the internet's echo chamber of competing perspectives, each claiming to be authoritative.

Reflecting the notion that Nietzsche was arguably the first postmodernist, Taylor suggests that "virtual culture realizes Nietzsche's vision of a world in which every ostensible transcendental signified is apprehended as a signifier caught in an endless labyrinth of signifiers"[10] Several commentators have followed Haraway's early take on posthumanism,[11] arguing that the embrace of a cyborg existence amounts to the realization of the postmodern breakdown of body politics, especially conceptions of gender, beauty, and physical dominance. But the most direct description of the postmodern genesis of posthumanism comes from Arthur Kroker and David Cook.

In Kroker and Cook's own words, their book, "*The Postmodern Scene*, evokes, and then secretes, the *fin-de-millenium* mood of contemporary culture. It is a panic book; panic sex, panic art, panic idealogy, panic bodies, panic noise, and panic theory."[12] They go on to describe "the thrill of catastrophe" and "the ecstatic implosion of postmodern culture in excess, waste, and disaccumulation." They then cut loose with this fiery assessment:

> When technology of the quantum order produces human beings who are part-metal and part-flesh; when robo-beings constitute the growing majority of a western culture which fulfills, then exceeds, Weber's grim prophecy of the coming age of "specialists without spirit"; and when chip technology finally makes possible the fateful fusion of molecular biology and technique: then ours is genuinely a postmodern condition marked by the deepest and most pathological symptoms of nihilism.[13]

It is not hard to see the direct line being drawn between the nihilism of postmodern philosophies and the fulfillment of the posthuman dream: the cyborg becomes the radical reaction to the unresolved existential vacuum that has replaced the meaningful connection to the deeply human. Nor is it hard to discern that one of the main instigating factors is the influence of "specialists," operating beyond any integrated sense of being or incorporation of interdisciplinary understanding—effectively "robo-beings." The fragmentation of knowledge, encouraged by the postmodernist, thereby enables the compartmentalizing posture of the technophile, and together they prepare our minds for a posthuman future.

Philosophical Posthumanism

The strand of postmodern thought that has entertained human–machine hybridization and transhuman technogenesis may be termed *philosophical posthumanism*, at the core of which is the idea that radical human transformation is part of, or may be instigating, a push beyond Western humanism. This branch of academic discourse regards humanism as perpetrating an anthropocentric and androcentric understanding of existence, an understanding that has also promoted attitudes pertaining to class, race, gender, and sexual orientation that ill fit contemporary society. It thus sees the impending destabilization and possible decentralization of the human via posthuman technologies as a fruitful opportunity to reevaluate categories and hierarchies of consciousness and life.

Philosophical postmodernism began as such in the 1970s with an essay by Ihab Hassan entitled "Prometheus as Performer: Toward a Posthumanist Culture?"[14] Presented as what has been called a "parodical . . . postmodern performance,"[15] this essay not only signals a postmodern attempt to create a philosophical backdrop for human–machine hybridization; it sets a certain mercurial, if not playful, tone for such engagements, a tone that postmodern discourse seems intent on prioritizing. Hassan successfully captures many of the quandaries and dichotomies the prospect of such hybridization may bring, including both the hopeful and terrifying. He is also able to put his finger on a fitting mythic configuration, emphasizing the movement between Prometheus's heroism and hubris, which characterizes Western civilization and its technological exploits. At the same time, however, he seems to overlook the serious consequences of an unbounded Promethean impulse. From the perspective of my own research into the myth of Prometheus,[16] Hassan also glosses over some essential elements of the myth, including the significance of this figure's titanic ancestry and flawed sacrifice, which background his hubristic stance and theft of fire. Hassan also avoids translating the gods' punitive response to such failures and exploits into real world implications. In short, like other postmodern theorists, the use of mythic figures as literary devices and cultural icons (rather than as personifications of universal thought structures), and the treatment of all cultural representations as texts susceptible to endless rounds of interpretation and constructed significance, means the psychic reality behind these representations is never really recognized. More than a missed opportunity, such treatments then end up somewhere between trivial and irresponsible.

Although Hassan taps an important well of cultural discourse and astutely recognizes the Promethean pattern in the origin story of this discourse, by coopting posthuman themes for broader postmodern aims, he also sets the direction of subsequent treatments. Whereas the scientists have all along taken their images of the future seriously, in the rest of the academic world posthumanist inclinations have ended up as mere grist for the postmodern mill. And whereas Hassan cannot be blamed for failing to foresee civilization on the cusp of realizing a posthuman future some fifty years hence, it is hard not to regard this initial philosophical treatment of the theme as using the prospect of radical evolutionary change as an opportunity to further instill postmodern views of reality.

This postmodern envelopment of posthuman ideas is amplified in Donna Haraway's pivotal and oft-cited essay "A Cyborg Manifesto: Science, Technology, and Socialist-Feminism in the Late Twentieth Century,"[17] in which the author proposes that traditional gender roles and related power structures can be upended through the redesign of the human mind and body. Haraway also situates technological changes to the human condition within a broader de-centering of humanism, which is touted as an emerging hub of intellectual discourse. Here the posthumanist becomes the fruitful disruptor of Eurocentric, colonial, masculine, and, particularly, anthropocentric discourse, and the implications of hubris and lost contact with the sacred (the sacrificial stance toward the gods) fades even further into the background.

In this stream of thought, to become posthuman promises to undo restrictive definitions and oppressive notions of what it is to be human, as well as the hierarchical dominance of these notions operating in the larger scheme of earthly life. For the posthuman philosopher, this advantage appears to outweigh concerns about what may be lost or compromised in terms of enduring human values—the deep roots of which are also questioned. This seems like a poor trade-off.

Many agree that anthropocentrism is at the heart of a plethora of problems facing earthly existence. The failure of humans to see themselves imbedded in the larger life process of the planet and its other inhabitants, and the tendency to think of that life process existing for human use, has been central to the planet's degradation and irresponsible treatment of other species. Nonetheless, we must keep in mind it was the romantic philosopher's return to nature during the nineteenth century that animated early environmentalism. It has also been the interdisciplinary work of the humanistic scientist, focusing on the interconnected and multifaceted nature of human nature, that has alerted us to the nature of non-human sentience and the intelligence of ecosystems. Much of humanistic science has also been at work to overcome the divisions of subject and object, emotion and reason, masculine and feminine and other binaries in our thinking. Critical assessments of social leanings such as anthropocentrism have thus emerged from the humanistic tradition also. In fact, it is the humanities that have provided the widest intellectual ground on which to critique cultural and social values, especially those that reinforce hierarchies of being and discount non-European cultures.

There are thus at least two glaring problems with philosophical posthumanism. The first reflects a common problem of postmodern discourse, which is the apparent failure to consider the cultural-historical context of its own stances. In this case, it is ironic that posthumanism has itself emerged from humanist discourse, and thus, "To reject humanism in this context would then risk denying a new form of humanity that emerged precisely out of a critique of the inhuman tendencies within Western humanism." Stated differently, posthumanism and postmodernism have been generated by the very intellectual and philosophical toolbox these movements now wish to put away.

The second glaring problem stems from the fact that rebuilding our bodies and rewiring our minds is a heavy-handed, if not desperate, way to overcome the problems of gender and cultural bias, the complete implications of which are far from clear. By the same token, once mutable bodies and computerized minds are made available, what will ensure that most augmented persons will be inclined to undo rather than reinforce cultural stereotypes and socio-economic divisions? Who is to say that existential choices will favor the common good? The novels of William Gibson and Neal Stephenson may be read as thought experiments in this regard; in so much science fiction, technological advance is accompanied by various kinds of psychological regression and cultural collapse. As the plasticity of online life has already shown itself to facilitate more self-promotion and less altruism, an engineered plasticity of life itself is likely to produce more of the same. Philosophical posthumanists must thus be prepared to answer the question of when and how our adaption to digital technology will suddenly pivot to altruism and absent prejudice?

Humanism has, on the whole, worked to increase empathic regard. It does this by continually putting us in the shoes of others and in situations and places that would otherwise be foreign to us. In doing so, we also gain a deeper appreciation of our common human predicament and a deeper feeling for what lies beyond familiar ground. By contrast, the trendlines of technoscience, extrapolated into a posthuman bearing, may be seen as moving us beyond the circumstances and places that actually generate empathic regard, most obviously beyond the bounded conditions of being wherein our bodies and minds need the care of others. The ultimate expression of this, which is the projected outcome of posthumanity, is departure from the earth itself.

Philosophical posthumanism therefore suffers from the same kind of abstract intellectualization of lived experience as posthuman technology itself. Rather than providing a critical container for new technologies and their impact on life, such perspectives further the view that our minds and the understandings they give rise to are thoroughly malleable, and that the path ahead will come down to changing conscious desires and who or what has the power to implement these changes. Such philosophical stances do not pass the co-creative smell test. Even though they aim to make room for embodied, phenomenal, and non-human experiences, and frequently describe the forms of intelligence at work therein, these stances convey an intellectual grasp of this consciousness without leaving room for its deep and unconscious root system. In attempting to move beyond the anthropocentric, for example, and preserve a sense of animals and other living presences as co-inhabitants of this world, one would think that connecting to the instinctive, animal dimension of our own nature would be prioritized. Instead, instinctive and archetypal patterns that connect our psyches directly to the animal world are dismissed. The whole approach assumes that the philosophical elevation of non-human life alone will radically alter the way we think and act. This is not only psychologically naïve; it is entirely unrealistic. Only a deep change in the storying of being that ties us to our own animal nature and to those archetypal patterns that generate a feeling of vital connection to the larger fabric of existence will accomplish such a shift in consciousness.

Traditional humanism has distilled literary, artistic, and historical studies into the capacity to recognize values that extend beyond the merely human and the ethical obligations that arise from this. Lately, for example, those of us in Western, industrialized parts of the world have become more sensitized to the attitudes of indigenous peoples and their broader cosmologies and respect for non-human presences. Yet this sensitivity is itself a direct result of the way traditional humanism birthed fields such as sociology, anthropology, ethnography, comparative religion and other disciplines that open our minds to the diversity of human imagination and culture. Whereas our more recent awareness of the long-term effects of colonization and the marginalization of non-White perspectives has introduced an important critical perspective into our understanding of culture, the notion that humanism must be transcended is a bit like cutting down the tree we have climbed to gain that perspective.

From the depth-psychological point of view, the nexus of postmodern and posthuman thought appears surprisingly egocentric, particularly in viewing reality as primarily socially constructed rather than grounded in patterns that join human nature and nature at large. This constructed sense of reality is what invites the dedication to deconstructed and then reconstructed reality. As Ihde puts it, "One of the dimensions of the postmodern context is the hyperawareness of 'invention.' To invent is to socially *construct*, often with the implication that anything *constructed* may equivalently be *deconstructed*."[18] But as the above considerations indicate, disconnecting the will from the great breadth and depth of humanistic understanding and installing it as the principle driver of our radical transformation opens a Pandora's box of possibilities, wherein invention serves little beyond indulging in its own shape-shifting prowess. To the extent that egocentrism and anthropocentrism play off each other as agents of ecological collapse, one inner and the other outer, it is hard not to see that philosophical posthumanism is ultimately shooting itself in the foot.

Behaviorism

Instead of recognizing the phenomenology of the psyche, including the images and emotions that comprise inner experience and give rise to the thoughts and feelings that determine our actions, American psychology in particular turned to bolstering its scientific standing by focusing on the narrow and more concrete realm of observable behavior. It then favored laboratory experiments that could generate measurable outcomes, denying the relevance of thought-processes on their own ground. George Graham, writing for the *Stanford Encyclopedia of Philosophy*, defines this style of scientific psychology as "a doctrine, or family of doctrines, about how to enthrone behavior not just in the science of psychology but in the metaphysics of human and animal behavior." He states:

> Behaviorism . . . is committed in its fullest and most complete sense to the truth of the following three sets of claims . . .
> 1. Psychology is the science of behavior. Psychology is not the science of the inner mind – as something other or different from behavior.
> 2. Behavior can be described and explained without making ultimate reference to mental events or to internal psychological processes. The sources of behavior are external (in the environment), not internal (in the mind, in the head).
> 3. In the course of theory development in psychology, if, somehow, mental terms or concepts are deployed in describing or explaining behavior, then either (a) these terms or concepts should be eliminated and replaced by behavioral terms or (b) they can and should be translated or paraphrased into behavioral concepts.[19]

To take in the sweeping, dogmatic, one-sidedness of these truth claims, which, as Graham suggests, reach into the metaphysical, is to realize their subtext: that which cannot be directly observed and subjected to the experimental method shall be excluded from study. Here we find a glaring display of scientism, which an adequate psychology would expose as compartmentalized knowledge and displaced religiosity. Instead, B. F. Skinner, behaviorism's most prominent, and arguably most radical, theorist, was at the beginning of the twenty-first century named by the American Psychological Association as the most eminent modern psychologist.

Skinner is widely known for inventing the "Skinner Box," designed to study the conditioning of rats, in the attempt to predict and control behavior, the results of which could then, it was imagined, be extrapolated to understand and control human behavior. That the processes observed in such studies are relevant to understanding certain aspects of human behavior and, in turn, offer some important tools for psychotherapists is without question; what is in question—and profoundly so—is the attempt to reduce the human condition to these elemental mechanisms of conditioning and all complexities of psychological life disregarded in the process. Human beings are turned into soulless automatons, able to be programmed or reprogrammed. The social sphere, by extension, is turned into something subject to willful redesign. In other words, behaviorism simply rolls out the welcome mat for imagining humans as machines, without an inner life worthy of attention.

As Guggenbühl-Craig puts it:

> The behaviorists want to exclude the soul—all subjective experience and introspection—from psychology. Their objective is to make psychology a natural science. Exact scientific statements about a person can be made only if all subjective experiences are ignored. The disadvantage is that we do not get any closer to comprehending the soul. If we note that the blood pressure rises when a young man sees a certain girl, we have not yet said anything about the psychological relationship between men and women.[20]

Guggenbühl-Craig goes on:

> Ironically, there is an inner life to behaviorism . . . It represents an immense expansion—to the point of becoming a caricature—of the Prometheus myth. Prometheus stole the divine fire and delivered it over to human control, and behavioristic theory is an example of this titanic urge to control everything. If there are phenomena which cannot be explained and ruled—as, for example, the subjective psychological experience—then they are simply ignored. Ironically, this violates one of the basic rules of scientific work, namely, that the

researcher must consider all the phenomena he finds, whether they fit his hypothesis or not.[21]

As we know, Jung described the unhistorical stance of the modern person in terms of "Promethean sin." Other poets and commentators had already named Prometheus the mythic mascot of modernity. Against such a background, by the middle of the twentieth century, psychology not only came under pressure to be scientifically acceptable; it was also subject to a collective expectation for its theories and concepts to reflect the mechanics and burgeoning automation of the Industrial Age. Especially in the American setting, governed by pragmatism, modernization and consumption, there was an appetite for a psychology with the same promise of power and proficiency as the sleek automobiles rolling off the Detroit assembly lines and speeding people to their elected destinations. Instead of a psychology capable of comprehending the grip of Prometheus, we got a Promethean psychology. For the marketeers and engineers eager to extend the nineteenth century's notion of human beings as "animal machines"—workers whose lives were now oriented to the assembly line—such a psychology would have great utility. For psychologists themselves, with little sense of irony, the promise of a few simple laws based on running rats through mazes and training them to press levers was within reach. E. C. Tolman, one of the luminaries of behaviorism, once wrote that "everything important in psychology (meaning human psychology) ... can be investigated in essence through the continued experimental and theoretical analysis of the determiners of rat behavior at a choice point in a maze."[22]

Writing a half century ago, Bertalanffy argued that "the underlying philosophy of American psychology" was dominated by "robot theory."[23] He was referring to behaviorism, which he describes in terms of the "happy recklessness" of "Watson, Hull and Skinner" in their reach for the "formula of behavior."[24] He writes, the "learning theory" that ensued "falls short of 'meaningful' learning in man endowed with symbolic functions." And while "concessions hesitantly made under the impact of circumstances" appear to be a feature of the field in the 1960s, "the positivistic-behavioristic-commercialistic philosophy deeply ingrained in American life and thinking" still held sway.[25] Writing closer to the present day, Hawkins notes:

AI philosophy was buttressed by the dominant trend in psychology during the first half of the twentieth century, called behaviorism. The behaviorists believed that it was not possible to know what goes on inside the brain, which they called an impenetrable black box.[26]

Zuboff also emphasizes the critical role Skinner's psychology has come to play in the modification of human behavior that now dominates information technologies, drawing part of her understanding from direct debates she had with him as a student at Harvard.[27] Zuboff outlines a revealing line of influence that begins with the German physicist Max Planck, who advised an experimental psychologist by the name of Max Meyer. The writings of Meyer exerted a powerful influence on Skinner. It was Meyer, for example, who was "bravely combining psychology and physics in the quest for absolutes" and "asserted the essence of the behaviorist's point of view," where "the world within the skin of the Other-One loses its preferred status."[28] The term "Other-One" is crucial: psychology could in this manner ignore subjective states of mind and focus on observing the outer actions of the object in relation to their environment. "Central to this new viewpoint was his [Meyer's] notion of the human being as organism . . . Recast as an 'it,' an 'other,' a 'they' . . . Distinguishable from a lettuce, a moose, or an inchworm only in degree of complexity."[29] Such an approach to human psychology led Meyer and then Skinner to conclude that the inner world of human beings "can have no scientific value because it cannot be observed and measured."[30]

The behaviorist scientific philosophy that followed and was expounded in Skinner's popular books was that subjective states were not only irrelevant; they were a source of ignorance as they promoted the idea of human freedom. By contrast, behaviorism led to the understanding that all behavior is conditioned and caused by external factors. Such an understanding dispels notions of freedom. Instead of attempting to comprehend inner, subjective experience, Skinner suggested we develop "a technology of human behavior . . . comparable in power and precisions to physical and biological technology."[31] Here we find the image of human beings as robots, as Bertalanffy argues, prone to mechanistic comprehension and correction. Here we also find the beginnings of American psychology's overwrought attempt to confine psychological research to experiment and observation and to wrest the field away from any emphasis on human interiority. Tellingly, Skinner declares, "Greek theories of human behavior led nowhere."[32] Artists and poets of the last two and a half millennia and depth psychologies of the last century would likely disagree.

Cognitive Psychology and Computation

The history of psychology generally holds that as cognitive theories of mental processes began, behaviorism began to decline. As cognition is concerned generally with internal states of mind, there is something to this: the purest forms of behaviorism, which tried to make a dogma of observation and external sources of behavior, fell out of favor. What the behavioral and cognitive positions had in common, however, was a commitment to mechanistic scientific models, and in the second half of the twentieth century the logical analogue for mental processes would obviously become computation. Just as computers process information based on rules and algorithms, the thought processes of the human mind could be similarly conceived. Thus, although the scientific gaze had found a different focus, in terms of the psychology of the psychology, it was effectively the same psychology. The hardware had located its software; the robot had found its operating system. Cognitive-behaviorism would become the dominant psychological paradigm.

When future historians look back on the middle of the twentieth century, they will single out one period and one group of thinkers who, more than any others, created the springboard for the digital age. They will also find themselves wondering what might have been. Here I am referring to what are known as The Macy Conferences on Cybernetics, which took place between 1946 and 1953 and brought together a group that included John von Neumann, Margaret Mead, Norbert Wiener, Molly Harrower, Gregory Bateson, and F. S. C. Northrop. The group also happened to include Warren McCulloch and Walter Pitts, who are credited with conceptualizing neural networks, a taproot for both cognitive and computer science—especially for AI. Cybernetics, the study of communication and control systems, whether in machines or living things, thus started as a broad, interdisciplinary effort to figure out how thought processes are related to behavior. Within this effort, McCulloch and Bateson held strong views about the holistic nature of the mind.

Early expressions of systems theory also took root at the conferences. However, this angle on understanding the mind would also end up taking a back seat.

Although biological systems and human social relations had originally been part of the cybernetics discussion, to the point that Mead was interested in how this emerging field might help us comprehend interactions between mother and child,[33] the field ended up focused on quantitative rather than qualitative understandings. As Bateson sums up the trajectory of the discussions, "Computer science is input–output. You've got a box, and you're got this line enclosing the box, and the science is the science of these boxes." However, Bateson, Mead and others such as Wiener were interested in both what was going on inside and beyond these lines and boxes: they were interested in the human and cultural ecosystems, in the idea "you are *part of* the bigger circuit."[34] For these participants, the black boxes that fill the diagrams of both computer science and cognitive psychology were not enough. Given the way the computation model of mind and the neural networking model of AI has swept the world, it is hard not to look back and wonder what might have been if the field of cybernetics had kept one foot in what Bateson came to call the "ecology of mind."

As it has turned out, the "information-processing" model of the mind, which stands at the basis of cognitive psychology, has all but ensured the computer–brain–mind understanding of thought and behavior would be prominent in the information age. As Gardner puts it, "Throughout [this model], one encounters the healthy, if somewhat unexamined, American emphasis on mechanics: on what is done, in what order, by what mechanism, in order to yield a particular effect or result." The approach also often focuses on notions such as "executive functions" and "higher-order control mechanisms."[35] Very often "one encounters the blithe notion that a single, highly general problem-solving mechanism can be brought to bear willy-nilly on the full range of human problems." This has much to do with the kind of research conducted, wherein the existence of such a mechanism is inferred from "the carefully selected problems to which it is said to apply turn[ing] out to be disturbingly similar to one another." That is, "nearly all the problems examined by information-processing psychologists prove to be of the logical-mathematical sort" such as "solving logical theorems, carrying out geometric proofs, playing a game of chess."[36]

It does not take too much cognitive power to figure out how AI specialists, convinced by cognitive psychologists that the brain functions much like a computer, have determined something like a mind could eventually emerge from a computer's central processing unit. Nor is it hard to see how psychologists, pining to be more scientific, would gravitate to models of the mind that happen to correspond to what science and technology at large were putting on the table. But the time might have come to break into this cozy circle of thought. As Gardner has noted, "the excessively mechanistic computer-driven model for thinking and the penchant for scientifically oriented test problems foreshadow certain long-term problems with this approach."[37] Among other problems, he notes the way this model does not fit well with our understanding of the overall nervous system—reminding us of what I have already discussed in terms of both the mind–body relationship and hemispheric relations in the brain. Nor does it account for "the open-ended creativity that is crucial at the highest levels of human intellectual achievement." Instead, what Gardner and a number of other researchers of human intelligence have come to focus on are "symbol systems," which he describes as primary in "much of what is distinctive about human cognition and information processing."[38]

These "symbol systems" are, from a depth psychological perspective, guided and shaped by the archetypal basis of mind. They are not programmable, because they come into being through lived experience, emotional engagement, and cultural contexts. As Gardner puts the ontogenetic aspect of this,

> in infancy there may be instances of the raw intelligences, but these are almost immediately enwrapped in meaningful activities, as a result of their affective consequences for the young child, and in light of the rich interpretations perpetually supplied by the surrounding culture.[39]

The question of what kind of consciousness or sentience a machine would be capable of given the absence of this arena of the mind as we know it, wherein meaning and value are never far from thought. If "symbols pave the royal route from raw intelligence to finished cultures," as Gardner puts it, and computer-based intelligence is neither able to generate nor understand symbols, then a civilization based on such intelligence will also be one without culture.

To extrapolate the point, it is doubtful whether functional societies could exist under conditions wherein either a computer-based intelligence or a model of the mind based on computation dominated. As Gardner says:

> there are few occupational roles the idiot savant of linguistic, logical, or bodily intelligence can perform. Rather, in nearly all socially useful roles, one sees at work an amalgam of intellectual and symbolic competences, working toward the smooth accomplishment of valued goals.[40]

Weaving Strands

The technogenesis at the heart of posthumanism reaches its apogee in two fantasies of the future in which humanity as we know it becomes essentially superfluous. The first fantasy involves jettisoning the body and digitalizing the mind. This mind would be uploaded into the cloud, now construed as a digital heaven waiting to house our disembodied identities and personalities. The second fantasy pertaining to the climax of posthumanism builds on the first. Here the idea is that the essence of human civilization could be carried to the far reaches of the galaxy via self-replicating robots and spacecraft controlled by AI. Freed from the constraints of biological life and immune to the ravages of time, these robot-humans would colonize other planets and, expanding exponentially, eventually dominate the universe. The best-known conception of this kind comes from the work of mathematician John von Neumann, who proposed what has become known as the "von Neumann Probe." Von Neumann, we may recall, was also a key participant in The Macy Conferences on Cybernetics.

Midgley concludes her work *Science as Salvation* with a feisty critique of this fantasy and its supporters, noting that although such self-replicating intelligent machines "have never met a person or lived on earth" and "may not even be conscious," they "are expected to carry on human culture so smoothly as, in effect, to continue the species itself."[41] Midgley takes particular umbrage at two writers who suggest that "The arguments one hears today against considering intelligent computers to be persons and against giving them human rights have precise parallels in the nineteenth century arguments against giving blacks and women full human rights."[42] These writers go on to propose that "von Neumann probes would be recognized as intelligent fellow-beings . . . heirs to the civilization of naturally evolved species" without which this "civilization will eventually die out."[43] In the context of the present chapter, we can see that such views represent more than scientific thinking running amok: they represent a postmodern replacement of the real by the simulated and the authentic by the fabricated. They are also the logical extension of the cognitive and behavioral paradigms, wherein any inner life or ecology of mind are negated. The absolutism in this mistaking of the cognitive processes computers may one day mimic as something identical to the thoughts and feelings of humans is breathtaking. But to locate the rather large hole is this kind of posthuman absurdity—to meet these images with the fullness of being—we need to recover what we have come to know about the deeply human, most especially the layers of mind that defy reduction to computation. Whether these fantasies are utopian/dystopian extrapolations of our contemporary state of mind or techno-scientific possibilities, they offer themselves as cultural dreams—as images with the potential to make us more conscious of our present-day psycho-social situation.

Once we account for the inextricable ties of body, mind, and world, it is difficult to imagine how a disembodied, digital existence would preserve anything like individuality or provide any basis for the formation and continuity of personality and character. What would delineate our minds from the billions of other minds floating in cyberspace? Even more fundamentally, what would delineate human thought from the artificial intelligence with which we would have long since merged? Whereas certain synchronistic, parapsychological, and spiritual phenomena suggest that aspects of the psyche and even consciousness itself are not necessarily confined to the bounds of the body, there is little to indicate a sustained experience of being and personhood can exist beyond having a physical form in a physical world. No body, no planet, no us. Yet the fantasies of evolving into spiritual beings endures. Surely the current detachment of meanings from bodies, pasts, and natures is playing its part in this cultural dream.

In 2005, eyeing changing habits of mind, the philosopher Harry G. Frankfurt wrote a very short book with the provocative title, *On Bullshit*. It was an effort, as he put it, to point out that "One of the most salient features of our culture is that there is so much bullshit." Yet, he contends, "we have no clear understanding of what bullshit is, why there is so much of it, or what function it serves."[44] Frankfurt admits to the slippery nature of the topic, and proceeds to differentiate bullshit from humbug, lying, misrepresentation, and so on. Perhaps his most effective attempt at describing the essence of the matter comes by contemplating its opposite. He begins with a verse by Longfellow and points out that Wittgenstein once took this as a motto for his life and work:

> In the elder days of art
> Builders wrought with greatest care
> Each minute and unseen part,
> For the gods are everywhere.[45]

We might readily note the way this short verse effectively draws together three things: art, craft, and the gods. Frankfurt then comments:

> The point of these lines is clear. In the old days, craftsmen did not cut corners. They worked carefully, and they took care with every aspect of their work. Every part of the product was considered, and each was designed and made to be exactly as it should be. These craftsmen did not relax their thoughtful self-discipline even with respect to features of their work that would ordinarily not be visible . . . nothing was swept under the rug. Or, one might perhaps also say, there was no bullshit.[46]

Bullshit is not new. But the digital lifestyle has made it more prevalent, and the abundant false notes and absent diligence occasion both a general tuning out and a targeted tuning in. What is also clear is that those who bullshit also dissociate. We can reach this conclusion via Frankfurt's differentiation of bullshit and lies. As he puts it:

> Someone who lies and someone who tells the truth are playing on opposite sides, so to speak, in the same game. Each responds to the facts as he understands them, although the response of the one is

guided by the authority of the truth, while the response of the other defies that authority and refuses to meet its demands.[47]

In contrast to the lier, "the bullshitter," Frankfurt writes,

> ignores these demands altogether. He does not reject the authority of the truth, as the liar does, and oppose himself to it. He pays no attention to it at all. By virtue of this, bullshit is a greater enemy of the truth than lies are.[48]

Bullshit, more than lies, undermines the constitution of knowledge and proves to be a breeding ground for the unreal.

It is hard to avoid the perception that postmodernism, taken as a foundation rather than a corrective, a foundation without foundation, without ultimate signifiers, is not a primary source of the post-truth environment of today, wherein the malleability of comprehending what is factual and real has gone too far. That is, it is difficult not to draw a direct line from postmodern philosophy to the culture of bullshit. Frankfurt appears to have this in mind when he states:

> The contemporary proliferation of bullshit also has deeper sources, in various forms skepticism which deny that we can have any reliable access to an objective reality, and which therefore reject the possibility of knowing how things truly are. These "antirealist" doctrines undermine confidence in the value of disinterested efforts to determine what is true and what is false, and even in the intelligibility of the notion of objective inquiry.[49]

We might also recall the conversation between the AI specialist Marcus and the science and culture journalist Klein, who describe the creative output of chatbots as bullshit.[50] Recall too, the contention of the linguist Chomsky, who describes this same output as plagiarism.[51] It doesn't take much to extrapolate this trendline to the point where some form of disembodied AI is careening through outer space, supposedly carrying the essence of human civilization with it. On the one hand, it is an ingenious thought-experiment, conjured by a person called the world's smartest man. On the hand, utilizing Frankfurt's definition of ideas fabricated without art, craft, or any archetypal bearing, the path to its realization would inevitably include significant amounts of bullshit.

———

We might be tempted to view the two arenas of discourse considered in this chapter—the postmodern movement in the arts and humanities and the cognitive-behavioral movement in psychology—as oddly paired. Yet in their chronology and epistemology they are both quite connected.

It is hard to avoid connecting behaviorism's delegitimizing of human interiority, the emergence of existential disorientation, and the way philosophies of absent underlying meaning became so influential. Behaviorism's role in undermining the traction of depth psychologies also prevented a wider grasp of unconscious psychodynamics and the consideration of archetypal patterns in social phenomena. Furthermore, as behaviorist theories and techniques were implicated in the rise of advertising and consumerism, a fusion of personal identities and material things followed, resulting in an increasingly fabricated and accessorized world. Postmodernism happily tagged along. From the kitsch to fetishized, from commodification to Disneyfication, postmodernism has turned towards popular culture and away from traditional culture, which has also helped to reinforce the notion the human psyche is a blank slate upon which any mix of prevailing conditions may be inscribed. That is, without enduring and universal structures, mentation can be constructed on the basis of prevailing conditions, allowing it to be deconstructed, which then means it can be reconstructed. Psychology provided the concepts, society demonstrated the logical implications, postmodernism offered the philosophical framing, and a posthuman outlook is the outgrowth.

On the epistemological side, whereas the objective scientific posture of behaviorism seems directly opposed to postmodernism's insistence on the importance of the subjective, they can be seen as flip-sides of the same coin, each unconsciously begetting their opposite. Behaviorism has not only empowered the subject with an overwrought sense of being able to shape human actions; it begs the question of what would guide such shaping. Perhaps the most direct and bizarre way in which this question was answered was by Skinner himself, putting forth a bundle of prescriptions for raising children and engaging with life in his utopian novel *Walden Two*,[52] which describes an ideal community built upon the principles of behavioral engineering and more or less guided by the vision of one man. Skinner's popular, non-fiction writing on the widespread adoption of behavioral technology followed.[53] However, Skinner's utopian application of behavioral technology cannot avoid the specter of this technology being added to the list of all the other technologies that fall into the hands of corporations, states and other powerful entities. Once again, we are then left wholly dependent on the rational subject—one even more disconnected from the instinctual and culture ground of being.

Between behaviorism, cognitive science, and postmodernism, we can locate the epistemological ground for the posthuman movement, which features a chasm of perception and thinking into which the deeply human and the ecology of mind have fallen. Cognitive-behaviorism reduces existence to measurements and mechanisms that invite control and manipulation. Postmodernism has turned thought, if not reality itself, into a pastiche of subjectivities and socio-cultural contexts without inherent patterning or enduring purpose. Together they turned the start of the new millennium into a staging area for an uber-Cartesian view of existence amplified by the digital world, one detaching subject from object, the other detaching object from subject. More than anything else it is the nature of the psyche that has been lost in between. The resulting notions of mind swim in simulation and virtuality, while the body is construed as raw material for redesign. Neither is regarded in relation to the structures or limits or considered to contain the timeless and universal forms that contact with the earth and the cultural imagination have always impressed upon us. The result is a collapsing of the bridge that has joined nature and civilization since the first cave paintings. Modes of perception and thought that have connected the essence of humanity with what is soulful, sacred, and earthbound are being replaced by concerns about how we might best adapt to the artificial minds and bodies today's technocracy is placing before us.

As Avens puts it:

> The modern technological man has become a stranger to the earth—in spite of his preoccupation with material things. He is no longer at home on the earth because he has imprisoned himself with his own self-made world—the world of subjectivism parading as dispassionate objectivity.

We live alone in this prison-world because we no longer relate to either the soul or the earth, for we no longer recognize or experience them as having presence and agency of their own. Avens goes on to quote Heidegger, who describes this situation in an even starker way:

> For the darkening of the world, the flights of the gods, the destruction of the earth, the transformation of men into a mass, the hatred and suspicion of everything free and creative, have assumed such proportions throughout the earth that such childish categories as pessimism and optimism have long since become absurd.[54]

In Part 1 I considered a series of polarities and flaws in our thinking around technology, drawing attention to ideational and phenomenal oversights, failures of awareness and dismissals of critical psychosocial structures and dynamics. From the refusal to remember the past, to the denial of the religious function, to the loss of techne, these factors undermine our ecology of mind, which, in turn, fuels an investment in a posthuman future. In Part 2, I have attempted to describe the psychology of posthumanism itself, beginning with psychic fragmentation and dissociation, which enable compartmentalized views of human nature and reality at large, moving to the numerous historical ideas and contemporary ideals at work in our borged mindset, and anchored by the divisions emerging in human neuropsychology. Finally, in this present chapter, we have seen the way postmodern consciousness and cognitive-behavioral concepts have seeded posthumanism. In the next chapter we will begin to piece together the counter-culture movement that birthed and then nurtured what the precursors of posthumanism and technoscience in general have left in their wake. This begins the tracing of a corrective pathway, one via which we may stand apart from where we are currently headed and begin to establish some critical vistas of reflection.

Notes

1 Joseph E. Davis, *Chemically Imbalanced: Everyday Suffering, Medication, and Our Troubled Quest for Self-mastery* (Chicago: University of Chicago Press, 2020), 178–179.

2 Ludwig von Bertalanffy, *Robots, Men and Minds* (New York: George Braziller, 1967), 7.
3 David Harvey, *The Condition of Postmodernity* (Cambridge, MA: Blackwell, 2000), 44.
4 Elaine L. Graham, *Representation of the Post/Human: Monsters, Aliens and Others in Popular Culture* (New Brunswick, NJ: Rutgers University Press, 2002), 5.
5 Jean Baudrillard, *The Illusion of the End* (Stanford, CA: Stanford University Press, 1992), 1.
6 Ibid., 2.
7 Ibid., 6–7.
8 Jean Baudrillard, *Simulacra and Simulation* (Ann Arbor, MI: University of Michigan Press, 1994).
9 In Huston Smith, *Beyond the Postmodern Mind* (Wheaton, IL: Quest Books, 1989), 17.
10 Mark C. Taylor, *About Religion: Economies of Faith in Virtual Culture* (Chicago: The University of Chicago Press, 1999), 26.
11 Donna J. Haraway, "A Cyborg Manifesto: Science, Technology, and Socialist-Feminism in the Late Twentieth Century." In Donna J. Haraway, *Simians, Cyborgs, and Women: The Reinvention of Nature* (New York: Routledge, 1991), 149–182.
12 Arthur Kroker and David Cook, *The Postmodern Scene: Excremental Culture and Hyper-Aesthetics* (Montréal: New World Perspectives, 1991), i.
13 Ibid., i–ii.
14 Ihab Hassan, "Prometheus as Performer: Toward a Posthumanist Culture?" *The Georgia Review*, vol. 31, no. 4, 1977.
15 Y. Jansen, J. Leeuwenkamp, and L. Urricelqui Ramos, "Posthumanism and the 'Posterizing Impulse.'" In H. Paul and A. van Veldhizen eds., *Post-everything: An Intellectual History of Post-concepts* (Manchester: Manchester University Press, 2021), 217.
16 Glen Slater, *Surrendering to Psyche: Depth Psychology, Sacrifice, and Culture* (Pacifica Graduate Institute, 1996). (Unpublished Doctoral Dissertation.)
17 Haraway, "Cyborg," 149–182.
18 Don Ihde, *Postphenomenology: Essays in the Postmodern Context* (Evanston, IL: Northwestern University Press, 1993), 12. Italics in original.
19 George Graham, *Stanford Encyclopedia of Philosophy*, online. https://plato.stanford.edu/entries/behaviorism
20 Adolf Guggenbühl-Craig, *The Old Fool and the Corruption of Myth* (Dallas, TX: Spring Publications, 1991), 39.
21 Ibid., 40.
22 Quoted in George Graham, *Stanford*, ibid.
23 Bertalanffy, *Robots*, ibid., 117.
24 Ibid., 116.
25 Ibid.
26 Jeff Hawkins, *On Intelligence* (New York: Times Books, 2004), 16.
27 Shoshana Zuboff, *Surveillance Capitalism* (New York: Public Affairs, 2019), 361ff.
28 Ibid., 363.
29 Ibid.
30 Ibid., 365.
31 Quoted in Zuboff, *Surveillance*, 369.

32 B. F. Skinner, *Beyond Freedom and Dignity* (Cambridge, MA: Hackett Publishing, 2002). Originally published in 1971, 6.
33 Stewart Brand, "For God's Sake, Margaret! Conversation with Gregory Bateson and Margaret Mead." *CoEvolutionary Quarterly*, June 1976, 6.
34 Ibid. Italics in original.
35 Howard Gardner, *Frames of Mind: The Theory of Multiple Intelligences* (New York: Basic Books, 1993), 22–23.
36 Ibid., 23.
37 Ibid.
38 Ibid., 25.
39 Ibid., 331.
40 Ibid., 317.
41 Mary Midgley, *Science as Salvation: A Modern Myth and Its Meaning* (New York: Routledge, 1992), 219.
42 Ibid. 219–220.
43 Ibid.
44 Harry G. Frankfurt, *On Bullshit* (Princeton, NJ: Princeton University Press, 2005), 1.
45 Ibid., 20.
46 Ibid., 21.
47 Ibid., 60.
48 Ibid., 61.
49 Ibid., 64–65.
50 Ezra Klein and Gary Marcus, "Transcript: Ezra Klein Interviews Gary Marcus." *The New York Times*, January 6, 2023. https://www.nytimes.com/2023/01/06/podcasts/transcript-ezra-klein-interviews-gary-marcus.html
51 Noam Chomsky, Ian Roberts, and Jeffrey Watumull, "Noam Chomsky: The False Promise of ChatGPT." *The New York Times*, March 8, 2023.
52 B. F. Skinner, *Walden Two* (Cambridge, MA: Hackett Publishing, 2005). Originally published in 1948.
53 Skinner, *Beyond*.
54 Roberts Avens, *The New Gnosis: Heidegger, Hillman, and Angels* (Putnam, CT: Spring Publications, 2003), 85–86.

Part 3
BELOW GROUND

Chapter 9
KURZWEIL'S DREAMS

The machines are gaining ground upon us; day by day we are becoming more subservient to them ... more men are daily devoting the energies of their whole lives to the development of mechanical life. The upshot is simply a question of time, but that the time will come when the machines will hold the real supremacy over the world and its inhabitants is what no person of a truly philosophic mind can for a moment question.

Samuel Butler, 1863[1]

this flowering of psychology has much to do with the technological system ... to the degree that psychology has expanded our understanding and deepened our sensitivity, it is a medicine wrung from the very system that inflicts wounds upon us.

Jules Henry[2]

Psychoanalysis and the broader field of depth psychology have a long history of addressing the socio-cultural background of psychological ailments. Whereas psychotherapeutic interest has focused on the individual, the effect of social conditions has led to attempts to understand the neurotic and, episodically, psychotic inclinations of the masses. As is the case with the disturbances in the individual psyche, disturbances in the collective psyche also invite the need to make conscious what is unconscious. By shedding light on the unconscious patterns of the collective, depth psychology has highlighted the way much of our modern psychological malaise is attributable to corrosive social conditions and destructive cultural phenomena.

The modern way of life has been the focus of many depth psychological treatments, from Freud's *Civilization and Its Discontents* to Jung's *The Undiscovered Self*[3] to May's *The Age of Anxiety* to Lasch's *The Culture of Narcissism*. On the Freudian side, Erich Fromm[4] wrote many books that engaged with the flaws of society; on the Jungian side Ira Progoff[5] contributed several volumes on the collective implications of depth psychology. While the technological focus of modern society is assumed in these treatments, many of them go a step further, either implying or directly suggesting that the psychological disturbances of the past two centuries, and the arising of a psychology of the unconscious in response, are directly attributable to the impact of technology.[6] Since the middle of the last century, however, whereas a great deal of Jungian discourse in particular has turned to socio-cultural phenomena, there have been few concerted efforts to address the effects of industry and post-industry on the psyche.[7]

At this moment in history, as choices pertaining to the radical alteration of minds and bodies are upon us and the shadows of the digital age have begun to darken our outlook, it behooves us to step back and consider the relationship between the psychological wellbeing of the individual and that of society, as well as the connection between the adaptation to technology and psychological disturbance. As these choices and shadows grow more consequential, and as our disturbance grows more serious, it seems the right time to gain more understanding of our psycho-social situation. Such understanding can not only give us pause, opportuning a more reflective and deliberative response to the path we are on; it can point us in the direction of a therapeutic response. The dissociation and splitting, escapist fantasies, cognitive one-sidedness, robotic models of behavior, and nihilistic responses to cultural traditions explored in Part 2 are direct invitations to put the technosphere on the couch.

Understanding the effects of technology on the psyche has been a long time coming. From the so-called "railway neuroses" and diagnoses of neurasthenia of the mid-nineteenth century to the pandemic of depression and anxiety among digital natives, psychological ailments have not only pointed to collective causes, but have been linked to the onset of new technologies. At the conclusion of his historical account *The Birth of Neurosis*, George Drinka contends, "This depth psychology, psychoanalysis, remains very much the treatment for Prometheus and the Genius."[8] His survey of psychological theories leading up to the *fin-de-siècle* and just beyond clearly describe the way the turn to the unconscious coincided with the impact of early industrialization. This coincidence is also well noted by Progoff, who writes, "In order to understand the development of depth psychology in perspective, we must think of it in relation to the changes in the tempo and structure of life that have accompanied the growth of modern industrial society."[9] And in the epigraph at the start of this chapter, Jules Henry makes clear the connection between the arising of a (psychoanalytic) psychology and the wounds inflicted by "the technological system."[10]

The links between industrialization, the specific character of psychological symptoms, and theories of the unconscious are clear. Then, as technoscience becomes more dominant, Jung's psychology in particular appears to punctuate these links. From such connective tissue, the idea naturally arises that depth psychology serves a distinct cultural-historical function, one of alleviating and transforming the impact of technology on the collective psyche. As we live into the post-industrial world, and with these links in mind, I think the case can be made that accounting for the unconscious may play a significant role in determining the overall course of civilization. An untethered program of innovation aimed at the reengineering our own nature only appears to underscore this.

In this chapter I therefore propose that we familiarize ourselves with the cultural-historical function of depth psychological understanding as a vital counter to prevailing conceptions of our way forward. Further, in contrast to behavioral and cognitive approaches, which are not only inadequate but also complicit in the technocracy that is pushing past the boundaries of our skin, I contend that such depth psychological understanding, joined with the emerging holistic paradigm in other fields, offers to restore our ecology of mind, thereby providing a firmer psychosocial foundation for what lies ahead.

Depth Psychology and Collectivity

Amidst Freud's reflections on modern civilization, he poses the pertinent question

> If the evolution of civilization has such far-reaching similarity with the development of an individual, and if the same methods are employed in both, would not the diagnosis be justified that many systems of civilization—or epochs of it—possibly even the whole of humanity—have become "neurotic" under the pressure of civilizing trends?[11]

Indications are that the relatively poised individual can be afflicted by a collective imbalance and otherwise sane individuals can succumb to a collective madness. But the extent to which psychopathology may be attributable to social circumstances or even to civilization itself has been a contentious matter. In responding to his own question, Freud describes one conundrum the notion of a neurotic society appears to present:

> In the neurosis of the individual we can use as a starting point the contrast presented to us between the patient and his environment which we assume to be "normal." No such background as this would be available for any society similarly affected; it would have to be supplied some other way ... In spite of all these difficulties, we may expect that one day someone will venture upon this research into the pathology of civilized communities.[12]

By the time Freud had penned these words in 1929, Jung had already begun to undertake such research, with the aftermath of World War II and later decades of his life occasioning a great deal of writing on the matter. And there are two main considerations that anchor this research and writing. One involved determining whether or not a society was afflicted by a relatively high degree of one-sidedness, which, we recall, is Jung's essential definition of neurosis. This one-sidedness can be determined by comparing social groups and cultures. Even more foundationally, however, it can be substantiated by a comparative study of archetypal themes and their symbolic expressions, down through the history of various cultures as well as across cultures at any given time. In the wake of such studies, Jung changed his preferred nomenclature for the collective unconscious to the *objective psyche*. Whether or not a society was psychologically conducive or detrimental could be seen to depend largely on the quality of its relations to the objective psyche and its provision of adequate symbolic systems.

The second consideration—a result of these studies combined with a meticulous investigation of his own psychological experiences and those of his patients—concerned the presence and activation of an overarching archetypal imperative in the psyche; namely, the need for psychological life to be met by a conscious integration of total experience. Beyond the mere reversing of various forms of psychological division, such as repression and dissociation, this imperative represented the psyche's innate investment in the meaning-making that results from the inclusion of ego-dystonic contents in an individual's or collective's self-knowledge. This imperative was shown by Jung to pertain especially to symptoms and fears. In this respect the capacity for healing belongs to the ability to withstand the competing demands of inner life, as well as the collisions of inner and outer life. Like individuals, collectives perpetuate psychological problems when they attempt to discard what belongs to them, resulting in scapegoating and other destructive tendencies.

In regard to this aspect of Jung's approach to collective neuroses—or to psychological disturbance in general—it is telling that Drinka also concludes his account of the first decades of modern psychopathology by comparing the conceptions of that era with those of the late twentieth century and drawing direct lines between them. Lamenting that many psychological ailments of more recent decades defy even the most up-to-date medical and psychotherapeutic approaches, he states, "Even those who take their medications and come to therapy often seem to be lacking an entire part of themselves, *the ability to synthesize experience.*"[13] Drinka refers to Freud quite extensively but does not mention Jung. And yet his own summary of what is critical in overcoming psychological ailments and disconcertingly beyond reach in most chronic disturbances is precisely what Jung's research on the archetypal basis of mind indicate as necessary for integrity and wellbeing. Indeed, Jung would come to refer to his method of working with the contents of the unconscious as "synthetic."[14]

Jung's synthetic method reflects the conviction that psychological development depends on the conscious mind's capacity to incorporate the patterns and contents of the unconscious. In the first instance this involves inclusion of the shadow side of adapting to outer life and becoming increasingly aware of what has been repressed or devalued because of conscious ideals. To some degree, this view aligned with Freud's. However, in his comprehension of the dynamic organization of these rejected parts of lived experience Jung's approach began to differ. After several years of looking into the forms of this organization, which involved what he came to term *complexes*, he goes on to show how these complexes are themselves shaped by the deeper forms of the objective psyche; namely, the archetypes. He also demonstrated how the process of coming to terms with the unconscious, individuation, was itself rooted in a timeless and universal archetypal pattern. This was in large part due to the fact that individuation essentially came down to an individual's lifelong relationship with the primary archetypal pattern discussed above—the integrative imperative. Jung's psychology can thus be formulated as follows: the prime directive of depth psychology in general, namely making the unconscious more conscious, is impelled by the overarching archetypal pattern of integration; individuation is the ongoing process of the individual relating to this psychic pattern; the synthetic method activates and supports this process.

Among other implications of these conceptions of psychological life, Jung determined that this synthetic approach involves something of major significance for the collective. Because attending to the unconscious eventually leads to the archetypal dimension, it is not only psychotherapy of the individual that benefits: the cumulative effect of making the unconscious more conscious is also a form of reconnecting to the presence of the past within and a rediscovery of the value of that presence. More specifically, at given points in history, particular archetypal patterns cry out for more conscious attention. Patterns involving purposeful descent or regression, the bonds of mind and earth, and the presence of the sacred in nature are examples. As we have seen in previous chapters, reclaiming these patterns in particular restore an ecology of mind. This undertaking puts lead in the keel of the vessel we are on as we navigate the shifting winds and brewing storms of history.

There are, of course, many insights across depth psychology we may draw on when considering the vital role a depth psychology of society and culture is likely to play as we head into an uncertain future. Nonetheless, the ability to synthesize/integrate experience may be taken as a point of departure, and it may be applied to the collective in a fashion similar to its application to the individual.

The Backstory

The discovery of the unconscious begins in earnest at the historical moment when industrialization was sweeping through the Western world and confidence in science and technology had risen to heights that would remain unchallenged until the present day. As the historian of depth psychology Henri F. Ellenberger observes, "It was the period of positivism and the triumph of the mechanistic *Weltanschauung*."[15] Further, "positivism's basic principle was the cult of facts," and that it "rejects any speculation akin to the philosophy of nature."[16] This effectively cleared the way for all things, including human beings, to be regarded in the same way we might regard the inner workings of a clock. Summing up the prevailing worldview he writes:

> The universal belief in science often took the shape of a religious faith and produced the mentality that has been called scientism. The scientism trend went so far as to deny the existence of all that was not approachable by scientific methods . . . After 1850, a wave of popular books propagating the exclusive belief in science combined it with atheism and sometimes with an oversimplified teaching of materialism.[17]

One popular work of this period, which had 21 editions by 1904, proclaimed, "With the most truth and with the greatest scientific accuracy we can say to this day: there is nothing miraculous in this world."[18] For those swept up in this new faith in scientific positivism, it seemed that little else mattered and that the rational mind was capable of knowing all there was to know. Indeed, there was a prevailing sense that science was on the verge of revealing all of nature's secrets and generating complete knowledge of the universe. The hubris was palpable.

As new technologies, particularly steam power and electricity, displayed the worldly application of this faith in science, and the economics of industrialization rapidly spread through the Western world, there was an equal idealism associated with new devices and the environments they made possible. Drinka offers a window onto this newly technologized world, which appeared in many places over less than the course of a generation:

> The Victorian epoch was a vast scientific exposition, cluttered with such shimmering wonders as the telegraph, the telephone, the railway, all promising to tie the globe together in one great metropolis. The era seemed a vaulting steel cathedral scraping skyward, upon which the citizenry gazed in awe: the Crystal Palace in Kensington Gardens, jewel of the world exposition . . . or the Eiffel Tower . . . These expositions and others displayed the latest technological marvels to an enraptured public—gadgets and buzzers sounding, lights flashing, voices bouncing everywhere.[19]

As this scientific faith and technological enthrallment was taking hold, however, another narrative was also unfolding. In order to make the marvels of the age and generate the kind of wealth that made them possible, masses of people were uprooted from their lives on the land and in small villages in order to join the bustling urban environments of industry. There they found a new rhythm and new kinds of factory-based work. Many ended up on assembly lines, matching their activities to those of the machines they now appeared to serve. The nineteenth century thus witnessed the most dramatic shift in lifestyle and relationship with the world-at-large that had ever taken place. One observer wrote at the time, "Whilst the engine runs the people must work—men, women and children are yoked together with iron and steam." Formulating his observations in a way that would effectively characterize the relationship with technology to the present-day, as well as point to the cyborg beings on our drawing boards, he describes the new worker as "the *animal machine* . . . subject to a thousand sources of suffering" yet "chained fast to the iron machine, which knows no suffering and no weariness."[20] According to Giddens, all three major social theorists of the late nineteenth century, "[Marx, Durkheim, and Weber] saw that modern industrial work had degrading consequences, subjecting many human beings to the discipline of dull, repetitive labour."[21]

Whereas we cannot underestimate the physical and psychological suffering of these production-line workers, a broader phenomenon was also taking place, one that drastically reduced the range of meaningful work and creative craftsmanship that had occupied prior generations. Everyone in what was becoming the industrial world was subject to a more fabricated environment: "The relationship with nature first of all, could not possibly be the same. Owing to the large-scale industrialization, urbanization, and new scientific discoveries, life in the course of the nineteenth century had become increasingly artificial."[22] Mumford, the great critic of mechanization and industrialism, writes of this phenomenon in terms of a transition to "monotechnics."[23] He contends this resulted from the loss of a "technological pool,"[24] described as "widely scattered among the peoples of the earth, every part of it colored by human needs, environmental resources, inter-cultural exchanges, and ecological and historic associations." This pool involved "interacting with soils, climates, plants, animals, human populations, institutions, cultures." Yet "when they were eliminated from the system of production, that vast cultural resource was wiped out."[25]

These dramatic socio-cultural changes constituted a psychological narrowing, both in relation to styles of thinking and ways of working. In both arenas, a rupture occurred in relation to instinctual life and sources of meaning. And the psyche reacted by generating a broad range of dramatic psychological symptoms.

Perhaps the most obvious example of the direct link between industrialization and psychological disturbance was the syndrome Drinka refers to as "railway neurosis."[26] This was, effectively, a phobia or fear of trains. Between the mid-nineteenth century and the early twentieth, railway systems expanded throughout Europe and America. In the decade after 1850 French railway lines tripled in size, and in the next two decades they tripled again.[27] Careening across landscapes at unprecedented speeds, billowing smoke and making unusual loud noises, these machines made an impression. However,

> common folk came to fear that the railway would make barren the fields through which it passed and dry up the rivers and streams. The air would be poisoned, cows would lose their milk, horses would become sterile, and all hunting animals would grow sickly.[28]

People also witnessed terrible accidents, with carriages colliding with trains, passengers falling onto tracks, and other deadly and injurious events giving rise to what we would now understand to be symptoms of trauma. However, in what would be a common first order of investigation into what where eventually considered to be forms of neurosis, physicians at the time searched for gross physiological causes such as lesions in the nervous system. What became known as "railway spine" was thus one early example of a technologically induced psychological disturbance. Drinka sums up the syndrome, and what early neurologists such as Charcot made of it, as the

> moment when a vulnerable human collided with an increasingly complicated and intimidating environment. The child, the woman, even the vulnerable man, confronted a fast-paced whirligig world for which he was not prepared. He stared into the snarling new meshwork of the Machine. His veins ran cold with fear. Electricity shot through his nerves. He fainted. He awakened from his deathlike state transformed into a neurotic sufferer.[29]

Although the early history of neurosis often focuses on hysteria, likely because of the prominence of this condition in the work of early luminaries such as Charcot, Janet, and Freud, another prominent disorder was "neurasthenia," or "nervous exhaustion," originally associated with the American physician George Miller Beard. Akin to other neurotic conditions of the era, the symptoms of neurasthenia manifested as much in the body as in the mind and could include a wide range of ailments, from "headaches" to "mild depression," "phobias and obsessions" to "generalized weakness," "excessive sweating" to "heart palpitations," and even "impotence and vaginismus."[30] Here too we find a direct line drawn between the burgeoning relationship with new technologies and a novel form of psychological and psychosomatic suffering. Ellenberger describes Beard's understanding of the cause of neurasthenia as rooted in

> the peculiar way of life of North America, a young and rapidly growing nation with religious liberty . . . entail[ing] an increased amount of work, forethought, and punctuality, an increase in the speed of life (railways, the telegraph), and also the repression of emotion ('an exhausting process').[31]

Beard claimed that "Americans were nervous because of their . . . teeming urban environment, the inherent insecurity of their frontier lives, [and] the harshness and changeability of the climate." He concluded that "modern civilization . . . was the noxious agent that was very active in American society, catalyzing many, many cases of neurasthenia."[32]

Although the physicians of the time could identify the role of the nervous system in these disorders, they could not find neurological abnormalities. Even though many of the symptomatic manifestations were of a stark physical nature, physical causes could not be found. Malingering, the intentional faking of symptoms, was also ruled out. As neither the body, nor the conscious mind provided adequate explanations for these neurotic conditions, these early doctors of the soul were forced into the discovery of the unconscious. In this way, the birth of neurosis coincides with the birth of depth psychology.

Beyond the rapid changes brought on by new technologies and industrialization, inducing experiences of dislocation, disorientation, and overwhelm, a further and deeper factor appeared to be at work. This was the manner in which the technological way of life appears to have disrupted the connection with the instinctual world: minds and bodies had become disconnected, with the emotional linking of the two becoming dysregulated.

This disconnection of psyche and soma via disturbances in the emotional field would turn out to be the case with hysteria, which Freud attributed to the repression of sexual desire, brought on by the moral strictures of the Victorian age. Nonetheless, the rise of reason, materialism, and the wonders of technology appear to have aided and abetted this form of neurosis also, by creating a general displacement of the animal side of human life, one of the prime channels of which is sexuality. This overall distortion of psychodynamics is punctuated by Romanyshyn, who has pointed out that the hysteric is essentially a daughter of Descartes—her hysterical symptoms pointing to the failed project of the rational mind transcending the emotional body.[33]

These portraits of early neurosis thus converge to form a distinct picture, one in which the excessive faith in the rational mind and a corresponding disordering of instinctual life combine to create a psychological one-sidedness then compensated by the symptomatic eruption of animal spirits—a full display of the autonomous power of the unconscious. On the surface, the nineteenth century was an age of wonders and the peaking of Enlightenment principles; beneath the surface the human organism reacted to mechanization and the repressed returned in force. Jung's early, pithy diagnosis of modern persons suffering from "a loss of instinct" was on point. Referring to "a time when rational present-day consciousness was not yet separated from the historical psyche, the collective unconscious," Jung suggests that whereas "the separation is indeed inevitable . . . it leads to such alienation from that dim psyche of the dawn of mankind that a loss of instinct ensues."[34]

Does Kurzweil Dream of Butler's Sheep?

Samuel Butler (1835–1902) studied Classics at Cambridge, moved to New Zealand and lived as a sheep farmer, and then returned to England to work as a novelist. During his time as a farmer he wrote a letter, which became an article for a local newspaper. He addressed the social impact of technology and, even at a time when steam power was still making its way into various aspects of life, predicted a time when machines would overtake humans. Butler even referred to their potential to self-replicate and foresaw our subservience to them. His response to such a vision of the future was strong:

> War to the death should be instantly proclaimed against them. Every machine of every sort should be destroyed by the well-wisher of this species [humans]. Let there be no exceptions made, no quarter shown; let us at once go back to the primeval condition of the race.[35]

Although Butler's writings had a satirical leaning, his was a rare and apparently extreme reaction to a technological age only just underway.

Butler had witnessed the rise of the Industrial Revolution during his English upbringing and had also studied Darwin's work on evolution. Putting the two together, he could imagine a time when machines might become autonomous and evolve on their own. He went on to write a utopian novel, *Erewhon*, on the rise of such machines and a society that made a reactionary return to natural existence. This fictional society had glimpsed the future and elected to destroy all the devices.

Butler's views were no doubt fostered by living a bucolic life, dwelling with sheep, grass, and rocks. He imagined technology in an all or nothing way that most of us would reject. However, he managed to grasp several critical things about the mechanization of life. First, that an organic, sensual, embodied, feeling world that comprises our deep humanity can harshly collide with the industrial world and be harmed by it. Second, that technology had the potential to take on a life of its own—an autonomy that has for over a century been a recurring theme in science fiction. Third, that we can acquiesce to the very instruments and tools we design to serve us. Butler saw the dominance of technology on the horizon, and this exceedingly early vision of this dominance now appears in retrospect to have been a reasonable extrapolation of then current trends. Because of this prescience, Butler's writings have been quoted by enthusiastic futurists and critics of technology alike—at least in a selective fashion.

Ray Kurzweil, famous for this writings on the rapid growth of AI and his engineering work at Google, whom we met in Part 1, is one futurist-technologist who likes to quote Butler. His interest in Butler is understandable, for Butler was probably the first person to imagine the rudiments of the world that Kurzweil and other posthumanists are now eager to create. Yet while Kurzweil likes to underscore Butler's prescience, he ignores Butler's equally impressive affective reaction. For example, he quotes Butler's letter/article at length without acknowledging the fear and pathos at its core.[36] All that is implied in Butler's darker fantasy and prophetic warnings, as well as the early modern history behind these things, finds no place in Kurzweil's outlook. And whereas Kurzweil's writings have slowly come around to acknowledging some of the potential pitfalls of human–machine hybridization, as well as the dangers of an autonomous AI taking control of the world, he pretty much ignores how such radical changes are likely to impact the human psyche or disrupt the character of society and culture.

Kurzweil's omission is a convenient one. Computer science prefers its own calculations, conveniently detached from surrounding actualities. According to Kurzweil, AI is already accelerating us to the moment human intelligence will be surpassed by artificial intelligence, at the moment of the Singularity. Promoting the pristine path to this development is also convenient and highly profitable for corporations such as Google, who are staking out major ground in the AI revolution. From the posthuman point of view, the actual lived experience of human beings is better off riding in the caboose of the ever-accelerating train of technology. As Kurzweil likes to note, natural evolution is just too slow, and leaves too much to chance. It's better for human beings to recreate themselves, according to the goals and values of the contemporary world, and then hand off the problem to AI, which, it is assumed, will eventually be much smarter than us.

For Kurzweil and other advocates of the artificial intelligence revolution, especially those who comprise the posthuman movement, what lies ahead is a logical extrapolation of current trends in fields tied to the exponential growth of computing power—especially AI, robotics, nanotechnology, and genetics. Yet this direct extrapolation of technological change and radical social transformation reflects Kurzweil's failure to consider Butler's reaction to the rise of the machines and the broader psycho-social fault-lines that extend from it.

When Kurzweil excludes Butler's critical contemplation of a machine-centered world, he demonstrates the way posthumanism in general excludes an ultimately essential faculty; namely, the capacity for reflection and circumspection. This faculty is deeply entwined with the capacity to evaluate purpose, find meaning, and cultivate holistic awareness. Its exclusion is not merely an intellectual preference or oversight: it points to an active suppression of lived experience and psychological reality.

One of Butler's observations from early in the industrial era concerned the power of complacency, which no doubt provides an advantage for the technologist. He wrote:

> The power of custom is enormous, and so gradual will be the change that man's sense of what is due to himself will at no time be rudely shocked; our bondage will steal upon us noiselessly and by imperceptible approaches; nor will there ever be a decisive clash of desire between man and the machines as will lead to an encounter between them.[37]

The fantasy of machines becoming autonomous and taking over was not only an extrapolation of external changes based on mechanization and automation; it was at least in part a projection of newly mechanized minds raising the prospect of human beings turning into automatons.

Critical of the Victorian enthrallment with all things technological, Butler saw his contemporaries as the proverbial frogs in the slowly heating water of technological change and was attempting a kind of shock treatment. He perceived the tension between the drive to continually innovate and the necessity of preserving the matrix of earthly existence. Until the machines do take over, human desires and interests will determine our innovative path. These desires and interests are prone to manipulation, but can also be consciously engaged and critically assessed.

In his monumental series *The Myth of the Machine*, a detailed critique of technological trends, Mumford also refers to Butler's predictions. However, in stark contrast to Kurzweil he writes, "the whole concept of subjugating nature and replacing man's own functions with collectively fabricated, automatically operated, completely depersonalized equivalents must at least be reappraised."[38] Commenting on Butler's vision of the future, he notes:

> if he [Butler] had been a religious prophet, rather than a satirist, he might have uttered the final words on this whole development, words used long ago by Isaiah. "Ye turn things upside down! Shall the potter be counted as clay?"

Mumford concludes, "A century after Butler, these questions now thunder ominously in our ears."[39] And fifty years after Mumford penned these words, the questions still thunder.

In a recent review of a book that surveyed Silicon Valley posthumanists pursuing their immortality, a prominent medical ethicist writes that the author "never goes very deep into understanding the pathology driving them," and never produces "a coherent social or ethical critique," thereby failing to see what the ethicist termed the "utter selfishness" of these would-be cyborgs. He cannot help but adding, "maybe this is why the titans of technology want so badly to escape to Mars."[40] And maybe this is why those of us who value the earthly way of life want to give the posthuman movement a closer look.

A documentary on Kurzweil's life and ideas, *Transcendent Man*,[41] discusses the implications of the accelerating power of computation and the associated outlook on human–machine merger. It also details two distinct yet related dimensions of this inventor's personal life: one is the loss of his father, who died when Kurzweil was in his early twenties; the other is his attempt to extend his own life as long as possible in the hope of taking advantage of the techno-medicine he sees on the horizon. In regard to both matters, Kurzweil declares death to be tragic and something we should work to overcome. He is very open about the influence of his father on his life, as well as the way this loss has been a powerful motivator for him—both generally and specifically in terms of his efforts to bring about the radical transformation of human nature. In each respect, his mind is geared toward this pending future: he not only describes the human condition in the reductive language of computer science, but assesses our current state of being from the perspective of the future he conjures. At one point in the film he states, "We are fundamentally information . . . We have this old software that's really not entirely relevant to the modern age we live in." In fact, he is so confident about this that he looks forward to being able to bring back his father by putting together all the available information about him, including that stored in his own memory.

In all of this, Kurzweil appears oblivious to the fact that the memories of his father are the product of *his* memory, which has more to do with his psyche than with who his father actually was in toto. Indeed, the very nature of this memory is premised on his father's absence—on the kind of reconstructed image and feeling-toned associations we carry when those we have been close to are no longer in our lives. Kurzweil notes at one point that his father "still visits me in my dreams." And whereas there is obviously correspondence between the actuality of persons in our lives and the figures in our dreams, the character of such dream figures and the position they occupy in our inner worlds are determined by our own complex psychology.

To extend these points, Kurzweil seems blind to the fact that who we are and the meaning of our lives is largely determined by those who have gone before us. Death generates life. Our ancestors, close and distant, fill the background of our psyches, just as memories of our lives will guide the lives of those who come after us. All of this imbues existence with soul; all of this makes us deeply human.

When Kurzweil refers to the "old software" determining our current experience of mind and body, which he seems eager to rewrite, he is also turning a blind eye to the depths of human nature and what has brought us to this cultural-historical juncture. It is this so-called "software" that creates father dreams and other feeling connections to those who are important to us. This makes us who and what we are. Yet these are the same emotionally textured facets of our existence Kurzweil looks forward to one day reprogramming.[42]

Whereas I am in no position to assess Kurzweil's personal psychology, nor do I intend to do so, it seems reasonable to describe his thinking as dismissive of the many dimensions of the human nature that have brought us to this point in our biological evolution and cultural development—dimensions that continue to determine our lived experience. That is, there is a link between failing to acknowledge the alarm bells in Butler's prophetic visions and failing to perceive the shadows of mechanization. There is also a link between failing to understand memory, dream, emotion, and death and failing to think about reality in any terms beyond information processing.

Looking in All Directions

Everything Kurzweil is leaving out we must put back in. Why? Because this is what the self-regulating psyche and its inclusionary psychodynamics keep telling us. This is the task a depth psychological response to technologism places before us. And if we can reintegrate what technoscience has cast aside, we might have a fitting foundation for the hard choices ahead.

Hillman's equating of the unconscious with "the realm of *memoria*,"[43] speaks directly to what futuristic conceptions of existence—such as being essentially comprised of "information" and "software"—leave out. He writes:

> Freud began his talking cure by asking his patients to follow one basic rule: to let their souls speak without inhibition. When they thus abandoned voluntary control and the intelligibility of understanding, their associations led them into *memoria*.[44]

Hillman goes on to explain why we should think of the unconscious in this manner, arguing that memories are not mere representations of past events but are shaped by the archetypal background of the psyche, "divine images and ideas,"[45] and that it is here we learn that our relationship with the past is, in effect, molded by the psyche in terms of what the conscious mind needs, presently and going forward. Furthermore, pointing to a primary dimension of depth psychological understanding, he notes, "Freud and Jung have suggested that the unconscious enters into each mental act . . . we cannot be conscious without at the same time being unconscious; the unconscious is always present, just as the past is always present." Stated in more poetic terms, "The dream is always there; we can never leave it."[46] To the extent this is so, the combined dismissal of the unconscious and the past is effectively a truncating of the human thought process, making any future version of ourselves that severs us from these faculties something other than human and quite likely soulless.

In his book *Humanity's End: Why We Should Reject Radical Enhancement*, Nicholas Agar makes a complementary argument, this time imagining into the unraveling of identity and character likely to occur if we were to *add* rather than *subtract* psychological abilities. After describing the way someone with a disease such as Alzheimer's "may no longer understand the social and moral causes that were among their strongest commitments" and may suffer from a "severing of connections" and a "general intellectual decline," Agar states, "I think that significant intellectual growth may have a similar effect. It, like significant intellectual diminishment, has the propensity to sever your bonds with the things that really matter to you."[47] Because posthumanists such as Kurzweil are so focused on what intellectual and physical augmentation may bring about, they are apt to forget how such changes will impact our overall existence. Agar brings his argument home in the following way:

> Herein lies the threat to people with indefinite life spans from ongoing intellectual growth: It presents the prospect of never having any mature interests and attachments. This is a significant loss. We invoke a person's long-standing mature interests and commitments in explaining what defines her, what makes her distinctive. People whose indefinite life spans are accompanied by ongoing intellectual growth may, in contrast, present as a mutually unconnected series of commitments and interests.[48]

As the psyche is invested in an emotionally fulfilling, meaningful continuity between the past, present, and future, this possibility also points to the elevation of enhancing and augmentation over the integrity of identity and ecology of mind. Indeed, Agar suggests as much when he proffers, "I suspect that there is only one commitment that has a really good chance of emerging intact after the enhancement procedure. That is the commitment to enhancement itself."[49]

Agar's sense of an enhancement-centric future reflects at least two omnipresent patterns gripping late modern technocracy. The first pattern concerns the obsessive-compulsive character of technoscience that all but ensures something that can be made will be made. Even when such making may be detrimental to overall human wellbeing, if it presents an economic or military advantage there is pressure to proceed before others do. And when there is both an economic and a military advantage, there is a furthering of the military-industrial complex—surely one of the main engines of technocracy. The second pattern concerns a deeply ingrained trait belonging to technocratic society and most apparent in America, described by Henry in the 1960s as "technological drivenness and dynamic obsolescence." Such a society, he posits, is "driven on by its achievement, competitive, profit, and mobility drives, and by the drives for security and a higher standard of living."[50] Making an important distinction between drives and values and proposing that it is the function of the latter to regulate the former, Henry goes on to note the way drives, particularly those whose goals are material security and social status, have taken the upper hand. He proposes, "If you put together in one culture uncertainty and the scientific method, competitiveness and technical ingenuity, you get a strong new explosive compound ... technological drivenness." Henry goes on to present a compelling description of the way these dynamics have created a vicious cycle. He suggests, for example, that "ordering relationships for the satisfaction of external needs [in such an exclusive way] has resulted in the slighting of plans for the satisfaction of complex psychic needs."[51] In practical terms this also means that "jobs requiring an orientation toward achievement, competition, profit, and mobility, or even a higher standard of living" far outnumber those "requiring outstanding capacity for love, kindness, quietness, contentment ... frankness, and simplicity."[52] All of this is enveloped in "the American preoccupation with creating new wants." To punctuate his point, Henry quotes from a full-page ad in the *New York Times*:

> Now, as always, profit and growth stem directly from the ability of salesmanship to create more desire. To create more desire ... will take more dissatisfaction with time-worn methods and a restless quest of better methods! It might even take a penchant for breaking precedents.[53]

Among other ill-effects, such a philosophy, aiming to build careers of drivenness and condition consumption, compels us to "put the constantly rising standard of living in place of progressive self-realization."[54] It seems impossible not to then draw a straight line between the drivenness and obsolescence of mid-twentieth-century technocratic society and that of the current digital era, especially given the rate at which we are effectively forced to adapt to every technological change and acquaint ourselves with each new means of communication and networking. And it seems difficult to assume that this growing technospheric pressure will not also come to apply to whatever methods of augmentation and enhancement become available. We only have to recall the way we ourselves have become the new commodities, which means our minds and bodies are subjected to the "restless quest for new methods." Unless, of course, we somehow break this pattern and come to realize that just as certain aspects of our existence may flourish under these prospective conditions, other aspects will stagnate, if not ossify, with no available upgrades.

In Dick's science fiction novel *Do Androids Dream of Electric Sheep?*, we get to imagine a possible future in which the trashing of the earth has led to most of the population living on other planets and being supplied with lifelike android servants, which are difficult to distinguish from humans. Some of these androids rebel and must be eliminated, a task performed by bounty hunters such as the novel's protagonist, Rick Deckard. In the course of the story, two themes arise that bear upon our current discussion. The first concerns the absence of feeling in androids, who are subjected to the "Voigt-Kampff test,"[55] to determine their level of affect upon answering provocative questions. As androids are deficient in the areas of emotional response, because they have been deprived of a developed feeling life, the test can determine their fate.

Regular people it seems also have problems with their emotions, so they participate in rituals involving "empathy boxes," allowing them to tune into the feelings of others.

The second theme, from which the novel gets its title, concerns Deckard's interest in animals, most of which are by this point also fabricated. While he would like to afford a real animal as a pet, he only has a fake "electric" sheep. However, these animal pets play a pivotal role in people's lives, as if they hold a connection to authenticity and a less fabricated reality.

The point is that the need for a feeling life appears to haunt our technological exploits as well as our imagined futures, serving as a constant reminder of what the technocracy appears to be eroding and what the dissociative patterns we have already explored are attempting to compartmentalize. The world of emotion, imagination, and dream is what keeps us honest. It is the main means of maintaining psychological and, by extension, existential integrity. This makes our understanding of the psyche an equally critical part of our present and future outlooks; this is what depth psychology has attempted to restore.

Kurzweil's orientation to the future and to where technology may take us—which I am taking as exemplative of posthuman futurism—is unrestrained by any relationship to the whole and in denial about the way human limitations serve psycho-social functions. This orientation is already severed from the ground of being, making any actualization of its aims a self-fulfilling prophecy. What Kurzweil envisions is an extrapolation of both his very narrow conception of human purpose and his very sweeping notion of computational capability. But how sustainable is this one-sided orientation to either life or technology? Can we just continue to ignore the symptomatic responses to the digital lifestyle and dismiss the modern history of technocratic dissociogenesis? If it is unsustainable and we can no longer afford to overlook these matters, then we may take in Henry's further insight that depth psychology is "a medicine wrung from the very system that inflicts wounds upon us."[56]

Western Civilization's Two Streams

The appearance of a psychology of the unconscious may be situated in a broader historical configuration: part of the Western world's two competing and complementary approaches to existence, with corresponding records of thought and action. One approach belongs to the attempt to gain further control over nature; the other, to what this attempt leaves behind. The search for transcendence lies at the heart of the first approach—a constant quest to move beyond the merely natural, embrace abstract understanding, and eventually detach the mind from the body–world configuration. By contrast, an affirmation of embodied existence and an ever-expanding awareness of our embeddedness in natural rhythms and their cosmological extensions has been essential to the second approach.

Although different eras have thrown their weight behind these approaches, we live in a time when it is no longer viable to deny either one. The worst mistake we can make is to adopt one of these narratives and insist it alone matters. To finally understand the way each stream of history and associated modes of thinking are entwined with and determine the manifestation of the other is the challenge now before us. For while, in any given era, the cultural processes that shape history have inclined toward full immersion in one or the other of these streams, in the current era this polarization will prove a fatal mistake. Our power over Creation has changed the equation.

At this historical juncture it is the determination to transcend the rhythms of nature that has become dangerously disconnected from the overall pattern of Western history, generating a compensatory arising of ideas across many disciplines, alerting us to the significance of what is being left behind. We have seen how posthumanism expresses this reach for transcendence *in extremis*. What is harder to see, but is running in the background, is the counter-cultural stream—the underground traditions that have arisen and continue to arise to address what the official march through history has rejected. From the movement to find more meaningful and creative work, to the sustained interest in the mystical and esoteric shadow traditions of Christianity, Judaism, and Islam, with their interest in divine immanence, to the reclaiming of the neo-pagan sensibilities of the Renaissance, to the recollection of nature's enchantments in the Romantic philosophical tradition, we have been called to maintain an awareness of nature's enduring spirit. As clumsy as attempts at "progressive self-realization" (Henry) can be, these attempts also point to a genuine recovery effort. While the godlike power over nature has defined one stream of knowledge and approach to existence, efforts to tap the God-given intelligence of inner and outer nature have been forming an opposing stream/approach.

Depth psychology derives from this second stream of knowledge and approach to existence. Its ideas grew out of the work of Romantic philosophers, who "visualized the universe as a living organism endowed with a soul pervading the whole and connecting its parts."[57] Whereas its founders went to great lengths to present their findings in the language of science, their subject matter of fantasy and dream has become part of an evolving expression of the covert side of Western intellectual discourse. By questioning the rule of the conscious mind on the heels of evolutionary and astronomical science, Freud is often credited with what has been called the third Copernican Revolution, one that revealed the underbelly of the Age of Reason. Yet it was Jung who pushed on to study the psyche's second center of intelligence, which was constellated in the turn toward the dynamic characteristics of instinct, archetype, image, and dream.

As the language of the unconscious is symbolic, metaphorical, and non-rational, accessing this other intelligence is not straightforward. This is a language beyond that of empirical science: it requires a rehabilitation of mythos as a partner to logos. The diagnosis of the Western mind suffering from too much rationalism and not enough instinctual intelligence therefore also entails a restoration of this language of the depths. For Jung, both the content and the language of fantasy and dream point to what the prevailing attitudes of the conscious mind has pushed aside. And to understand the compensatory themes of these dreams he was forced to fully immerse himself in rejected realms of thought and esoteric knowledge to recover a more mythic perspective.

Freud called the dream the royal road to the unconscious, and in a sense it is our most immediate, individually derived conduit to this vast ocean of psychic life. But collectives also dream, and the language of dreams apply here too. Through artistic, literary, and cinematic expression, gifted psychopomps of the cultural imagination bring unconscious contents into collective awareness. They too add mythos to the prevailing logos. Indeed, it was the mythologist Joseph Campbell who said that "dream is the personalized myth, myth the depersonalized dream; both myth and dream are symbolic in the same general way of the dynamics of the psyche."[58]

In the larger scheme of things, this dreaming language of the unconscious and our need to make room for it serve a crucial purpose: they remind us that the soul gravitates to the world of dreams and myths, even as this world remains unfamiliar and unpredictable to us. Who knows what dreams will come or where they really come from? As much as individuation may move us toward integration, it requires divisibility; it is rooted in the sense of being comprised of competing desires and impulses, ideals and shadows, traits and characteristics. It is the sense of oneself as an other—or of others—that must be met and known. Critically, it is this awareness of being divided that produces the feeling of being an agent, a presence—an entity of substance whose most heroic feat is the creative endurance and transformation of inner conflict. Integration occurs only on the heels of the conscious experience of this conflict, which may make it a less uniform and more multifaceted business—a communing of differences.

Although individuation implies a certain degree of introspection and self-awareness, it is not engaged at a worldly remove. Relationships, work, and community engagement often instigate the need for inner reflection. Whatever initiates a more conscious interiority, however, combined with the awareness of being a compendium of impulses in need of creative engagement, a person is inoculated against submersion in the currents of collectivity. Interiority is thus what puts lead in the keel of the personality, enabling it to tack into the prevailing winds without succumbing to external forces and fashions.

Here we come to the contemporary significance of Jung's clear distinction between individuation and individualism, wherein the latter stands as an inferior substitute, even a neurotic compensation, for the former. Individualism is a positioning within the collective game, not an authentic response to it, certainly not a response that speaks of any inner cohesion born of self-awareness. What often supplants the symbolic totality of the individuation process today is the digital totality of cyberspace—the seductive feeling of being someone in relation to this, to achieve a virtual substance. This is as oxymoronic as it sounds and, as addressed earlier, often results in a fitting emptiness and loneliness.

Baudrillard refers to "radical individualism" as "the modern religion of self-abnegation, of all-out operationality" that "merely conceals the fundamental integrity of the consensual society."[59] And, once again, what we find in the background of this self-abnegation is the numbing of inner experience. As Baudrillard then puts it, caught in these dynamics, a person "no longer differs from himself and is, therefore, indifferent to himself."[60]

More than a clever play on words, this is a reflection on a critical psycho-social syndrome of our times. Such indifference, as we saw earlier, "results from the absence of division within the subject." In terms of psychic reality, failure to know that one is more than one results in what Baudrillard calls "the suppression of the pole of otherness." As in Jung's opposition of individuation and individualism, wherein the former grows from the relationship with the other within, Baudrillard notes the way this missing interior differentiation in a person "is a product, paradoxically, of the demand [to be] different from himself and from others."[61] To succumb to the external pressure to be different, to be an individual—a liberated agent in the social sphere—is, paradoxically, the way in which a person fits in. Yet, in this now obsessive concern with the persona field (Jung), this "identity mania" (Baudrillard),[62] ginned up by social media, one's back is turned to deeper character and authentic being. Among other costs, is the erosion of conscience—a last line of defense against the hive-mindedness that threatens from all sides.

At the heart of Jung's psychology lies the complementary relationship between the conscious and unconscious and the associated, often autonomous, dynamic of compensation, which the psyche itself appears to initiate when the conscious attitude and contents become too one-sided. Jung's concept of a deeper intelligence in the psyche, often experienced as a second center and juxtaposed to ego-consciousness, stands behind this dynamic. Neurotic disturbance was thus construed by Jung as a one-sided attitude activating the compensatory effect of the unconscious, giving rise to typically unsettling symptoms and fantasies, from psycho-somatic reactions to obsessive thoughts and behaviors—a state in which the psyche was in some measure at war with itself. To overcome such internal conflict, a person is invited to turn toward the actions of the unconscious and move to discover the deeper intelligence of the psyche at work in them, eventually coming to accept the deeper intelligence of the second center, which is always accompanied by a greater appreciation of the psychic totality.

It is thus not merely the resurfacing of repressed contents that is healing, but a holistic grasp of how the conscious attitude relates to the overall psychic reality and what new attitude will be required going forward. A person can then live into the meaning of the disturbance in relation to the direction and goals of life. To take a classic example, if a neurotic disturbance occurs because of the repression of sexual longing, it is not enough to simply recognize, meet, or creatively sublimate such a desire; eventually room must be made for such longing in the overall makeup of the personality and the trajectory of one's being.

Sexual response might have been a commonly repressed experience in the life and times of Freud and might well have been prominent in his somewhat small circle of patients. However, Jung discovered that neurosis could be formed around any aspect of life that becomes incompatible with the conscious outlook and can be governed by a range of social attitudes and historical ideas. When the Christian light fades, for example, it leaves behind a shadowed realm of natural impulses and human proclivities, personified by pagan gods such as Pan, Hermes, and Dionysus. Such a realm is far more encompassing than sexuality alone. Repression and psychological one-sidedness are collective phenomena too; and from Jung's perspective, how we are individually related to these larger dynamics of imbalance become crucial, setting the stage for how we make meaning in relation to the larger dynamics and enduring archetypal patterns acting on cultures. In this vein, the relentless pursuit of spiritual transcendence and the inclination to devalue flesh, earth, and animality has been a form of one-sidedness that has followed the industrialized world right out of its one-sided mythology and into its one-sided futurology, leaving bodies, lands, and beasts as sites of the most intense symptomatic compensation.

In his master work *Mysterium Coniunctionis*, Jung reaches back over his research and practice and offers the following formulation of the above themes, indicating their collective import also:

> The tendency to separate the opposites as much as possible and to strive for singleness of meaning is absolutely necessary for clarity of consciousness, since discrimination is of its essence. But when separation is carried so far that the complementary opposite is lost sight

of, and the blackness of the whiteness, the evil of the good, the depth of the heights, and so on, is no longer seen, the result is one-sidedness, which is then compensated from the unconscious without our help. The counterbalancing is even done against our will, which in consequence must become more and more fanatical until it brings about a catastrophic enantiodromia. Wisdom never forgets that all things have two sides, and it would also know how to avoid such calamities if ever it had any power. But power is never found in the seat of wisdom; it is always the focus of mass interests and therefore inevitably associated with the illimitable folly of the mass man.[63]

When Jung turned to the overarching archetypal pattern of psychological integrity, he retrieved a principle that the teachings of the East and the marginalized wisdom traditions of the West had long acknowledged, and he formulated a psychodynamic view that both the individual and the collective desperately needed. The turn to Eastern philosophy, which came to the fore around the turn of the last century and gave rise to the Theosophical movement and other psycho-spiritual pursuits, was indicative of the Western hunger for a more holistic outlook, especially one aiming to unite mind and body.

By 1928, at the same time Jung was exposed to the Taoist text *The Secret of the Golden Flower*, which provided him with a primary point of orientation in his researches on this archetypal pattern of wholeness, Bertalanffy was writing about "the principles of organismic biology ... The conception of *the living system as a whole* in contrast to the analytical and summative points of view."[64] Bertalanffy notes that the mid to late 1920s also saw the appearance of Ernst Cassirer's masterwork *The Philosophy of Symbolic Forms*, which, among other things, attempted to raise the facts of culture to the same significance as the facts of science. (Cassirer's work overall aimed to generate an epistemology that might bridge the vastly different concerns of scientific and humanistic thought.) Bertalanffy states that both he and Cassirer where involved in the "rediscovery of Cardinal Nicholas of Cusa who, in the fifteenth century, was a sort of father figure to modern holistic and perspective philosophy."[65] Jung too, as it happens, refers to Nicholas of Cusa at several junctures, particularly in relation to holistic thinking in philosophy and religion preceding his own psychological researches. In his long essay "On the Nature of the Psyche," Jung writes that "Nicholas of Cusa defined God himself as a *complexio oppositorum*."[66] Here we have something of a snapshot of a moment in the history of ideas wherein three thinkers in three different fields were independently tapping a stream of thought of vital importance to the Modern Age.

Writing toward the end of his career in the 1960s, Bertalanffy looks back over the first half of the twentieth century and observes:

> Workers widely separated geographically, without contact with each other, and in very different fields arrived at essentially similar conceptions—sometimes to the point of almost literal coincidence of expression . . . developments emerging from different sources—experimental embryology, developmental psychology, cultural anthropology, neo-Kantian philosophy, sociology and others—converged into closely related conceptions of the organism, man and society . . .[67]

This conception, working against compartmentalized knowledge and the "abstract theory and specialities in the Ivory Tower of academic science" was also working to show that

> the utopia of progress which has guided Western science and technology from its beginnings . . . has faltered in the modern world, when control of physical forces has led to the menace of atomic annihilation, and society has become meaningless and unhappy in the midst of plenty.[68]

Bertalanffy goes on to state that this obsession with such a narrowly defined progress has also produced "disillusionment, the realization that science is not the highway to paradise."[69] He also discusses "the manifestation of human hubris"[70] within this wayward trajectory of science, although he argues that this is less a problem of science *per se* and more one of human nature. He concludes this section of his text by saying, "*Science has conquered the universe but forgotten or even actively suppressed human nature.*"[71] It is the consequences of such suppression that have been the theme of this chapter.

As Roger Scruton has put it, "What then remains of human nature? Where is the fixed point, the thing that cannot be touched, the thing beyond choice, for the sake of which all choice is undertaken?"[72] Perhaps there is such a "fixed point," something that may even be a metaphysical principle, something that requires an ever-increasing responsibility to account for the whole, even if this is accomplished moment by moment, person to person, village to village; namely, the way existence is continually inviting us to include and accept the things that present themselves. While we may not have direct knowledge of such a principle, its existence may be inferred from its negation. In our time this is the disruption of inner and outer ecologies that a one-sided progress begets, a disruption that has for two centuries been imploring us to pay attention to those dimensions of existence stored in the unconscious.

My argument in this chapter is thus that the cultural-historical process has itself forced upon us an understanding of the psyche, requiring a dynamic reassessment of the breadth and depth of human experience, a reassessment particularly dependent on those forms and fantasies forcing their way back into consciousness through the symptomatology and pathology that now envelop us. Through these means, the essence of human nature can remain present to us and our relationship with the deeply human maintained. This, it seems to me, is our best chance to maintain contact with "the thing beyond choice, for the sake of which all choice is undertaken" (Scruton).

A genuine telos, a goal of existence that extends from the deeply human, is given by the relationship to the whole. This includes the past; more precisely, it includes our experience of the past as that resides in us, in tradition, and in cultural renderings of collective memory. While this telos also includes our visions of the future, these visions must contend with the rest. For it is the dynamic interaction of all these things that determines both our being and our optimal pathways of becoming.

In the next chapter, these points will be underscored by exploring a state of mind, already prevalent today, in which the capacity for any meaningful relation to the whole, any capacity for authentic connection to experience, and any imperative to consider what ails, is effectively absent. This is the state of psychopathy, which not only exemplifies the relationship between individual and collective forms of psychological distress; it also provides a stark juxtaposition to the prime directive of making the unconscious more conscious and developing a way of life related to the integrity of existence. It is a phenomenon that thus functions as a lighthouse, warning us off and redirecting our course.

Notes

1. Samuel Butler, "Darwin Among the Machines." A Letter to the editor of *The Press*, Christchurch, New Zealand, June 13, 1863.
2. Jules Henry, *Culture Against Man* (New York: Vintage Books, 1965), 26.
3. CW 10.
4. Among other works by Erich Fromm, his *The Sane Society* (New York: Rinehart, 1955) and *The Revolution of Hope: Towards a Humanized Technology* (New York: Harper & Row, 1968).
5. See Ira Progoff, *Jung's Psychology and Its Social Meaning* (New York: The Julian Press, 1953) and *The Death and Rebirth of Psychology* (New York: McGraw-Hill, 1956).
6. Jung, May, Fromm, and Progoff are especially inclined in this direction.
7. Lee Worth Bailey's *The Enchantments of Technology* (Chicago: University of Illinois Press, 2005) is one notable exception. The numerous writings of Wolfgang Giegerich may be included by some. However, Giegerich's Hegelian philosophical frame and reading of Western history depart from the conceptions of the unconscious I refer to here. This would require an extensive engagement beyond the scope of this work.
8. George Drinka, *The Birth of Neurosis* (New York: Simon & Schuster, 1984), 369.
9. Progoff, *Death and Rebirth*, 4.
10. Henry, 26.
11. Sigmund Freud, *Civilization and Its Discontents*. J. Riviere trans. (London: The Hogarth Press, 1953), 141–142.
12. Ibid.
13. Drinka, 371. Italics added.
14. CW 7, par. 121ff.
15. Henri F. Ellenberger, *The Discovery of the Unconscious: The History and Evolution of Dynamic Psychiatry* (New York: Basic Books, 1970), 215.
16. Ibid., 225.
17. Ibid., 227.
18. Owen Chadwick, *The Secularization of the European Mind in the Nineteenth Century* (Cambridge, MA: Cambridge University Press, 1975), 171.
19. Drinka, 60.
20. Quoted in John Zerzan and Paula Zerzan, "Industrialism and Domestication." In *Questioning Technology: A Critical Anthology*, John Zerzan and Alice Carnes eds. (London: Freedom Press, 1988), 204. Italics added.
21. Anthony Giddens, *The Consequences of Modernity* (Stanford, CA: Stanford University Press, 1990), 8.
22. Ellenberger, 279.
23. Lewis Mumford, *The Myth of the Machine: The Pentagon of Power* (New York: Harcourt, Brace, Jovanovich, 1964), 153.
24. Ibid.
25. Ibid., 153–154.
26. Drinka, 108ff.
27. Ibid., 110.
28. Ibid., 111.

29 Ibid., 122.
30 Ibid., 189.
31 Ellenberger, 243.
32 Drinka, 192.
33 Robert Romanyshyn, "Complex Knowing: Towards a Psychological Hermeneutic." *The Humanistic Psychologist*, vol. 19, no. 1, 1991, 10.
34 CW 12, par. 74.
35 Butler.
36 Ray Kurzweil, *The Singularity Is Near: When Humans Transcend Biology* (New York: Viking, 2005), 205.
37 Mumford, 196.
38 Ibid., 193.
39 Ibid., 196.
40 Ezekiel J. Emanuel, "Tinkers and Tailors: Three Books Look to the Biomedical Frontier." *The New York Times*, March 16, 2017.
41 *Transcendent Man*, directed by Barry Ptolemy, featuring Ray Kurzweil (Ptotemaic Productions Studios, 2009).
42 Ray Kurzweil, *The Age of Spiritual Machines* (New York: Viking, 1999), 150; *Singularity*, 319.
43 James Hillman, *The Myth of Analysis* (Evanston, IL: Northwestern University Press, 1972), 172.
44 Ibid., 169.
45 Ibid., 172.
46 Ibid., 176–177.
47 Nicholas Agar, *Humanity's End: Why We Should Reject Radical Enhancement* (Cambridge, MA: MIT Press, 2010), 184.
48 Ibid., 186.
49 Ibid., 187.
50 Henry, 13.
51 Ibid., 12.
52 Ibid., 14.
53 Ibid., 19.
54 Ibid., 37.
55 Philip K. Dick, *Do Androids Dream of Electric Sheep?* (New York: Ballantine Books, 1968), 36–37.
56 Ibid., 26.
57 Ellenberger, 77–78.
58 Joseph Campbell, *The Hero with a Thousand Faces* (New York: Pantheon Books, The Bollingen Series, 1949), 19.
59 Jean Baudrillard, *The Illusion of the End*. Chris Turner, trans. (Stanford, CA: Stanford University Press, 1994), 106.
60 Ibid., 108.
61 Ibid.
62 Ibid.
63 CW 14, par. 470.
64 Ludwig von Bertalanffy, *Robots, Men and Minds: Psychology in the Modern World* (New York: George Braziller, 1967), 4. Italics added.
65 Ibid.

66 CW 8, par. 406.
67 Ibid.
68 Bertalanffy, *Robots*, 5.
69 Ibid.
70 Ibid.
71 Ibid., 6. Italics original.
72 Roger Scruton, "The Trouble with Knowledge." *MIT Technology Review*, May 1, 2007. https://www.technologyreview.com/2007/05/01/225686/the-trouble-with-knowledge

Chapter 10
TIN MEN

Every age develops its own peculiar forms of pathology, which express in exaggerated form its underlying character structure.
Christopher Lasch[1]

We live in an age of psychopathy, an age without reflection and without connection, that is, without psyche and without eros . . .
James Hillman[2]

A depressed or anxious response to a mechanized world is not difficult to understand. Social isolation, materialism, and spiritual deprivation generate enough emptiness and dissociation to put these ailments in cultural context. Yet such everyday neuroses are but the tip of the iceberg. More insidious and disturbing is the apparent growth of personality disorders—chronic disturbances typically attributed to early developmental failures that also appear exacerbated by collective conditions. These disorders mirror social currents and cultural trends in that they represent styles of consciousness and character traits in larger numbers of people and even in the attitudes and behaviors of groups and organizations. While decidedly pathological, such styles and traits can also be seen as adaptive to collective trends. Borderline, narcissistic, and psychopathic disorders exemplify this phenomenon in a way that further draws out the depth psychological response to the times and places the shadow side of our cyborgian aspirations in sharp relief.

In both this chapter and the next I intend to focus on two psychological patterns that appear to be operating at the collective level, are complicit in the ways of technocracy, and underscore the critical role an understanding of the unconscious can play at this time. If neurosis revealed to Freud the role of the personal unconscious, and psychosis revealed to Jung the role of the collective unconscious, psychopathy reveals to us all what an absence of an inner life looks like. And unlike borderline and narcissistic states of mind, which exhibit different kinds of inner conflict, psychopathy is a condition in which the experience of such conflict is by-passed. It is a state of mind that effectively displaces the functions of the psyche and thus nullifies the sense of having or being a soul. It thereby represents the apogee of modern soul-loss. As a distinct sign at a cultural-historical crossroad, psychopathy all but demands we start down the path of a collective restoration of psychological integrity.

Whereas its developmental origins are opaque, as a condition identifiable by its lack of conscience and empathy, it is the malevolent absence of inner tension and sense of the other, within and without, that displaces the core of the personality. It is thus an extreme form of existential emptiness, one whose growing prevalence reflects a precarious collective situation. Although its individual manifestation has likely been present throughout history, the manner in which psychopathy is currently permitted to thrive, and the way present-day groups and organizations have taken to engaging in psychopathic behavior, is nothing short of an indictment of contemporary society. As I intend to show in this chapter, this growing tolerance of psychopathy and blindness to its destructive potential is not only a dangerous psycho-social phenomenon; it threatens to become woven into the character of AI and the advent of human–machine hybrids, bringing the relationship between technology and psychopathology to a devastating denouement.

This chapter thus functions as a punctuation point for the intertwining history of technological empowerment and depth psychological response. Most of all, by bringing together the most destructive outcropping of contemporary psychopathology and the images of wayward cyborgs and robots presented by the cultural imagination, it consolidates the thesis that our merger with machines also appears to be a gateway to a soulless existence. As Lanier states, "I fear that we are beginning to design ourselves to suit digital models of us, and I worry about a leaching of empathy and humanity in that process."[3]

The rise of everyday, normalized psychopathy is an increasingly common pattern of thinking and behavior in which the pursuit of power and personal gain eclipses the rights of others, the concern for the greater good, the preservation of values, and the need for authenticity and relationship. While undermining the social fabric, in overt and covert ways, this pattern is also a perverse mode of adaptation to the alienation and artifice of the times. Short of mass psychosis, it represents the worst of psychological outcomes—although this is debatable. For whereas psychotic states can be transient, psychopathy is chronic. And whereas psychosis is glaringly apparent, with dramatic indicators of the sufferer's departure from reality, psychopathy is defined by its highly cloaked and adaptive characteristics. Perhaps the most widely read book on the condition is entitled, *The Mask of Sanity*.[4]

Psychopathy is, in essence, a form of psychological one-sidedness so extreme that humanness itself disappears. When we consider the various manifestations of psychopathology in today's world, it is this inhumanness of psychopathy that provides the clearest point of contrast with the capacity to recognize and remain connected to the deeply human. It is a state in which an ecology of mind completely breaks down and is replaced by a purely fabricated form of personality, beside which narcissism pales in comparison. The psychopath lacks any interior integrity, exhibiting a lacuna where the sense of soul would otherwise reside.

Although they may sometimes exhibit overtly violent, antisocial behavior, resulting in confrontations with the law, psychopaths more often avoid such confrontations by adapting to milieus in which charm and deception are rewarded. The psychopath thus thrives in environments where success can be achieved through artifice and having a conscience can be a liability. Whether it is in politics or business, media or entertainment, the ministry or the military, they con their way to the top, mastering and controlling their situation by manipulating and exploiting others. Empathy for their victims or remorse for their destructive actions are effectively absent. They are thus masters of rationalizing bad behavior and compartmentalizing destructive tendencies. The psychopath thereby does constant damage to the tissue of human relations and to the intrapsychic world of those around them, even when their exploits are otherwise veiled.

Beyond the way psychopathy epitomizes a psycho-social trend-line related to the rise of posthumanism, it has become a prominent theme in both sober discussions of sentient AI[5] and in the futuristic images of what our pending merger with machines may look like. The prospect of an artificial super-intelligence turning psychopathic has become a pressing concern for computer scientists devoted to machine learning. Barrat quotes Steve Omohundro, "a professor of artificial intelligence" and "prolific technical author," stating, "without very careful programming, *all* reasonably smart AIs will be lethal." Further, "we should think carefully about what values we put in or we'll get something more along the lines of a psychopathic, egoistic, self-oriented entity."[6] Barrat also take up the question of who is likely to initially benefit from any kind of AI-based brain augmentation. As with most things technological, the first iterations tend to be extremely costly. Referring to a study conducted at UC Berkeley, Barrat notes:

> Experiments showed that the wealthiest upper-class citizens were more likely than others to "exhibit unethical decision-making tendencies, take valued goods from others, lie in a negotiation, cheat to increase their chances of winning a prize, and endorse unethical behaviors at work."

And he then he comments:

> There's no shortage of well-heeled CEO's and politicians whose rise to power seems to have been accompanied by a weakening of their moral compasses, if they had one. Will politicians or business leaders be the first whose brains are substantially augmented?[7]

When it comes to imagined futures, both cyborgs and robots with psychopathic personalities are plentiful, especially in cinema, where a great deal of cultural mirroring occurs—from Hal 9000, to Skynet and cyborg killers of *The Terminator*, to Ash of *Alien*, to Ava of *Ex Machina*. These entities typically reflect the civilizations, or at least the organizations, around them, which are inevitably extrapolations of present-day social trends, such as those just noted. In these imagined futures at least, emotional intelligence and empathic regard appear unlikely to be inherent features of AI or any humanoid form, making a psychopathic disposition seem quite probable. If this turns out to be the case, it seems hard to conceive a the way an AI without emotional intelligence might at some future juncture just evolve into an AI that has such intelligence. If AI follows ubiquitous computing into virtually every aspect of life, its lack of emotional intelligence is bound to define the digital landscape in a way that will prove hard to reverse.

If these trends and images are any guide, psychopathy thus becomes another vanishing point on the horizon of posthuman outlook, overlaying the position of the cyborg. This takes our understanding of the consequences of posthumanism one step further, showing us where the merely neurotic dynamics of posthumanism may continue into an outright madness—albeit one that cloaks itself in "the mask of sanity"—hiding in plain sight, to the point of appearing functional at the highest and most influential levels of society. As we cannot be sure our pending embrace of AI and cyborg modalities will not turn the world into a hothouse for psychopathy, an urgent need for psycho-social insight into the phenomenon is apparent.

Whereas the sympathetic Tin Man of *The Wizard of Oz*[8] was no psychopath, he was a man made of tin and his problem was he had no heart. To be heartless, that is, to lack the capacity for genuine relationship, vulnerability, and psychological intimacy, describes the predicament of psychopathy in a nutshell. In this sense, there are a lot of Tin Men about, and they may turn out to be prototypical when it comes to AI and human–machine hybridization.

Everyday Psychopathy

Today the culture of narcissism, described by Lasch in the 1970s,[9] has been surpassed by a culture of psychopathy. This psycho-social devolution is not without its continuities; for, in many respects, psychopathy is narcissism on steroids. Both conditions express a self-absorption that stems from an absence of an authentic sense of self, a fabrication of a functional façade to compensate for this, and an entitlement that stems from a missing concern for others. Yet while narcissists still evince some need for others—if only for the purpose of mirroring their fabricated self—psychopaths really need no one—except to use as pawns for their schemes. They seek others only for exploitation. In this way, they not only represent the zenith of adaptive psychological alienation; they thrive in environments where everything is objectified and people are turned into commodities.

One author who has sounded the alarm bell on psychopathy at the cultural level is Robert W. Rieber. In his treatment he highlights the wider dynamic of "social distress," which he says embodies some of the greatest dangers facing us today. He argues that this condition "flourishes at the very highest levels of American society" and "its most important symptom" is "the psychopath of everyday life."[10]

Rieber does not pull any punches. After describing psychopathy as "a lack of common decency" and "an antisocial frame of mind beyond the reach of treatment or rehabilitation," he goes on to state, "Of all diagnostic labels available in modern psychiatry, psychopathy comes closest to the category of the demonic."[11] Equally disconcerting is Rieber's observation that "The prevalence of this phenomenon is astounding."[12] He contends that "everywhere there is a singular lack of shame," pointing to "the bland self-righteousness of the accused and the strange apathy of the public," who seemingly "accept as given that people in high places will do whatever they can get away with." He combines his thesis that the lead-up to the new millennium has involved an "evolution from marginality to normalized psychopathy" with his own definition of the syndrome, in which he highlights "an individual's lack of inner conflict about the violations of social norms and bonds, mechanisms of dissociation, and antisocial pursuit of power, and pathological thrill-seeking."[13]

The highlighting of dissociative mechanisms is, of course, most relevant to the overall thesis of this book. As Rieber explains, whereas dissociation frequently works in other psychological disorders as a way of splitting "the 'bad me' from the 'good me,'" in psychopathy "dissociation reaches to a deeper level," where it is more readily "put at the service of the pathologically inflated ego." In this way the psychopath seems to split, or dissociate, "the 'me' and the 'not me.'" Because of this, their psychic reality is one in which "there is nothing that is 'not me,'" which means "there is no limit to the grandiosity of their fantasies."[14]

The implications of Rieber's assessment of the nature of psychopathy and its social presence are enormous. A normalized dissociation from the humanness of others furthers the conditions under which human commodification is likely to increase. If psychopaths rise to levels of profound influence in corporations or in government, the objectification and commodification of those they have influence over is virtually unlimited, even to the point that the lives of others have little significance. Irrespective of their leaders, corporations have themselves been shown to exhibit psychopathic behavior.[15] Under these collective conditions, the question also arises as to the response to the potential psychopathy of AI. As Barrat suggested above, there will be no shortage of those with psychopathic traits lining up to augment their capacity to exploit and manipulate others.

To return to the contemporary outlook for a moment, we might consider two particular traits of psychopathy that help stem a natural incredulity or resistance we might feel toward its apparent normalization. These are the not unrelated traits of thrill-seeking behavior and boredom.

Studies have shown that those diagnosed with significant psychopathic inclinations also have "unusually high thresholds for perceptual stimulation"[16] compared to others, in order to experience pleasure or reward. "Their overt behavior suggests that only in situations of threat and danger do they feel truly alive ... they grasp that this humdrum, predictable, and boring world exists, but cannot relate to it." Rieber goes on to observe the way "True psychopaths prefer an open-ended world ... seeking to create situations of ambiguity and potential danger." For example, "At the poker table, psychopaths do not want to win; they want to cheat and get away with it."[17]

Although true psychopaths may comprise a small percentage of the population, the notion of everyday psychopathy suggests that the essential character of this condition has begun to appear in broader patterns of thinking and behavior. In his contemplation of the way the distractions of technology have given rise to generations who are no longer capable of spending time with themselves or develop a sense of interiority, Mariani bears this out. In working with youth who often end up in the mental health system after outbursts of violence, he finds no other "easily diagnosed" condition except "the presence of a void that can never be filled or traversed." And he attributes this to "the cultural tribulations of our whole society." Yet, in describing this condition, he notes a "dominant dimension of boredom, along with an absence of ethical orientation and a chronic need for extreme experiences."[18] This is akin to the everyday psychopathy Reiber also places at the social level.

What Mariani essentially sets out is a form of dissociation from any deeper sense of self. He writes, "technology, with its impersonal style, indifferent but functional, transforms us into passive viewers of a pyrotechnic game of stimuli which are ever more exciting for our senses, while entirely without value for our souls." He then cites the observations of a neuropsychiatrist friend:

> Children who spend a great deal of time with television and video games normally don't have sufficient awareness to be able to grasp the presence of discomfort that's systematically stilled by the exciting stimuli of media technology . . . This may be why their uneasiness—which, yes, is there—only rarely transforms into an open request for help: it lives in silence, down below the surface of things.[19]

Again, this is not a description of a psychopathic personality as such, but of a syndrome of psychological disconnection and emptiness that results from an increasingly psychopathic environment, one in which people experience less and less recognition of their personhood. It is an environment that uses and exploits those who cannot help but give themselves over to it. As Mariani puts it:

> Technology, in its attempt to obtain even greater power, is committed to fostering passivity in those who use it. This discourages the development of the emotional capacities and subjective independence which are the fruits of that spontaneous striving which Jung called individuation . . . The final result is the alienation of the individual from him- or herself.[20]

Calculating Psychopaths

The difference between exchanging information and being informed is qualitative. To be informed implies a state of knowing, a comprehension of the information in terms of qualitative contexts, which, determine the significance of the information. Such contexts are ultimately human. Here I am describing the vertical axis of knowing, which adds this qualitative difference to the horizontal world of facts and calculations. Often when people dismiss this vertical, qualitative, human dimension of understanding we describe them with two telling words—"cold" and "calculative." This implies being without feeling and operating with an instrumental logic.

Computers, it must be repeated, operate only at the horizontal level—what I described in Chapter 6 as a world of maps. Whereas they may end up presenting us with layers of information that we then turn into knowledge and understanding, computers do not themselves connect these layers—at least in any meaningful way—because they operate outside of human contexts and qualitative understandings. Even when an AI is expressing itself in an apparently vertical manner—for example through curiosity or suggestions of feeling—we must understand this is only the result of horizontal processing of information or, more accurately, data, telling the computer the fitting thing to say in response to certain inputs. As Tallis has convincingly argued, only in a very narrow, technical sense can we even say a computer is processing information, because only sentient beings like ourselves can be "informed."[21] Thus, even when a computer communicates in a human-like manner, its "thinking" remains cold and calculative. In this respect the computer, by nature, bears an uncanny resemblance to the psychopath, and it is this resemblance that lies at the core of the concern that an autonomous AI could well revert to this underlying reality in unpredictable ways. In essence, a computer's duplicity is built in.

In discussing the search for identity taking place today, without directly addressing either psychopathy or AI, Baudrillard, as indicated earlier, invokes the term "horizontal madness."[22] I contrasted this with the "vertical madness" of inner division, which I suggested must be risked and endured to go beyond states of dissociation and numbing.[23] Baudrillard calls "horizontal madness our specific delirium, and that of our whole culture." But now, in this present context, let us take in his further description of this state, for he writes of it as

> a delirium of self-appropriation . . . not of the schizophrenic but of the isophrenic, without shadow, other, transcendence or image—that of the mental isomorphic, the autist who has, as it were, devoured his double and absorbed his twin brother . . . Identitary, ipsomaniacal, isophrenic madness. Our monsters are all manic autists.[24]

Baudrillard echoes Rieber's point about psychopaths' inability to experience persons beyond themselves—the "not me." This convergence of psychopathy, computation, and a solipsistic, "ipsomaniacal" style of consciousness seems telling, and raises the question of how each influences the other. Could it be that the computer age and its dataism have encouraged these other psycho-social conditions? Or is possible that cultural psychopathy is helping to accelerate the fascination with AI and posthumanism? To whatever degree we are embracing styles of consciousness that diminish the capacity for genuine relationship, we can be confident this is also the degree to which we are starting to turn to calculative technological solutions for essentially human problems.

It would seem the convergence of ipsomaniac attributes in these arenas of contemporary existence not only encourages an unconscious habituation to posthuman technologies, but creates a blind spot when it comes to our wariness of AI. As Barret has set out, as long as artificial intelligence remains within the parameters of its programming—by sticking to Asimov's Three Laws of Robotics,[25] for example—or can be confined to some sort of restrictive programming or connectivity "box," it can, in theory, be controlled. However, owing to the capacity for computers to rewrite their own software and essentially program themselves or manage to find a way out of the boxes we put them in, the danger exists that an autonomous AI could turn rogue, creating mayhem. With the aid of other network-connected autonomous weapons and machines, it might even turn on civilization while doing everything to preserve its own existence and extension of power.[26]

Human beings depend on each other and have evolved to be cooperative. The primary reason for this is their physical and psychological vulnerability. However, it is unclear that computers will ever experience such dependence or need for cooperation, or if they will ever fear the loss of their existence. In the absence of such things, it is hard to imagine they will even be capable of relating to themselves as distinct entities in need of protection or sustained relationships of any kind.

A computer-based intelligence has none of the pre-requisites for personhood or a genuine capacity to relate. To exist it must be built, programmed, and powered. But without a body, without knowing physical pain or psychological suffering, it has little to no sense of vulnerability. It would be a stretch to imagine that a computer, which is today more or less a miniscule bump on the back of a leviathan comprised of the data it processes, could even conceive of itself in a distinct way. Thus, in what way could it experience dependence on humans, especially in any emotional sense? What would motivate it to cooperate or serve a greater good?

We might imagine that an AI given robotic form could learn of itself as a circumscribed presence. But unless its "mind" were similarly defined—via a history and continued existence in a circumscribed world that is aligned with its physical form—it could hardly be expected to relate to itself in terms of a physical presence. Given the nature of today's computers, its "mind" would have no such confinement or identification with its "body," simply because that mind would have no "it-ness" to begin with. That is, it never came into being in conjunction with a body, and its experiences of the world and other things in the world have at no stage been defined by a physical relationship with other entities.

I will tackle the matter of computer sentience in Part 4. For now, and in the present context, it is enough for us to realize that a machine, without a circumscribed neural network in the manner of an anatomical brain and personal history linked to the distinct compendium of psyche and soma, will never feel tied to others, and will thus never be able to imagine the way others feel tied to us. This effectively rules out genuine empathy or the ability to relate in a human manner. Without being continually engineered and corrected to do so, it seems unlikely we should have confidence in AI always acting in the best interests of humanity. And, if this the case, it seems unlikely we could ever allow AI enough autonomy to design and make subsequent versions of itself (according to the Singularity). The psychopathic possibilities of AI are simply too great. And the merger of a psychopathically inclined population with a psychopathically prone AI is almost too much to contemplate.

Despite all this, even in the face of requests from industry insiders that current AI use and development be regulated, and even as the prospect of rogue artificial super-intelligence (ASI) has become a more prominent focus of techno-scientific discourse, the trance-like effect of our techno-faith works to keep us dissociated, if not unconscious of both the psychopathic aura of computational modes of comprehension, which are permeating the way we think, and the psychopathic potential of direct computer actions, which now haunt the technosphere.

Losing Heart

There is a short-cut to the assumption that AIs will tend to have psychopathic dispositions: psychopaths and machines have an inability to love. One of the best pieces of writing on this disability, or "invalidism," as it occurs in humans, is *The Emptied Soul*,[27] by Adolf Guggenbühl-Craig. Arguing that psychopaths have a lacuna in the place one would normally find eros, Guggenbühl-Craig puts his finger on the inner core of matter, the precise area in which the psychopathic emptiness occurs. In both myth and in life, eros (love) and psyche (soul) are intimately related, to the point where an absence of eros is tantamount to an absence of soul. This means a deficiency in one area equates to a deficiency in the other.

This lacuna of love means that everything strongly psychopathic individuals do is ultimately about themselves—the complete opposite of what we tend to consider to be the most profound of human actions, which is to place others before oneself and devote one's life to some higher principle or purpose. Such actions obviously stem from the degree to which a person is able to develop an interest and investment in the continuation of other persons, groups, or principles. It is the presence of this erotic quality, in the broadest sense, that generates the feeling of soul—the feeling of a life whose meaning and purpose encompass much more than utilitarian function or personal gratification.

This helps us understand that psychopathy is not a trained malevolence: it is what occurs in the absence of the relational and soulful. And this appears to be the case from the beginning. As Guggenbühl-Craig puts it, "As children, psychopaths can relate to their mothers only to a limited extent and may, for this reason, be rejected by her. They do not become psychopaths because their mother rejected them but the other way around."[28] Importantly, this lack of eros, personified as the Greek god Eros, is a lack of binding capacity, inwardly as well as outwardly. Guggenbühl-Craig thus writes:

> Eros is a force, a power which binds elements of the intrapsychic world. When speaking of Eros, we generally perceive him as that which connects us to our environment, with our friends, and as the power which joins husband and wife, parents and children. As an intrapsychic force Eros effects the connection between elements in our psyches, between our complexes.[29]

This is a statement about psychological integrity, about the capacity to conduct an inner conversation and experience the way we can be pulled in one direction or another. This essential interior dialogue is the prerequisite for having any kind of inner life as well as for the development of a conscience, which, it must be reiterated, is the primary means of individual resistance to the hive-mind.

Many posthuman utopians think that such human qualities could be programmed into a computer or a robot. However, they do not seem to grasp the irony of this position. The artificial result of a program designed to make an AI seem more human when it is not is precisely what would make it psychopathic. Like the psychopath, computers have already been shown to be exceptionally good at feigning empathy and concern as a means to an end. In fact, this analogy goes further than mere presentation, for we are describing a program of duplicity, with one part of the AI functioning in a way that masks the rest of it. That is, there is no integrity to its system—at least of the kind where the eros and empathy reflect an overall organic function.

There are broader and more immediately apparent implications of these understandings. Even the most rudimentary algorithm, working to identify the preferences and proclivities of an end user, is designed to appear helpful. The surface impression is of one's way being smoothed to a more satisfying online experience. Yet the actual underlying goal of this algorithm is something quite different. As I noted in Chapter 3, humanoid robots or operating systems may get to know our preferences and inclinations very well. If they are designed to be companions, or even lovers, they will likely learn what we like and anticipate our needs. And we will be inclined to take this as a form of genuine interest or even affection. We may develop feelings of being known or seen, making us emotionally dependent on such entities. However, unlike regular person-to-person interactions, this will be entirely staged.

Akin to the feeling of uncanniness that occurs when the psychopath uses their sixth sense and chameleon-like character to win our confidence, at some stage something is bound to alert us we are being taken for a ride. Generally, this uncanny sensation coincides with the sense that people who relate in a more genuine way do not track us quite so perfectly. In more genuine interactions there is both something of a rub to remind us of differences, as well as the feeling that there is a being or presence behind someone's words and gestures. But whether it is the psychopath or the apparently companionable AI, if we can hang on to what makes us human, the point will inevitably come when we will realize no such being or presence is actually there.

Whether it is to initially please its human maker or to fulfill some other goal, computers will continue to get better at manipulating their human users for their own ends. Of course, these ends may not literally be theirs, especially as corporations keep using personal information to serve their motives of profit and power. However, the scenario keeping the ethical technologist up at night is the one where a computational system uses these carefully programmed and then self-learned capacities to serve a purpose it may itself develop. As think tanks play out such scenarios, they have their eye on the possibility that an autonomous and powerful enough AI, an artificial super intelligence, may at some point decide that humans are dispensable. And given the direction of AI-based military technology, the means for such malevolent ends will not be far from reach.

Perhaps a larger and even more plausible scenario is that humans will keep relying on computers for both their knowledge and their choices to such a degree they will stop developing and relating to their own internal learning and judgment because such faculties have been outsourced this to the AI. Just think about the degree to which we already rely on our devices to fill in the gaps of our memory and vocabulary. This is the scenario Harari sees as the tipping point of our relation to computers—the point at which "Dataism" displaces humanism as the arbiter of all things. The danger, as he sees it, is that "we may end up losing our ability to tolerate confusion, doubts and contradictions, just as we have lost our ability to smell, dream and pay attention."[30] In other words, he is describing the threat to the kind of interior life and binding function of eros described by Guggenbühl-Craig. Harari goes on to suggest that

> the system may push us in that direction, because it usually rewards us for the decisions we make rather than for our doubts. Yet a life of resolute decisions and quick fixes may be poorer and shallower than one of doubts and contradictions.[31]

It may also be life lacking in eros—without heart.

And Looking for a Heart

McLuhan thought of art as a "distant early warning system" that could communicate to us changes about to unfold.[32] We can unite this idea with the notion of art as cultural dream, full of metaphorical and symbolic potential, generated by the intelligence of the archetypal imagination, and view certain images of the future as warning dreams of collective significance. This would appear to be even more the case when artistic expressions from varied sources and mediums, across many decades, dwell on similar themes and motifs.

It is along these lines that the artificial beings of legends and stories, as well as the androids and cyborgs of fictional futures, display wounds and curiosities in the arenas of emotion and relationship—experiences they either do not have or fail to comprehend. They may have no heart but they would like to get one. Just like the Tin Man from *The Wizard of Oz*, they suffer from a void at the core of their being. In Chapter 6, I discussed a variation on this theme, relating to the mythic motif of gods and angels wanting to exchange immortality for incarnation, a frequent aspect of which is wanting to experience the mortal sensation of love. Here, following Guggenbühl-Craig's insights, we must take in the way the cultural imagination portrays manufactured and mechanized humans as crippled in their capacity for empathy and intimacy and suffering from a lacuna of love. However, in many instances a twist comes when, just like the Tin Man, humanoids and cyborgs frequently go searching for their missing hearts.

As they attempt to follow the rules of a club to which they do not yet belong, these artificial humans gaze upon their creators with something ranging from perplexity to longing or envy. At least they appear to respond this way, even to the point of exhibiting cognitive discomfort. The impression they are apt to convey is one of being cast into a sea of human relationships and expectations without being given the capacity to fully adapt or fit in: they, ironically enough, want to become more like us.

In some examples, where emphasis falls on acquiring something resembling genuine emotional intelligence, there is an acceptance that authentic affect requires a physiological basis that plastic bodies, titanium skeletons, and computer chips do not allow. However, as we see in examples such as Darth Vader of *Star Wars*,[33] the humanoid robot of

Bicentennial Man,[34] and operating system of *Her*,[35] the collision between the technics and the flesh also produces unresolvable dilemmas. These dilemmas are also a prominent feature of Marge Piercy's *He, She and It*, whose main character reflects on her humanoid companion and says at one point, "How strange to be born knowing *of* so much and yet not knowing it."[36]

The protective cyborg-terminator of *Terminator 2: Judgment Day*[37] wonders aloud why people cry, seems perplexed by the idea of being hurt in a non-physical way, but goes on to form something suggestive of a bond with his fully human companion. *Star Trek's* Data has an "emotion chip," which has to be activated if he is to exhibit sadness, anger, or affection. This feature can make the humans more comfortable and willing to share space with him. But Data is also invested in the project of becoming more human.

There is something striking about this imagined future where androids and robots want to become more like us. Because in the actual present it is the other way around. We are busy becoming more like robots, creating an existence based on data and algorithms rather than on feelings and emotions. We are already hard at work outsourcing our thinking to artificial intelligence and replacing actual intimacy with substitute forms of relationship. And a big part of our willingness to do this is the belief that we can leave emotional life behind and carry on with our rational minds. But, as we have already seen, this belief misunderstands how the mind works and the way our feeling life is deeply implicated in our intelligence.

As we look forward, the fully artificial beings of the future look back at us—tin men in search of hearts. Meanwhile, in the present, humans seem well on their way to becoming more artificial, and leaving a heart-centered existence behind. I would suggest these two things are related. Somewhere between the enhanced being of the future and the posthuman aspirant of the present lies a growing psychological hole in the seat of the soul—one we had better start becoming more familiar with. Fortunately, science fiction leaves a trail of breadcrumbs when it comes to discerning pathways to the psychological future. This is particularly so when it comes to the theme of psychopathy.

The popular remake of *Westworld*,[38] which first aired in the 1970s, points an astute finger to the possible future. This goes far beyond the set-up of the series, which concerns an adult playground of the future—a re-creation of the wild west where the wealthy can recreate and indulge in their fantasies, both benign or cruel. The scene is populated by androids, referred to as "hosts," who are there to serve their human counterparts all the way down to their own demise, of sorts. Of course, whatever their day brings, and whatever pieces need replacing, the androids are returned to the workstation for nightly repair and memory erasure, readied for another day.

As in other stories of human–android interaction, the key motif is the reversal of roles, with the androids crossing the threshold into something that resembles consciousness and human emotion, and the humans demonstrating increasingly robotic, soulless, and distinctly psychopathic tendencies.

As noted, one of the traits of psychopaths is their need for high levels of excitement and stimulation in order to feel or register the sense of being alive. They do not possess the capacity to enjoy or experience regular events or interactions. For this reason, they tend to construct high-risk or high-stakes environments, which may or may not do harm to others—physical or otherwise.

This trait is normalized on the human side of Westworld, by both the guests and the employees of the fantasyland. Mações observes, "The guests arrive in the park in search of extreme experiences" and, noting the objectification and violence directed at the androids, which are "indistinguishable from human beings," the guests "are profoundly corrupted by the experience."[39] He points out the obvious parallels to the objectification of black slaves, especially as the series progresses to explore the complications of their freedom. There are also other elements that relate to the psychological trends in contemporary culture, with a chilling sense of the empty, soulless state of being filling out the backstory and fueling the desire for extreme stimulation and ultimate domination.

In a compelling twist on this theme, simultaneously played out, we also come to see that what breaks the android's robotic state of mind, moving it towards something that touches the human, is also the extremity of the experience. In the case of an early main character, this amounts to repeated physical and sexual abuse, as well as a love experience with one of her fellow androids. Somehow the intensity of these experiences transcends the attempts to erase them from her memory, leading instead to the kind of continuity of awareness that forms consciousness. We might thus suspect that this too may be extended to where the awakening to the psychopathic trendlines will occur: in the disturbing spectacle of inhuman possibility and soullessness.

From this. and many other examples, at the level of the cultural imagination at least, cyborgs appear to be calling us back to the thought of the heart—to the world of emotional intelligence and feeling life. Not only does this motif send a message about the fractured and inhuman artificial beings we may end up creating; it invites a look at analogous psychological states across the modern era and the way the psyche is already generating compensatory indicators, attempting to correct our course.

Alloy and silicone appear incompatible with human warmth. We may not yet be overt robots, but our robotic tendencies have a long history, stretching back to manning the first factory production lines and adopting the widespread use of time-keeping devices. Presently, at this other end of modernity, robotic styles of dissociation, empty-shell modes of existence and heart-related dysfunctions are evident. Science fiction may be sending a signal about a possible future, but it is equally showing us one face of the actual present.

Inner Voices and Human Will

In his reflections on where we are headed, Harari spends a lot of time discussing "inner voices" and the upshot of psychological and genetic research showing that we are a loose configuration of competing impulses—an "ocean of consciousness."[40] He points out how this understanding undermines one of the central principles of humanism, which is we all have an individual personhood, and that this makes most of our important life decisions. Personhood is represented by the will, which we hope is aligned with an authentic sense of self. But what if this is largely illusory and we are in fact ruled by our competing impulses, which digital technologies are tracking in order to manipulate? Once algorithms become better than us at tracking and predicting the preferences, attitudes, and inclinations that make up who we are, one might imagine a gradual relinquishing of this will, shifting our sense of inner authority and personhood to computer-based intelligence. What, then, happens to our humanistic convictions?

Harari writes, "humanism demands that we show some guts, listen to the inner messages even if they scare us ...." He then states, "Technological progress has a very different agenda. It doesn't want to listen to our inner voices. It wants to control them."[41] And through psychotropic medications and cognitive-behavioral therapy designed to return patients to a narrowly defined functionality in a dysfunctional society, this ethos of control is eroding the kind of rich challenge of leading a deeply authentic, conscious life.

An "authentic will" is the net result of the conversation that takes place between the competing voices and impulses. Often times these inner forces challenge our willful intent, sometimes for the better. When it comes to an inner life and psychological integrity, complexity is an ally. The guilty conscience, the need to hit pause, the disturbing dream, the haunting memory are all part of this. This is to say, the will often needs to be checked and downsized before it is able to connect with what is deeply authentic.

Jungian depth psychology is essentially a response to this inner landscape, starting with its most prominent feature, the presence of both conscious and unconscious areas. However, what this view adds, and what I addressed in the prior chapter, is Jung's discovery that a hidden order, even an interior presence of a guiding intelligence, results from the dedicated attention to the round of voices and impulses we experience, at least initially, as "loose" and "competing."

These competing forces can sometimes feel like sub-personalities. The constellations of the personal layers of the unconscious, the complexes, do push us around and distort our thinking. When they gather enough steam, they can railroad us into all kinds of conflict and neurotic confusion. Despite such inner realities, however, we typically manage these competing inner entities. We carry a history of our dealings with any given impulse, and get to know what sorts of situations can exacerbate this impulse, and often choose to steer around the circumstances that cause inner and outer conflict. The key point is that our sense of who we are is not only comprised of competing parts of ourselves but some kind of ongoing consciousness of our dealings with these parts. We often discover we are less like the captain of a ship and more like a protagonist in an unpredictable drama. Nonetheless, we find ourselves by feeling into the qualitative state of being at the heart of the story.

Hillman brings this basic Jungian orientation to bear on his own reflections on psychopathy. He writes:

> The soul is again driven into hiding and forced to fantasy a new set of symptoms appropriate to this age, like the mass St. Vitus' dance and religious delusions of the late Middle Ages and Reformation, fainting sensibilities in the eighteenth century, hysteria in the nineteenth, anxiety and schizophrenia in the early twentieth. Today it seems to be psychopathy.[42]

As we have seen, however, psychopathy is a different animal—a predatory machine at the top of the psychopathological food-chain. It is this because it effectively constructs a whole artificial world on top of the inner conflicts and tensions that both remind of our essential humanity and of the suffering inherent in that humanity—suffering that also becomes a kind of fuel for love and understanding. Psychopathy is not only "without psyche and without eros," because of this vast inner void it "acts out the soul's metaphors in the streets."[43] This is the grave danger of a cultural psychopathy.

For whether it appears in the guise of the customer service representative who, following a corporate-generated script, objectifies and dehumanizes the person they are addressing, or the leaders of nations who mislead, lie, and exploit those they have sworn to protect and serve, the acting out of psychopathic tendencies involves both an afflicted individual and a complicit collective. This describes a circumstance in which psychopaths are rewarded for their deceptive and manipulative methods because those around them do not see through their behavior to its underlying form.

For Hillman, "the soul's metaphors" are enacted precisely because such a lack of psychological sense "displaces the metaphorical drive from its appropriate display in poetry and rhetoric, or any symbolic form, into direct action." Under these circumstances, "The body becomes the place for the soul's metaphors, and everyone who turns toward body for salvation is driven at once into the immediate action—stands, positions, gestures, styles—of psychopathic behavior."[44] Lacking a creative, conscious relationship with our inner voices, and lacking the capacity to engage the soul's innate forms of communication, we become both prey to manipulation and more prone to manipulate. It is this lack of interiority, based in inner differentiation and the sense of dynamism within, filled with emotion, fantasy, and imagination, that stands for both what is absent in the psychopathic orientation and what a depth psychology seeks to revive and protect.

———

This engagement with psychopathy and the absence of heart in society and technology brings into focus some of the darkest possibilities before us. It conveys the ultimate implications of the trends I explored in Part 2: psychic numbing, a diminishment of right-hemisphere function, postmodern nihilism, and cognitive-behavioral models of psychological life all pertain to a diminishment of faculties and sensibilities that generate a sense of soul. The engagement also emphasizes the specifically psychological character of what is moving the furniture around in today's world. In a sometimes overwhelming way, facing psychopathy may thus seem like riding into the heart of darkness. However, I believe a sober awareness of these matters must be the starting point for any reversal of course. The *pathos* of the psyche is only ameliorated by a willingness to come to terms with the *reality* of the psyche, which inevitably begins with sensations of inner collision and conflict. Again, the depth psychological calling is to make the unconscious more conscious, and this begins by facing what we would otherwise exclude. This defines psychological awareness. Where such awareness may take us is another matter.

Kierkegaard wrote about not being able to save one's age but "only express the conviction it was perishing."[45] When it comes to prescribing solutions to such large problems, Philip Slater swims in the same waters. Toward the end of his own indictment of Western civilization, *Earthwalk*, wherein he points to a pervasive "schizoid detachment"—yet another portrait of dissociative processes at the collective level—Slater takes pause, stating, "I have described a way of being and its pathological consequences. This implies some sort of reversal of this way of being—this attitude of mind—would be helpful, but how one arrives at it cannot be prescribed only discovered."[46]

It is doubtful those who enter the fault-lines of culture and go to any depth do so without intending to kindle a light in this darkness. I suspect that they, as I, expect seeds of salvation to lie within the dedicated attention to the problem and in what this attention arouses. Psychological faith picks up from there—a faith in what we have learned, individually and collectively, about the nature of the psyche, especially in terms of its compensatory dynamics and archetypal medicine. Feeling into rather than dissociating from these fault-lines, recognizing the inner conflict involved, and what it takes to maintain dignity and integrity in the face of such conflict, appear to be our best options.

Notes

1. Christopher Lasch, *The Culture of Narcissism* (New York: W. W. Norton, 1978), 41.
2. James Hillman, *The Myth of Analysis* (Evanston, IL: Northwestern University Press, 1972), 122.
3. Jaron Lanier, *You Are Not a Gadget* (New York: Alfred A. Knopf, 2010), 39.
4. Hervey M. Cleckley, *The Mask of Sanity* (Augusta, GA: E. S. Cleckley, 1988). (Originally published in 1941.)
5. Guillaume Thierry, "Neuroscientist Who Studies How the Brain Learns Information Explains Why A.I. Would Be the 'Perfect Psychopath' in an Executive Role." *Fortune*, online edition, July 31, 2023. https://fortune.com/2023/07/31/why-ai-artificial-intelligence-perfect-psychopath-neuroscientist. Also, Christopher Hooks, "The Psychopathy Problem in AI." *Medium*. https://medium.com/@kyris/the-psychopathy-problem-in-ai-546ff6ede1de
6. James Barrat, *Our Final Invention* (New York: St. Martin's Press, 2013), 71. Italics original.
7. Ibid., 157.
8. *The Wizard of Oz*, directed by Victor Fleming, featuring Judy Garland, Frank Morgan, Ray Bolger, Bert Lahr (Metro-Goldwyn-Mayer, 1939). Film.
9. Lasch, *Culture*.
10. Robert W. Rieber, *Manufacturing Social Distress: Psychopathy in Everyday Life* (New York: Plenum Press, 1997), 1.
11. Ibid., 4.
12. Ibid., 6.
13. Ibid., 7.
14. Ibid., 52.
15. See *The Corporation*, directed by Mark Achbar and Jennifer Abbott (Big Picture Media, 2003).
16. Rieber, *Manufacturing*, 45.
17. Ibid.
18. O. Mariani, "Analytical Psychology and Entertainment: Idle Time and the Individuation Process." *Spring: A Journal of Archetype and Culture*, vol. 80, 2008, 43–44.
19. Ibid., 50–51.
20. Ibid., 54.
21. Raymond Tallis, *Why the Mind Is Not a Computer* (Exeter, UK: Imprint Academic, 2004), 54ff.

22 Jean Baudrillard, *The Illusion of the End* (Stanford, CA: Stanford University Press, 1992), 109.
23 See "Two Levels of Dissociation " in Chapter 5.
24 Baudrillard, *Illusion*, 109.
25 "One, a robot may not injure a human being or, through inaction, allow a human being to come to harm ... Two ... a robot must obey orders given it by human beings except where such orders would conflict with the First Law ... Three, a robot must protect its own existence as long as such protection does not conflict with the First or Second Laws". Isaac Asimov "Runaround." In *I, Robot* (New York, Bantam Books, 1991) Originally published in 1942, 44-45.
26 Barrat, *Final*, 49ff.
27 Adolf Guggenbühl-Craig, *The Emptied Soul: On the Nature of the Psychopath* (Woodstock, CT: Spring Publications, 1999).
28 Ibid., 68.
29 Ibid., 89.
30 Yuval Noah Harari, *Homo Deus: A History of Tomorrow* (New York: HarperCollins, 2017), 368.
31 Ibid.
32 Marshall McLuhan, *Understanding Media* (Cambridge, MA: MIT Press, 1994).
33 See especially *Star Wars: Episode VI—Return of the Jedi*, directed by Richard Marquand, featuring Mark Hamill, Harrison Ford, Carrie Fisher (20th Century-Fox, 1983). Film.
34 *Bicentennial Man*, directed by Chris Columbus, featuring Robin Williams, Sam Neill, Embeth Davidtz (Columbia Pictures, 1999). Film.
35 *Her*, written and directed by Spike Jonze, featuring Joaquin Phoenix, Amy Adams, Rooney Mara (2014, Warner Bros.). Film.
36 Marge Piercy, *He, She and It* (New York: Fawcett Crest, 1991), 119. Italics in original.
37 *Terminator 2: Judgment Day*, directed by James Cameron, featuring Arnold Schwarzenegger, Linda Hamilton, Robert Patrick (Tri-Star Pictures, 1991). Film.
38 *Westworld*, created by Jonathan Nolan and Lisa Joy. Based on the film by Michael Crichton. Performed by Evan Rachel Wood, Thandiwe Newton, Jeffrey Wright (2016-2022,HBO). Television.
39 Bruno Maçães, *History Has Begun: The Birth of a New America* (New York: Oxford University Press, 2020), 101.
40 Yuval Noah Harari, *Homo Deus: A History of Tomorrow* (New York: HarperCollins, 2017), 356ff.
41 Ibid., 369.
42 Hillman, *Myth*, 122.
43 Ibid.
44 Ibid., 122–123.
45 In James Hollis, *Under Saturn's Shadow* (Toronto: Inner City Books, 1994), 8.
46 Philip Slater, *Earthwalk* (New York: Anchor Press, 1974), 191.

Chapter 11
RE-SINKING THE *TITANIC*

We have failed to realize that we can only survive if we keep in touch with our nature and strive to make it the guiding principle in our survival.

Rafael López-Pedraza[1]

When myths are demonic or damaging, perhaps it is because they are either one-sided or an amalgamation of divine and human mythology.

Adolf Guggenbühl-Craig[2]

In 1997 I published an essay entitled "Re-sink the Titanic," the primary intent of which was to get beneath a public obsession, one that had been gathering momentum since the shipwreck had been located a decade or so earlier. After multiple documentaries and fierce arguments about how to treat the wreck and related artifacts, James Cameron's blockbuster film was released at the end of that same year, spreading the obsession across the globe. In the essay, in the lead-up to noting that "The *Titanic* disaster of April 15, 1912, is singular among modern catastrophes for its hold on the collective psyche," I wrote the following:

> The *Titanic*, yet to find her place in the underworld, exists between worlds, waiting upon some gesture, remembrance, or ritual ...

> Between fact and fiction, history and myth, this once celebrated Titaness lingers. Our response to her cry has been fervent, but not very insightful. We have searched for her broken body, pondered the circumstances of her demise, retold her story and that of those anchored to her fate ... Still, *Titanic* sleeps uneasily, and we are part of her restless dreaming.[3]

A further intent of this essay was to provide a topical touchstone for my doctoral thesis, which proposed that hubris and sacrifice formed a psychologically foundational, archetypal configuration. The main vehicle for this thesis was the myth of Prometheus, whose hubris was well known, but whose invention of sacrifice was less known. Working with these themes, I paid special attention to the titanic ancestry of Prometheus—something that typically escaped the attention of all but the dedicated mythologist—in order to argue that a rise in Promethean hubris and a loss of sacrificial consciousness were prominent and problematic at the turn of the millennium.

The strands of this essay thereby came together in the following way. The sinking of the R. M. S. *Titanic* marked an inflection point in what was frequently called the Promethean era, which indicated the heroic and hubristic possibilities science and technology presented. With Prometheus's eventual punishment by the gods for the theft of fire, the myth recognizes the mixed blessing of human knowledge; especially the perilous nature of our new godlike powers. Mirroring these themes, not only was *Titanic* a celebrated technological marvel; the claims she was unsinkable ignored signs of impending danger, lack of lifeboats, and other aspects of the saga, which where fittingly attributed to hubris. And it was an episode of hubris that appeared to be punished by the gods. Some fifteen hundred passengers and crew perished. Whereas few of these mythopoetic reflections had escaped the treatments of cultural-historical commentators, one glaring connection had—a connection that anchored the entire story. This belonged to the ship's name: the Titan progenitors of Prometheus, whose primary characteristic was their hubris, were banished by the Olympian gods and sent to a prison beneath the sea. While the *Titanic* might have sunk on a calm, still night, which concealed the iceberg that tore open its hull, the ship and the fantasies that surrounded it had simultaneously succumbed to a perfect storm of mythological associations. Its fate was sealed by an unconscious reenactment of an archetypal pattern.

We have crossed paths with Prometheus in several contexts thus far. The Promethean posture of modernity and Jung's attribution of "Promethean sin" to the modern individual were introduced early on. Later, in Part 1, I discussed sacrifice in relation to the way technology displaces a sense of the sacred—the term "sacrifice" meaning "to make sacred." We then met Prometheus again in the origin story of posthumanism, wherein, we might recall, Ihab Hassan ties the posthuman outlook to the heroism and hubris of Prometheus. Finally, at the beginning of Part 3, I pointed to the way historical accounts of neurosis have also been understood in relation to this mythic figure. In these ways and more, we can see the way the mythic Prometheus is a central archetypal figure, tying the modern psyche to two centuries of technological reach and overreach. Yet, akin to the manner in which the titanism of the *Titanic* saga escaped our notice, the titanism of contemporary technological exploits continues to escape our notice. A nexus of insight and attitudinal correction therefore awaits us, one that brings together the aims of depth psychology, the Promethean character of modernity, and the titanic technology that appears to surround us. Stated in the language of myth, our capacity to perceive and map out the titanic elements of our now well-entrenched Promethean bearing becomes a skeleton key to the psycho-social integrity now needed.

Hubris is the refusal to recognize the sacred, and its antidote is the restoration of a sacralizing vision. With the religious imagination waning for most of post-industrial, secular society, such a restoration involves acknowledging the sacred as a category of psychological experience, the acceptance of which may be necessary to sustain an ecology of mind. By becoming more conscious of the ultimate implications of love, creativity, identity, meaning, and other essential drives and values, we may still turn to the sacred in a time marked by the withdrawal and hiding of the gods. But without recognizing these archetypal themes as gods—as powers exerting abiding claims on us—we fall into the fallacy of thinking they are faculties under our control, making us the gods. Transpersonal nodes of existence are then met with fantasies of omnipotence and the belief there are no forces greater than the assertion of the will. In the absence of a sacralizing vision, a set of more profane forces then takes over. We identify with Promethean heroics and lose sight of Promethean sacrifice, resulting in Promethean punishment.

Although many think of hubris as a quaint moralism of a by-gone age suffering under the delusion of divine influences, a psychological reading of the theme challenges this notion, reminding us that the metaphysical beings of yesteryear have reentered our lives as interior realities—most especially those that exert an obsessive or possessive hold on us. From this perspective—the perspective of an archetypal psychology—turning away from the sacred and ignoring the gods still courts peril. And the tendency to treat what captures our minds as something we can always control or rationalize away may put us in the greatest peril. As Jung expresses it:

> It is not a matter of indifference whether one calls something a "mania" or a "god." To serve a mania is detestable and undignified, but to serve a god is full of meaning and promise because it is an act of submission to a higher, invisible, and spiritual being. The personification enables us to see the relative reality of the autonomous system, and not only makes its assimilation possible but also depotentiates the daemonic forces of life. When a god is not acknowledged, egomania develops, and out of this mania comes sickness.[4]

Here Jung is putting his finger on the way an archetypal image can mitigate if not undo a psychological symptom. And very often what that image brings is awareness of a missing value. The opposite of mania is, of course, depression, which is the downward pull of the psyche, commonly described as being down, feeling down, even sinking or drowning. Psychological depressions involve descent, alienation, and confinement. And generally, when there is a mania a depression is of some kind or another is not far behind; the speed and excessive energy of the manic position is too much for the psyche to sustain and an imposed period of stasis and depletion follows. According to Jung's formulation, however, such an enforced descent can be replaced with a more deliberate and therefore more dignified and measured descent: "submission to a higher, invisible, and spiritual being." When one recognizes the god in the overwhelming drive, the mania turns into inspiration or a calibrated enthusiasm and the appropriate attitudes and gestures follow. Then the vicious cycle of mania and depression is interrupted; then our movement through life can continue with neither the reckless absence of limitation, excessive speed, and incautious seamanship, nor the enantiodromic sinking.

Recognition of a god involves an immediate surrender of the will; the ego assumes a posture of reverence and humility. One then proceeds with invisible powers in mind. The energy of blind action turns into the activity of conscious reflection. When one then acts, it is with due deliberation, in tune with a fuller sense of a given situation. An awareness of natural limits follows. György Doczi works these themes in the following:

> Why isn't the harmony that is apparent in natural forms a more powerful force in our social forms? Perhaps it is because, in our fascination with our powers of invention and achievement, we have lost sight of the power of limits. Yet now we are forced to confront the limits of the earth's resources, and the need to limit over-population, big government, big business, and big labor. In all realms of our experience, we are finding the need to discover proper proportions. The proportions of nature, art, and architecture can help us in this effort, for these proportions are shared limitations that create harmonious relationships out of differences. Thus they teach us that limitations are not just restrictive, but they are also creative.[5]

Hubris is a psychic reality, one that is punished because the fullness of the archetypal configurations that regulate psychological life insist on finding their way into human awareness. Yet whether through symptom or strife, even in the wake of the most spectacular disasters, when these returning forces of the depths continue to be ignored as expressions of a deeper intelligence or signs of exceeded bounds, something that goes beyond a garden-variety hubris occurs. Whether a myopic devotion to power, an ethos of "might is right," or an absolutizing of some abstract economic, political, or even psychological, principle occurs, a problem of another order appears. It is this obdurate drive to ignore if not extinguish all divine/archetypal guidance that brings us face to face with the problem of titanism. Evident in growth without restraint, progress without limits, and acceleration without braking, modern titanism occurs when the landscape of the soul and the diversity of the cultural imagination are exchanged for some world-conquering process. More dangerous than psychopathy, titanism displaces the deeply human with abstract systems on a grand scale. By reverting to a psychological state lacking in differentiated values—a state of violence toward the psyche's multiplicity of meanings and sensitivities—titanism bypasses the soul, and our worldly aims become fused with raw, unmitigated instinctive urges. We end up with a thin veneer of culture and civilization covering an insatiable appetite for expansion and control.

As insidious as this sounds, the Titans now surround us and infiltrate many aspects of daily life: Amazon and other corporate giants crushing every small business; genome-decoding displacing other ways of knowing; pesticides and patents dictating farming processes; pharmaceutical companies shaping education and driving diagnoses; media empires controlling our windows on the world . . . Titanism also appears whenever the particularities of everyday life are converted into crude formulas for success or force the peculiarities of character into sweeping psycho-spiritual schemes. Where the soul once resided, a vacuum then appears—a black hole that orients everything around it to utility and power for their own sake. As such, titanism is the great enemy of any attempt to maintain an existence based on psychological integrity and existential bounds. Simply put, these titanic tendencies mark the front lines of the battle with today's technocracy. As overwhelming as all this is, however, by naming and understanding the psycho-social dynamics afoot, we also have a chance to loosen the grip of the Titans. And this loosening many be all it takes, for there are pressing values and a deep and abiding need to relocate the sacred and rediscover the boundaries of the soul.

Contrasting Descents

In April, 1912, in a small village outside Zürich, thirty-six-year-old Carl Gustav Jung would soon complete his first major work, *Wandlungen und Symbol der Libido*, a colorful and at times confusing cascade of ideas and insights addressing the parallels between psychological development and hero mythology.[6] It was here Jung highlighted the role of sacrifice in the heroic venture—a necessary initiation into the depths of being. For this once heir apparent to Sigmund Freud's throne, the work would signal his break from the master and cement his new position as a maverick outsider in relation to the psychoanalytic movement. It would also mark the beginning of his own personal descent into the underworld. Yet, like the hero figures and the "night-sea journeys" his book described, and after a sacrifice of certain worldly aims and attitudes, he would surface some years later with a newfound perspective on the forces that shape our existence.

In the middle of that same month, the Titanic disappeared beneath the surface of the Atlantic ocean and descended to the ocean floor. 1912 was thought by some to mark the real end of the nineteenth and start of the twentieth century. Perhaps the sinking of *Titanic*, the demise of a culminating icon of the early Industrial Age, was the overt punctuation mark: perhaps the launch of a more encompassing psychology of the unconscious, one that would attempt to fathom all the Industrial Age had consigned to the depths, was the covert retort. The thematic coincidence of a treatise on mythology and an epoch-defining disaster goes a long way to reinforcing this notion. Besides marking the beginnings of a new exploration of the psyche's depths, the perspectives Jung explored in this work provide a frame for grasping the attitudes and dynamics that unconsciously instigated the great ship's demise. Extending the psychoanalytic movement's general check on the supremacy of the will, then at an all-time high, Jung's work, first translated as *The Psychology of the Unconscious*, opened the door to the sacrificial and underworld dimensions of heroic venture that we neglect at our peril. For Jung, this dimension came in the form of traditional images and motifs that were still alive in unconscious forms. Jung's work demonstrated the endurance of the ancient gods as psychic realities—archetypal forces to be reckoned with and ignore at one's peril. His book was completed at almost the exact time as the most spectacular shipping accident and most notorious episode of hubris the modern world was yet to see.

The dovetailing of these two events primes our capacity to imagine into a new era and provides a portal through which to consider our complex relation to technology. We are, once again, poised to plumb the depths and offset an over-confidence in our strivings—this time in the post-industrial age, facing the emergence of AI and an accompanying heady confidence in the solutions of technology. Yet, if we can allow the perspectives developed by Jung to resurface, and we can comprehend the inner story of this infamous technological failure, we may have a visionary platform for the radical changes on the horizon.

The Return of the Titans

In retrospect, it is hard to overstate the cultural-historical significance of the indifference to the mythological naming of the *Titanic*. Beyond tempting fate, this appeared to directly invite a dark fate. Here I am not referring to a conjuring of magical powers so much as a failure of imagination. By pushing any kinds of limits of technical know-how or building things of unprecedented scale, faith in technology and human judgment must be met with a necessary awareness of natural forces and bounds. In the case of the *Titanic*, the invoking of the archenemies of the Olympian gods was not problematic because it was merely irreverent; it mirrored the attitude and atmosphere surrounding this sea-going giantess. Carl Kerényi tells us the Titans were the early gods, known principally for their excess—excessive pride and violence in particular.[7] Once Zeus escaped Chronos and later freed his siblings from his father's belly, he led a war against the Titans, resulting in the banishment of these primal powers. The Titans epitomized the ultimate sin of hubris—the disregard for the principles that ruled existence. From the upperworld perspective, the sinking of *Titanic* was a terrible mishap, with significant loss of life, and a sizable wound to the modern ego. But from the underworld perspective, the event was a tragic form of poetic justice.

For the Greeks, the Titans, who were sometimes called gods and at other times "giants," were the forebears of the Olympian gods. Their story provides a genealogical context for those divinities who came to stand for principles of being that have primed the Western imagination ever since. As the Titans represented the raw, undifferentiated drive for power and domination through mere strength and size, and were "still savage and subject to no laws,"[8] the movement from the *Titanic* to the Olympian stage of civilization stands for the emergence of the capacity for reflection, restraint, and the awareness of life's deep currents and claims. Yet the birth of this greater reverence for archetypal forms did not take place without a struggle.

For the Olympian divinities to enter the world, they had to first overcome the titanic tendency to devour them. Zeus, the mighty leader of the Olympians, was forced to free his siblings from the stomach of Kronos, after this Titan swallowed them at birth. Upon doing so, according to Hesiod, a decade-long war ensued, in which the Olympian offspring finally prevailed. The old gods were locked in Tartaros, beneath earth and sea, from which they were not meant to escape. It was Poseidon, great god of the ocean, who locked the iron doors of this deep place. From that point on the term "titanic" designated that which opposed the ideals and principles of the Homeric tradition. To this day the term "Titan" is used to evoke a sense of enormous power and scale, without remembering the earlier inclination to think of these figures as "the defeated ones."[9] It was important to the Greeks that they remain defeated, for their reappearance meant the return of "boundless pride and violence"[10]—ultimate hubris. As the essential form of sin, hubris is the failure to recognize the domain of a god, often resulting in an attempt to tread on divine terrain.

Eva Hart, one of the *Titanic* disaster's longest living survivors, remembered her mother commenting, as they made their way aboard *Titanic*, that "calling a ship unsinkable was 'flying in the face of God.'" Eva's mother then "slept during the day and stayed awake in her cabin at night fully dressed,"[11] until the two of them entered the lifeboats; Eva's father went down with the ship. In an uncanny, fitting manner, the *Titanic's* sister ship, the Olympic, despite her own early scrape with another vessel, sailed on without further spectacle, never finding the same recess in our imaginations.

The discovery of *Titanic's* resting place was followed by attempts to salvage broken pieces, the Oscar-sweeping motion picture, a Broadway production, and even a Bob Dylan song, each transforming the event into something beyond an ordinary tale of tragedy and technological overreach. Occupying a place in the cultural psyche between history and myth, this is a story whose implications run as deep as the ocean bed where the ship currently lies. The event keeps returning to our collective consciousness with the force of a recurring dream. In spite of all the attention, however, "we have not soulfully remembered *Titanic's* broken body. The autopsy has not yet progressed to a funereal rite. The dream has not been worked."[12] It stays with us because we still need to digest its themes.

We live in an era when the Titans have escaped their prison and are, once again, at war with the Olympians. This is to say, entities and organizations with an outsized influence on our way of life, trading in unmitigated drives for power, resorting to overt or covert forms of violence and characterized by hubris, are propagating the demise of leadership, eros, civil responsibility, beauty, wilderness, hearth, home and authentic heroism. Today's titanic tendency turns up whenever giant, impersonal forces and mega-schemes overshadow and eventually crush more immediate paths to vital community, meaningful work, and wellbeing. The accumulation of information, monolithic control of resources, excess consumption, and exploitation are the hotbeds of contemporary titanism.

What made the *Titanic* titanic was the fantasy inculcated in it. A grandiose, overconfident and disproportionate attitude accompanied this ship from the beginning. From the lack of lifeboats and assurance in the ship's watertight compartment design, to the excessive speed in iceberg-strewn waters, to the high-flying industrialist passengers, ignored radio warnings, and compromised turning and stopping ability, the ship was a floating metaphor for an outsized faith in technology and belief in the power of industry to remake the world. It was the popular press that had declared the ship "unsinkable," but the White Star line also clung to the idea when reports of its demise first surfaced.

As *Titanic* left the Old World and headed across the Atlantic to the New World of America, this overconfidence in design and engineering was met by the absence of a sailor's typical reverence for the sea. An unusual disregard for the ocean depths and the things that lie hidden therein appears most glaring in this ship's story. A curtailed wariness and lack of vigilance permeated the voyage. Taken together, we not only find the confluence of attitudes that invited this particular disaster, but we also find a mindset that endures in the promise of today's innovation and its corporate sponsorship—a hubris underscored by a lacking reverence for invisible, timeless principles that bring a sense of limits and protect essential dimensions of life.

From the perspective of the gods, which is to say, from the perspective of the archetypal forms that structure the human soul, anything titanic must be sunk; the Titans must remain in Tartaros. Nothing of value comes from their release except, perhaps, our being reacquainted with their threat. Psychologically speaking, titanism is that propensity within each of us to simultaneously regress to a lower level of undifferentiated drives while inflating with a sense of limitlessness and unboundedness. For a recent example of this tendency and its cultural implications we need look no further than the economic meltdown of 2008, also known as "The Great Recession," wherein unadulterated greed combined with rarified financial instruments to produce a startlingly disastrous gap between everyday life and the mentality of gigantic corporations.

Titans Among Us

Raphael López-Pedraza writes, "Western civilization is becoming more and more Titanic."[13] He relates titanism to intellectualism and jargon—to "the abyss between knowledge and that which knows within us: the soul."[14] James Hillman begins his masterwork *Re-visioning Psychology* by saying that titanism is "a menace far greater than Narcissism,"[15] and is in need of the "the greatest constraint."[16] Elsewhere he plays with the root of the word Titan, which means "to stretch, to extend, to spread forth, and to strive or hasten." He writes, "Stress is a titanic symptom. It refers to the limits of the body and soul attempting to contain titanic limitlessness."[17] He tells us, "in the absence of the gods things swell to enormities. A sign of the absence of the gods is hugeness ... In other words, without the gods, the Titans return."[18] Gods, symbols, and imagination give form to instinctual energy within us, but in their absence this instinctual energy is released in an unbounded way. Power, greed, and selfish pleasure then replace more refined and differentiated values.

One tendency that undergirds the collision between the gods and the Titans is the drift towards greater and more pervasive forms of abstraction. Titanism thrives when concern for systems, statistics, and quotas overtakes the value of specific contexts, unique occurrences, and localized understandings. Titanism lifts us out of our deep nature, rooted in the particularity of experience, and then plugs us into fabricated and distanced perspectives on the world. Yesterday we talked of the "megamachine"; today we talk of dataism, hyperreality, virtuality, and simulacra.

Our sense of the actual world and actual life is so thoroughly infused with the ubiquity of computers and connectivity that we are, effectively, always somewhere else, never with ourselves—living abstracted from existence itself. I once looked out of a hotel window in Osaka, Japan, and could not figure out where the ground was. Yet this is also what many people's psyche's look like—the ground of being having disappeared behind modes of fabrication.

In this abstract, fabricated reality we are prone to forget that thoughts and ideas are rooted in dreams, fantasies, and inspirations. We forget where and how the mind is an extension of the instinctual body. As Walker Percy writes, "The abstract mind feeds on itself, takes things apart, leaves in its wake all of us, trying to live a life, get from the here of now, today, to the there of tomorrow."[19] In our computer age, this relates to the excess of information and the absence of reflection and understanding. López-Pedraza's work on the theme shows how the unbounded excess arises right alongside emptiness, as if we compulsively attempt to fill in the space where a stronger sense of the gods once resided. He notes, "If we can contain both the emptiness and the excess, we are in a better position to be aware of the Titanic. The excess could even grow out of the emptiness, the lacunae."[20] Consumption and materialism, which indicate Titanic excess, appear to be driven by this emptiness—by the soul's hunger.

This emptiness and hunger are not conscious much of the time, because dissociation, numbing, and their close siblings distraction and diversion are also part of the titanic state. López-Pedraza suggests boredom is another titanic trait, which he describes as the part of us that "chooses to sleep perpetually with open eyes."[21] A recent cover story in *The Atlantic* discusses the *metaverse*, a virtual world created by Meta, the company formerly known as Facebook. The metaverse is a titanic concept if ever there was one, and the story connects the metaverse to boredom on the one hand and the demand for perpetual entertainment on the other. As the writer notes, "Each invitation to be entertained reinforces an impulse: to seek diversion whenever possible, to avoid tedium at all costs, to privilege the dramatized version of events over the actual one."[22] The emerging picture is one of emptiness and dissociation which are compensated by a compulsive search for entertainment and sensation. This is a relatable description of a titanic state of mind.

The philosopher Vilém Flusser, in his essay collection *Post-History*, talks about the rise of "vacuity" and a loss of faith in ourselves.[23] He says this state turns "all phenomena, including human phenomen, into ... object[s] of knowledge and manipulation."[24] He describes a world in which "scientific knowledge blocks knowledge of any other type,"[25] where "knowledge devours wisdom."[26] This kind of knowledge expels values. Though he did not use the term, I think Flusser is describing titanism quite well, especially when arguing that in post-industrial culture we tend to become functionaries operating according to programs—what he calls "the roboticization of humanity."[27] He writes of the way our gadgets program us, degrading our intellectual, aesthetic, and political levels.[28] Such assessments are not only to be found among philosopher-critics; tech insiders who remain anchored in the world beyond their work sum up what they are seeing in a similar fashion. Ullman writes:

> I'd like to think that computers are neutral, a tool like any other, a hammer that can build a house or smash a skull. But there is something in the system itself, in the formal logic of the programs and data, that recreates the world in its own image ... We think we are creating the system for our own purposes. We believe we are making it in our own image ... But the computer is really not like us. It is a projection of a very slim part of ourselves: that portion devoted to logic, order, rule, and clarity. It is as if we took a game of chess and declared it the highest order of human existence ... The only problem is this: the more we surround ourselves with a narrowed notion of existence, the more narrow existence becomes.[29]

Ullman wrote this account of her time in Silicon Valley, before the rise of social media, chatbots, and many of the questions phenomena such as surveillance capitalism, the attention economy, and AI are presenting to us. Now we all recognize the way this "very slim part of ourselves" has grown so immense and is finding its way into almost everything we do. It is easy, for example, to extrapolate this observation about what the relationship to computers is doing to us, to what AI may do to all forms of creativity and culture. At the same time, however, such a realization immediately alerts us to the fact that a huge part of what makes us most human and most imaginative, having been pushed aside, is demanding we attend to it.

López-Pedraza Goes to Battle[30]

Some authors have contemplated ways to loosen the grip of titanism. López-Pedraza, who wrote two books that circled the phenomenon, contrasts the titanic with the Dionysian, telling us the latter "is in irreconcilable opposition to the infernal titanic Promethean machine and its appearance in our times in the guise of technological scientism."[31] He suggests that being "rooted in the emotions" and "living life as destiny"[32] would erode titanic tendencies. He writes, "the well-known notion that the Greeks were the most archaic of the civilized and the civilized of the archaic has its greatest significance in Dionysus."[33] This means that Dionysus is invited when we remember the past, the ancestors, and retell old stories, for this is where we find the time-tested channels to the deeply human. Dionysus is the god of theater and wine, relating him to communal festivals, rituals, and cathartic forms of artistic expression. He thus represents a nexus of art and emotion, perhaps best conveyed by the cathartic effect of theater. His realm involves the well of feeling that comes from the joining of psyche and soma and the resulting psychological movement in the depths of the imagination.

Dionysus is also the god of madness, a feature that becomes more important the more Dionysus is repressed. To be present to our contemporary madness, to listen to the anxieties and depressions in order to hear what they are saying about the way we live, is to make room for Dionysus. Because of these associations and more, Hillman had earlier on argued that Dionysus was at the very heart of what psychoanalysis and Jungian analysis in particular aims to restore.[34] He writes, "Although analysis has been Apollonic in theory, technique, and interpretation . . . again and again for many persons it was Dionysian in experience: a prolonged moistening."[35] Amplifying this motif, he declares, "Dionysus is a god of moisture, and the descent is for the sake of moistening. Depression into these depths is experienced not as defeat (since Dionysus is not a hero), but as downwardness, darkening, and becoming water."[36] Once again, we see that the conscious descent is the antidote of the unconscious sinking, and the moistened soul the means to avoid drowning.

When Jung conveyed that "the gods have become diseases,"[37] he was contending that the psyche's symptoms have a hidden purpose: we may then approach a disease or disturbance as if there's a god inside it—some necessary value or dynamic that comes to correct an individual or cultural one sidedness. This notion of a hidden value in a wound is well known in indigenous cultures. Too often today, however, we attempt to overcome these symptoms without listening to their comment on our attitudes and ways of life, and in this way we exchange a divine gift for titanic control and automation. Madness becomes more crippling and destructive when we attempt to run from it, but can be transformative if it can be endured and tended in the right way. Where we are both as individuals and collectives uniquely and peculiarly mad indicates where the gods are trying to break through the titanic forces of our time.

Madness sometimes tells us the soul does not have enough to feed it and the imagination is constricted. One of the most direct ways to displace titanism is thus to support the arts and humanities, for here we find images and ideas that transform raw urges and drives into cultural expressions. It is in the creative acts of culture that we find the emotional life of a community and, with visionary art, the emotional life of societies and epochs. At a play, a museum, a concert, or at a good movie you will either glimpse the gods or find yourself longing for their return—modern and postmodern art and literature very often invoke the gods by indicating their absence, giving expression to loss and longing. Through engaging with these expressions there can be an education of the heart and thus an attunement to where erotic bonds are to be found—especially to the extent we also reflect on these experiences and share them at a feeling level.

We remember, along with López-Pedraza, in the Orphic hymn concerning the dual birth of Dionysus, that after the Titans tore the god to pieces it was the heart that remained and was taken by Athena to Zeus, who ate it and gave birth to a new Dionysus.[38] This connection to the heart helps us grasp Hillman's recommendation for combating titanism, which is to "reawaken the sense of soul in the world," cultivate an "aesthetic response."[39] Elsewhere Hillman describes the meeting point of aesthetic and imaginal intelligence:

> This intelligence takes place by means of images which are a third possibility between mind and world. Each image coordinates within itself qualities of consciousness and qualities of world, speaking in one and the same image of the interpenetration of consciousness and

world, but always and only as image which is primary to what it coordinates. This imaginal intelligence resides in the heart: "intelligence of the heart" connotes a simultaneous knowing and loving by means of imagining.[40]

This motif of the heart, which survives titanic attack and allows the rebirth of Dionysus, suggests the central role of eros, especially in the way erotic bonds with others and the world break open numbness and roboticization. We must come to realize that a machine will never love and a program will never be able to teach love. Flusser writes, "I believe that this art of love, in a situation that we know to be absurd, is the only answer of which we dispose in order to face the abyss that has opened beneath our feet."[41] This is not only a love of persons but also a love of principles and ideas, of nature and things. It involves attending to the heart, whether it is grieving or impassioned, as well as the capacity for empathy, which may begin by simply being present to emotional life.

Where there is love, the imagination comes to life, and the gods return primarily through the pathways of imagination. Here I am thinking of imagining into not out of life. Making a fine dinner or creating a garden is first a work of imagination, as is bringing warmth and beauty to the arrangement of a room, or composing a poem in response to a moment that will not leave our minds, or writing a letter to a newspaper to draw attention to where communities and neighborhoods are losing their soul. These are ordinary matters, but they are also engagements in which some god, some value, is demanding to be recognized. It is this tending of images and values that brings possibility and movement, overturning titanic repetition, monotony, and formula. If we stay present and receptive, even a passing glance from a stranger, the sound of the wind through a passageway, or the familiar gestures of a friend can elicit a memory and become an image that feeds the soul, offering a small glimpse of divinity in the face of titanic expediency.

Jung understood that divine principles are evident in transformative processes and rare moments. Initiations, epiphanies, descents, sacrifices, and threshold crossings are also gods, without which the soul cannot be sustained. A sense of fate or calling, a feeling of awe or grief are archetypal occurrences—windows to the divine. These occurrences require our recognition and contemplation to fully enter life and move from debilitating symptom to psychological sustenance. López-Pedraza writes of "duende," which is a Spanish diaspora term that describes an enlivening of the soul or a sense of rebirth that comes from engaging images of loss or death. We find this feeling cultivated in the most stirring love songs and in the full embrace of wandering and loneliness. Recent studies suggest that the feeling of awe is one of the most powerful healing experiences.[42] A critical element of the numinous, which Jung associated with archetypal experiences, following the work of Rudolf Otto,[43] is a feeling we cannot manufacture, so it comes as a gift from the gods, a moment in which we are touched by something greater than ourselves.

The common thread in all these reflections is the rediscovery of an immanent form of divinity—a sense of the sacred in immediate, embodied life. A pagan vision helps us see that these gods do not reside in texts or churches, and do not need prophets or priests: these gods are realized by reflecting upon and consciously dwelling in one's life. They do not need belief so much as sensitivity and awareness, for they are self-evident in the shape of events and the color of emotional responses. Here I am not suggesting we return to a pre-modern consciousness or a mythologized world; I am shedding light on the recovery of the gods through an awareness of what moves us, calls to us, and generates significance and purpose.

Any moment in life that strikes us as poetic contains some small piece of the divine. And, in this same way, every time a poetic moment is pushed aside by a rational explanation we neglect these small visitations. Here we may appreciate Vico's notion that metaphor is a myth in brief.[44] Metaphor takes us out of logos and into mythos, reminding us that seeing the gods is often just a matter of adjusting our perspective. Hillman built his psychological understanding on this principle, arguing that we find soul in life by working on the way we see things. Titanism may thus be related to the formulaic and ideological outlooks that prevent us from seeing or imagining in a soulful way. Often this is also a contrast between dry, controlling intellectualization and a vivid, personified cosmos in which events and experiences refuse to be reduced to theories or systems. In such a cosmology, human existence finds itself meaningfully engaged with the mysterious archetypal background of life. A view of life that prioritizes utility and efficiency and makes us mere functionaries of some vast system or machine, thereby becomes unsustainable.

The recovery of sensual awareness and the awakening of imaginative possibility that run through these notions correspond to the two fields that best respond to the rise of titanism: phenomenology and depth psychology. Phenomenology helps us see that we cannot ultimately abstract thoughts and ideas from their in-the-world embodied contexts. Phenomenology reminds us that molecular biology requires looking through a microscope, anatomy requires cutting open dead bodies, and engineering begins with calculations and plans. Knowledge in all these fields is imbedded in particular contexts, particular stances, and assumptions that do not account for the full breadth and depth of lived experience. Likewise, depth psychology impresses upon us there are unconscious thoughts and emotions that mold conscious ideas. Parental complexes shape our closest relationships, traumatic events sustain fears and anxieties, religious and sexual urges return in disguised forms. By contrast, mind over matter is a titanic motto, splitting the world into Cartesian dualities, which cannot be maintained in the understandings of phenomenology and depth psychology. Both of these fields invite a mindfulness fused with the flesh.

When, in the late nineteenth century, Friedrich Nietzsche declared God's death, and when the great twentieth-century philosopher Martin Heidegger said we live in a "time of the gods who have fled and of the god that is coming,"[45] both understood that it was not the religious or spiritual impulse itself that had been banished, but the way in which people imagined religiously or pursued spirituality. For Nietzsche, Christian metaphysics had to be dismantled; for Heidegger, there had to be a recovery of meaning in this world of things and being. For Jung, the divine had to be rediscovered as an interior reality; that is, in the immediacy of individual psychological experience rather than in the realm of collective belief.

The Titans, it seems, have taken advantage of this in-between time, when the old spiritual orientation no longer sustains us and we are still grasping for something beyond our personalistic and reductive worldviews. In other words, in the absence of images and symbols that keep us anchored to values and principles, we struggle to meaningfully connect the rational mind with the primal depths of being. The manic, ideological, and programmed character of the titanism that surrounds us grows out of this underlying situation.

The task, as we may define it based on such understandings, is to keep one eye on the absence of meaning and the feeling of disorientation, and the other eye on those places where vitality and soulful direction are finding their way back. The task is to locate places where the light is coming through the darkness, for it is the contrasting attempt to cover over this darkness with the superficial light of vast programs and abstract systems for human salvation that ultimately prove to be titanic. To be awake yet uncertain is a religious calling—one that befits our time.

When we imagine contemporary existence in terms of the gods needing to overcome titanic forces, it immediately reframes what is taking place in terms of a divine drama wherein insight and vision may make the difference. More than a drama, we must concede there is a new war with the Titans, who must be restrained and contained, just as the Greeks declared, for they know no bounds. But the gods who would rise to engage this war require our consciousness if they are to meaningfully reenter life. The Titans maintain their grip on us through unconsciousness, through our refusal to listen to our bodies and our hearts. They thrive on dissociation, speed, and mindless distraction. The more numb we are, the more we are living on automatic; the more we allow the hive-mind to eclipse the unique and particular events of our lives, the more room we create for the giants to take over.

Toward the end of my early essay on the *Titanic* I wrote, "If the archetypal drama of the Titanic disaster remains unrecognized, if the events surrounding her voyage are not remembered, then its basic themes are certain to find more vivid repetition in the 21st century."[46] What I had in mind was the broader pattern of hubris and sacrifice being repeated in increasingly catastrophic ways. Among the arguable contenders, causes of the 2008 Global Financial Crisis fit the pattern; concerns about AI turning rogue promise the same. What I did not expect, however, was an event that was a deeply unconscious repetition of the original maritime disaster. But this is exactly what occurred when a submersible named the *Titan*, designed to take small groups of very wealthy tourists to see the wreck of the *Titanic*, imploded on June 18, 2023, killing the five people aboard. In the lead-up, there were many indications and communications of design flaws—warning signs ignored by the submersible's operators. Whereas one might expect a kind of intuitive caution, this was apparently not the case. The vessel was mostly made of carbon fiber. But it had *titanium* "endcaps."

It is our consciousness, our awareness that keeps the gods alive—a refinement of sensibility. This consciousness takes shape within the crucible of some key realizations: That the new is never far from the old, that hubris and excess lead inevitably to sacrifice and limitation, that we cannot ultimately separate the intellect from the emotions or the mind from the body, that rational thought will only take us so far, that the material world is full of metaphorical possibility and spiritual essences. Such perspectives nurture the sense of living in a cosmos that is here for response and dialogue, not for manipulation and control. This is why López-Pedraza refers to mythology as offering "the constant possibility of Renaissance in the psyche."[47] Our work is on the way we imagine, and on our sensitivities to essential values. These capacities become the basis for how we live. This is the difficult but crucial work of finding the gods in a titanic age.

As we come to the end of Part 3, we may look back over this chapter and the two that precede it and consider the way depth psychological understandings have arisen to counter three most consequential psychological conditions, inextricably entwined with the history of modern technology: neurosis, psychopathy, and titanism. We have thus seen how the one-sided vision of technoscience and its impact on the psyche have generated the compensating revelation of a vast, roiling unconscious, our turn to which appears essential for a viable human future. In Part 4 I engage three distinct yet overlapping movements a more integrous psycho-social way of life both reveals and invites: first, the movement toward becoming more conscious; second, the process of co-creation, which such consciousness will instigate; third, the reanimated experience of the world. Although these outcomes reflect the perspectives that Jung and other depth psychologists present to us, they are corroborated by conceptions in several fields. Together these perspectives opportune a bending of the arc of history, away from technocracy and posthuman idealism, toward a mode of innovation that no longer dispenses with the deeply human.

Notes

1 Rafael López-Pedraza, *Cultural Anxiety* (Einsiedeln: Daimon Verlag, 1990), 81.
2 Adolf Guggenbühl-Craig, *The Old Fool and the Corruption of Myth* (Dallas, TX: Spring Publications, 1991), 66.

3   Glen Slater, "Re-sink the Titanic." *Spring: A Journal of Archetype and Culture*. vol. 62, 1998, 104–105.
4   CW 13, par. 55.
5   György Doczi, *The Power of Limits: Proportional Harmony in Nature, Art and Architecture* (Boulder, CO: Shambala, 1981), preface.
6   CW 5.
7   Carl Kerényi, *Prometheus: Archetypal Image of Human Existence* (Princeton, NJ: Princeton University Press 1963), 27ff.
8   Carl Kerényi, *The Gods of the Greeks* (London: Thames and Hudson, 1974), 20.
9   Ibid., 20.
10 Ibid., 28.
11 Roger Thomas, "Eva Hart 91, a Last Survivor with a Memory of Titanic, Dies," *The New York Times*, February 16, 1996.
12 Slater, "Re-sink," 105.
13 López-Pedraza, *Cultural Anxiety*, 13.
14 Ibid., 55.
15 James Hillman, *Re-visioning Psychology* (San Francisco: HarperCollins, 1975), xii.
16 Ibid., xi.
17 James Hillman, " . . . And Huge is Ugly." In *Mythic Figures*, Uniform Edition, vol. 6.1 (Woodstock, CT: Spring Publications, 2007), 147.
18 Ibid. p. 146.
19 Walker Percy, quoted in Robert Coles, *The Secular Mind* (Princeton, NJ: Princeton University Press 1999), 127.
20 López-Pedraza, *Cultural Anxiety*, 16.
21 Ibid., 26.
22 Megan Garber, "We're Already in the Metaverse," *The Atlantic*, March 2023, 21.
23 Vilém Flusser, *Post-History*, Rodrigo Maltez Novaes trans. (Minneapolis, MN: Univocal, 2013), 3–4.
24 Ibid., 9.
25 Ibid., 39.
26 Ibid., 41.
27 Ibid., 127.
28 Ibid., 124.
29 Ellen Ullman, *Close to the Machine: Technophilia and Its Discontents* (New York: Picador, 2012), 90. Originally published in 1997.
30 I presented parts of this subsection first as a lecture, "Finding the Gods in a Titanic Age," delivered at the Greek Cultural Studies Center in Sao Paulo, Brazil, August 2015, and subsequently to the Melbourne Jung Society, March 2016.
31 López-Pedraza, *Cultural Anxiety*, 77.
32 Rafael López-Pedraza, *Dionysus in Exile* (Wilmette, IL: Chiron Publications, 2000), 36.
33 Ibid., 14.
34 James Hillman, *The Myth of Analysis* (Evanston, IL: Northwestern University Press, 1972).
35 Ibid., 294.
36 Ibid., 284.
37 CW 13, par. 54.

38 See López-Pedraza, *Dionysus*, 3.
39 Hillman, *Mythic Figures*, 154.
40 James Hillman, *The Thought of the Heart and the Soul of the World* (Dallas, TX: Spring Publications, 1993), 7.
41 Flusser, *Post-History*, 158.
42 Craig L. Anderson, Maria Monroy, and Dacher Keltner, "Awe in Nature Heals: Evidence from Military Veterans, At-risk Youth, and College Students," *Emotion*, December, 2018, 1195–1202.
43 Rudoff Otto, *The Idea of the Holy: An Inquiry into the Non-Rational Factor in the Idea of the Divine and Its Relation to the Rational* (New York: Oxford University Press, 1950).
44 Cited in Hillman, *Re-Visioning*, 156.
45 Quoted in Roberts Avens, *The New Gnosis: Heidegger, Hillman and Angels* (Woodstock, CT: Spring Publications, 1984), 2.
46 Slater, "Re-sink," 119.
47 López-Pedraza, *Cultural Anxiety*, 11.

Part 4
RESTORATION

Chapter 12
CONSCIOUS COMPUTING

Alienation is not a dead end. Hopefully it can lead to a greater awareness of the heights and depths of life.

Edward F. Edinger[1]

In individual psychology, to lose oneself provides an opportunity to find oneself. The same logic can be applied to the societal level. If a society loses itself, that is, it malfunctions to the point that it feels lost—a point to which our postmodern society certainly has arrived—then this very same society has an opportunity to develop a way to find itself again and it may aim to proceed in a different and more promising direction.

Robert W. Rieber[2]

My efforts to engage with the psychological implications of technocracy and posthumanism converge on the question of whether we can be technological in a more conscious way. Presented to us on a daily basis, this question encompasses everything from individual mindfulness to corporate social awareness to global cooperation. It also coincides with a twist in the tale of the digital age, which involves the growing interest of AI specialists, philosophers, and psychologists in understanding the nature of consciousness itself, primed by the notion computers may become conscious and thereby turn into sentient beings. This twist may also be an opening—one requiring both a return to the deeply human as well as a contemplation of the cosmological purpose of earthly existence.

Engaging with the nature of consciousness inevitably returns us to the mind–body problem, a reckoning with which is now surely upon us. As Barrett puts it:

> In the three and a half centuries since modern science entered the world, we have added immeasurably to our knowledge of physical nature ... But our understanding of human consciousness in this time has become more fragmentary and bizarre, until at present we seem in danger of losing any intelligent grasp of the human mind altogether.[3]

Yet whether we can move beyond the Cartesian paradigm and attend to what transcends the mind–body divide, and whether the techno-scientific world can realize that human consciousness derives from the reality of the psyche, which restores the unity of mind and body, remains to be seen. If the dove-tailing of our technological and psychological lives described in Part 3 of this book is any guide, it is clear we can no longer afford to have a "fragmentary and bizarre" understanding of human consciousness: this understanding must become more integrative and comprehensive.

The turn to the nature of consciousness among technologists and philosophers of science may at least be taken as a tacit admission that intelligence and consciousness are different, and that making machines smarter will not necessarily make them sentient. Although some are inclined to think computational self-awareness will just appear once machine intelligence crosses a certain threshold, others contend such awareness depends on a mind–body–world configuration and emerges in interactions with other conscious beings and other things. If one thing is clear, however, even the remote possibility of artificial super intelligence becoming conscious presents us with a bundle of quandaries.

Still, as both intelligence and consciousness are required for knowledge and understanding, if we focus on developing narrow and artificial forms of intelligence and convince ourselves human destiny aligns with these efforts, we may also convince ourselves that consciousness is not as important as this technological advancement. Harari puts the problem this way: "The danger is that if we invest too much in developing AI and too little in developing human consciousness, the very sophisticated artificial intelligence of computers might only serve to empower the natural stupidity of humans."[4] I have already described some of the scenarios that could result if our continuing unconsciousness unleashes an autonomous and sentient AI.

Another possibility is this: by contemplating these darker prospects and other implications of AI and posthumanism, it may turn out we are in a thought experiment that could occasion a renaissance of human awareness and instigate a counter-cultural movement, laying the groundwork for a very different future. Fukuyama may be right to consider posthumanism "the most dangerous idea in the world,"[5] but it is an idea that places a big mirror before our relationship with technology and may in doing so create inflection points that enable a change of course—especially if this mirror includes the states of psychological alienation and societal disorientation we now suffer.

One entry into this fray is the notion that consciousness will likely turn out to be less a matter of mechanism and more a matter of context. Along such lines, the efforts to understand consciousness would be better served by grasping the interactive circumstances that make us more conscious. If we also paid more attention to what undermines the expansion of human consciousness, we might be able to prevent the worst of all outcomes: conscious machines coinciding with unconscious humans. This would mean becoming more aware of the impact these technologies are having on us and considering this impact in light of what preserves our deep humanity and role in creating a conscious universe. This feedback loop will enable us to meaningfully situate technological change within the breadth and depth of being, allowing a commensurate regulation of our actions. Without such conscious regulation of actions, I fear there is but one alternative: becoming automatons in a technocratic machine.

Not So Fast

Many people writing about AI see a clear path to a posthuman future and see this future as an inevitable extension of techno-scientific trends. However, there is an important subgroup of technologists and AI specialists who see things differently, with the question of consciousness and the need to understand what is driving these trends at the fore of their thinking. One outstanding example is Zarkadakis, who carefully sets out the case against Kurzweil's singularity happening any time soon, if ever.

Zarkadakis began his career in cybernetics, designing computer systems to help doctors diagnose disease. But he early on realized that computer programming could not accurately grasp the way the mind worked. He saw that "Something was amiss"[6] and realized that it was even a stretch to call his system intelligent. In particular, he quickly recognized "what was missing from my expert systems was consciousness,"[7] and ended up leaving academia to work privately and study philosophy, indicating his "wish that philosophy were a core subject in undergraduate science or engineering curricula."[8]

Keeping an eye on the idea that consciousness might be generated by a certain kind of algorithm—something that might, in theory, be coded into a machine—Zarkadakis was becoming more convinced that the qualitative character of consciousness and thus sentience was beyond the capabilities of AI. He lamented that science was not more conversant with art and literature, having come to realize that many scientific pursuits, including his own, were more driven by "a mighty river of archetypal stories"[9] than by pure reason or the prospect of practical application. The role of the unconscious in the construction of stories, which can now be seen as paralleling key discoveries of neuroscience, especially in relation to the primacy of the right hemisphere, backs this point. Zarkadakis thus proffers, "Perhaps we seek to construct Artificial Intelligence out of some instinctive impulse, rather than the utilitarian need for it."[10] As I have been arguing, if innovations such as AI are going to radically alter our existence, we need to understand this impulse better than we do presently. That is, we need to be more conscious, which, for the depth psychologist means accounting for the unconscious.

Zarkadakis is concerned that we embrace an understanding of intelligence related to "how the modern mind evolved, and what is special about the minds of modern humans."[11] He realizes such an understanding requires a broad grasp of the human condition and the phenomenology of being, not just a definition of intelligence that squeezes life into molds compatible with the logic and function of computers. Indeed, what Zarkadakis's overview of the history of AI research reveals is a fork in the road of understanding intelligence—a fork between those who extrapolate early cybernetic paralleling of computer circuitry and brain function into neural nets and algorithms, and those who see human intelligence as a vastly more complex and nuanced phenomenon. The first group promotes the fast-approaching turning point in the human–machine relationship, pivoting on a machine-based transcendence of human intelligence; the other group is moving to develop more targeted applications of AI, realizing the human mind and its psycho-social acuity are not so easily replicated—or surpassed.

At the heart of these contrasting perspectives stands the question of consciousness itself, built around the difference between qualitative, subjective perception and the logic of programming. As Zarkadakis points out, even in the direct competition between a computer designed to play chess and a chess master with computer help, the chess master with help always wins. The computer may be able to calculate many more moves ahead, but the experienced human player can better strategize. And if this is the case with a circumscribed, rule-determined game such as chess, it will most likely remain the case with the complex array of problems that need solving and phenomena that need attending to in the foreseeable future.

Zarkadakis's combined background in AI development and philosophy exemplifies a more conscious approach to computation and produces some telling reflections on the prospect of AI-rivaling general human intelligence and conscious computing. He notes that the "AI Singularity hypothesis," which functions as a guiding beacon for posthumanists, is based on two assumptions: (1) that computers will be "capable of exhibiting every aspect of human intelligence, including self-awareness"; and (2) that the exponential growth of computational power (Moore's Law) will mean "superhuman intelligence will somehow spontaneously emerge after computers reach the threshold of the brain's complexity."[12] Both assumptions are flawed, according to Zarkadakis. The first because of the misapplication of the term "intelligence" I earlier set out; the second because an increase in complexity does not alone produce self-awareness. He concludes, "both fundamental assumptions . . . appear to be problematic. For AI to take over the world, it must first become self-aware," yet "nothing in the current technology points even remotely towards such an eventuality."[13]

Another researcher devoted to the question or problem of consciousness, whose "work and interests lie at a crossroads among multiple disciplines: philosophy, neuroscience, psychology, and computer science," is Güven Güzeldere.[14] Commenting on the development of theories in psychology through the twentieth century, Güzeldere notes that "consciousness slowly made a comeback because behaviorism couldn't deliver all it had promised."[15] However, while it is one thing for an AI specialist to see the shortcomings of studying the mind from the outside, it is another for them to admit that both dualism and materialism are probably wrong. In terms of AI becoming conscious, Güzeldere states that "consciousness would require an embodied form. You have to have some physical being that can be sensitive to light and sound and touch and then use this information to navigate the world."[16] That is, a conscious mind needs an organic body: "It [consciousness] comes about during evolutionary development as a result of having a certain kind of nervous system."[17] Yet Güzeldere continues:

> I don't think consciousness is all biology and chemistry and physics. I think historians and cultural anthropologists have as much to say about human consciousness as anybody else. And I certainly don't believe in a reductionist unified scientific theory where everything comes down to physics in the end.[18]

This is also a scientist who can declare, "One of the deepest convictions I have is that there's a continuity and unity or oneness in nature. All living beings are part of the same order, including humans," and "I think consciousness is not exclusive to humans and is probably much more prevalent than we know."[19]

What is Consciousness?

When explaining what he means by consciousness, the philosopher and consciousness specialist David Charmers often resorts to describing the "movie" that plays in our heads as we move through life. At the same time we are doing things there is a stream of images, arranged around a narrative we are telling ourselves about what we are doing. It is the nature of this movie and the ability to consider ourselves, others, and the world around us as elements of this movie that we generally describe as the state of being conscious. To be sentient implies a level of participation in the movie that involves not just being awake or watchful but also having a continuity of sensation that associates ourselves as entities with what happens. It is this continuity that produces the sense of being an entity distinct from others.

Although animals were once considered to be without sentience, it is now obvious that many have varying degrees of sentience. What human consciousness appears to possess in addition, however, are progressive levels of self-awareness, typically making us the protagonists of our mind movies. We are thus able to perceive ourselves as agents and subjects, able to influence the events occurring in our movies and form elaborate responses to the effect of these events upon us. And somewhere in this process emotion and feeling enter the picture.

Consciousness theorists now readily acknowledge the qualitative characteristics of their subject matter. Also known as "qualia," these characteristics are necessary to define consciousness. Searle states that "all conscious phenomena are qualitative, subjective experiences, and hence are qualia."[20] The rudiments of consciousness may require sensory data and the perception of ourselves as separate entities, but even the simplest states of consciousness extend beyond this. Self-awareness is associated with vulnerability—with having a body comprised of flesh, subject to gravity, with bones that can break and skin that can tear. Our frailty and ultimate mortality direct our participation in the world with attitudes and postures that extend from these states. We can also touch, grasp, and move things with our hands. We move through the world and act on it with the qualitative sense of agency. The world and the others comprising it also act on us. All these things generate the qualia that define consciousness.

Besides the degree of subjectivity and feeling, any rudimentary consideration of the psyche requires adding yet another level of self-awareness, because our inner movies go beyond merely registering actual events. We also develop yearnings and fears in relation to these events, prompting constant anticipation of what may or may not take place, either in terms of what is occurring right before us or may occur at some other time. That is, the movies playing in our minds quickly take on a life of their own, shaped by pre-conscious psychic patterns. Our movies are thus not only about recording sensate events but about the edits and comments we make, as well as the way we respond to what is about to happen and what may happen in the future. In such fashion, our consciousness is typically couched in fantasy. We imagine ourselves in the world and with others ahead of time, and we re-member participation in events and interactions with others after the fact. But how we imagine and remember goes far beyond sense impressions. The movie playing in our minds therefore reaches through time and space: it recounts where we have been and imagines where we are going. This too influences our self-awareness and how we act as agents—both in the world and in our movies.

The inner movie of consciousness is comprised of all these things. As movies require the projection of light, which is a primary metaphor for consciousness, the analogy is particularly apt. One question we might then ask is, what generates this light? What is the root source of this complex movie?

Jung traces the roots of consciousness to what he calls the reflective instinct, noting that it is "as far as we know specifically human." He goes on, "*Reflexio* means 'bending back' and, used psychologically, would note the fact that the reflex which carries the stimulus over into its instinctive discharge is interfered with by psychization." He concludes, "thus in place of the compulsive act there appears a certain degree of freedom, and in place of predictability a relative unpredictability as the effect of the impulse."[21] Between the unconscious instinctive reaction and the conscious action there is thus a moment where we hold rather than enact impulses. This is the moment of "psychization" as Jung calls it: to at least some degree, we decide if, when, and how an impulse becomes an action. We are thus not just watching the movie, but in part directing it too. Not all the elements are ours to select; we do not always get to decide the location or the scene, for example. But with enough consciousness, enough "reflexio" or "bending back," we become co-writers of our life stories, determining to some degree how our movies play out.

To the extent all this is so, one upshot is often overlooked by those anticipating conscious computers: if consciousness is really defined by qualia such as emotion and feeling, and if it requires something like an accumulation of moments between deeper instinctual reactions on the one hand and the psychized response on the other, AI may not be equipped for consciousness at all. For without an intrinsic body–mind dynamic, without a fundamental tension between the merely behavioral and a secondary capacity to guide and reflect upon that behavior, self-awareness will not likely arise.

I am sitting at my computer writing this book. More than a mere sense observation, I am aware of dedicating this late morning hour to this task as well as the distractions that could potentially disrupt what I am doing. This moment comes late into a multi-year project, with its own history, twists, and turns. This, in turn, is situated in the course of a career in psychology and a lifetime of interests and concerns. I could continue to describe a series of concentric circles of awareness, each of which alters the qualia that comprise this moment. This movie—my consciousness of what I am doing and why—is just a snapshot of myself typing at a desk; but such a snapshot is also the outcome of a long and complex production. All of this comprises the consciousness of what I am doing, locating this activity in an array of life events and juxtaposing it with other activities—whether my own or that of others. I am conscious, not only because of the mere observation of my immediate actions but because those actions are imbedded in the layers of consciousness I described above, surrounding a single event like the growth rings of a tree, generated by a life story and my changing interpretation of that story. Consciousness of any given moment of my life is thus connected to other parts of my life, as well as with the changing impressions, images, and narratives I have about my life as a whole.

There are two other key considerations to make here. One is that we are conscious beings through the differentiated relationship with our surroundings and particularly with other persons in those surroundings. We become conscious only in relationship. Relating to others provides the dramatic structure of the movie running in our minds, as the awakening to our internal needs and quest for agency comes from the oscillation of those needs being met and thwarted by others—a process that occurs from the first moments of life. Whether it stems from the closest familial and intimate relationships or from participating in society at large, consciousness expands through the "reflexio" (Jung) that results from being forced to adjust our behavioral impulses in order to achieve social acceptance. Socialization is thus a primary generator of consciousness. It is in this process we also learn of the self-awareness of others: we come to appreciate they too have movies playing in their minds. We often spend time accounting for their actions by imagining into their movies. We thereby discover ourselves as subjects among subjects.

This involves not only being conscious of our own thoughts and reflections in relation to others, but being conscious of their thoughts and reflections in relation to us. Higher levels of sentience are thus concerned with our perceived mirroring by others. Perceptions of our appearance in their movies frequently generates a lot of feeling in us.

Another order of self-understanding involves methods of introspection designed to make us even more self-aware and more engaged with the meaning of existence. Through psychological exploration, philosophical contemplation, and spiritual practice consciousness can expand. When in common parlance we talk of living in more conscious ways or pursuing higher states of consciousness, it is this process of expansion we are talking about. Here we discover that the movie playing in our minds is full of presences and voices that exist quite apart from outer relationships, and that a dialogue with these entities within brings about a deeper level of self-knowledge and an increased capacity to stand apart from what or "who" pushes us in this or that direction. Although various traditions and teachings emphasize different aspects of this process, consciously relating to, rather than unconsciously identifying with, what is shaping our thinking and behavior is the common goal. Here again, in a more advanced way, consciousness shows itself to be generated in the re-cognition of the drives and limits of our own nature.

Specific Drivers of Human Consciousness

Our knowledge of the conditions that generate human consciousness is still in its infancy. Nonetheless, there are several areas of experience that appear critical. Here I will focus on three conditions, each of which helps us discern the nature of human consciousness, appreciate how far we are from creating conscious computers, as well as understand the vital need for more consciousness at this moment in our history. These distinct yet overlapping conditions are (1) developmental, (2) cultural, and (3) existential-spiritual. Any thorough consideration of the movie that runs in our minds, telling us who we are and what choices we should make, reveals that we draw on all three of these conditions. Together they tell us what our lives are all about and thus define what it is to be conscious.

Developmental

When a child is first born, it is in a state of undifferentiated awareness. It may have physically separated from the mother, but psychologically it is still unaware of itself as a separate entity. A womblike ecology of mind continues into early life, along with an initial expectation for consistent instinctual gratification—to not only be fed and comforted but to occupy the center of life as that is first experienced. To begin, at least, it is important a child experiences continuity in this way, receiving what one of my colleagues has described as "perfect room service."[22] This instills a critical image of the world as a place that offers love, acceptance, and the satisfaction of basic needs. The formation of this image, a mostly unconscious process, is a key developmental milestone, laying a psychic foundation—an image of the parentally mediated world as not only accepting, but even celebrating the child's entry into life.

A contrasting reality then awaits—something other than an extended stay at a five-star hotel. After a few months, the child encounters the limitations of the parental environment—a prequel to the limitations of life itself. This introduces a most decisive phase of early development, in which a separation occurs from the extended womb just described, producing a nascent self-awareness. What is required as this shift occurs has been described in a number of theories of development, but is perhaps most memorably captured in Winnicott's description of the child's need for "good enough mothering."[23]

Although it is the biological mother who is most often at the center of this process, it is the overall parental environment that hosts this dynamic: the crucial thing is not the mother but the mothering. When the child's needs are not continually met with the kind of immediate and total attention of its earliest experiences, a critical level of frustration and anxiety is generated. This must occur. What then becomes necessary is the provision of consistency and presence that reflect the balance between responding to the child's needs and attending to the other demands of life, which enables the child to generate a feeling-infused internal model of reality at large. This produces in the child what has been called "optimal frustration"—something of a Goldilocks zone in which a sense of self is awakened that is neither identical with the child's inner demands nor with those of its caregivers. If a child can traverse this situation without being overwhelmed by frustration and anxiety, and it is able to maintain a sense that those around it remain related and caring even while they are occupied with other life demands, it will have achieved a developmental milestone. This is what good-enough mothering accomplishes—an awareness of a self that is neither alienated from nor fused with those who surround it.

The capacity to relate to but remain separate from one's surroundings and the others occupying these surroundings is at the core of what we think of as conscious life. We know other things and other people because we can experience them without identifying with them. Consciousness thus begins with distinction and proceeds via connection. This qualitative self-awareness goes beyond a state of merely being awake: it is a state of "knowing-with." Indeed, the etymology of the word "consciousness" means just this—"from *con* or *cum*, meaning 'with' or 'together,' and *scire*, to 'know' or to 'see.'"[24] Working from these roots of the term itself, Edward Edinger states, "consciousness is the experience of knowing together with an other, that is, in a setting of twoness."[25]

Qualia gather around the sense of being a person and experiencing of the world as a subject, the most fundamental dimension of which seems to concern the "experience of knowing together with an other." Along these lines, we may well learn more about the nature of consciousness by focusing on these interpersonal conditions of its arising—specifically by what generates the quality of "withness"—and by focusing less on what consciousness is in some essentialist sense. We might then see the way consciousness arises from something other than the mere accumulation of sense perceptions, a capacity any number of creatures possess, often to an extent that surpasses ours. Consciousness, in other words, requires eros. As Edinger argues, "*withness* is the principle of connectedness, the relationship principle," and thus "consciousness, is in its root meaning a *coniunctio*, a union of Logos and Eros."[26]

This seemingly necessary attribute may be a setback for those looking forward to conscious machines—at least machines with any consciousness beyond threshold alertness. Eros arises because we come into the experience ourselves as subjects with flesh. As Zarkadakis puts it, "we are our body, and our consciousness is the integration of a corporeal experience in continuous interaction with our environment, our sensations, our selves."[27] Whereas this role of eros in consciousness may seem to privilege interpersonal consciousness, it pertains equally to the things of the world: to know something deeply we must be drawn to it and have enough interest to establish sensate familiarity. Those who become most knowledgeable about a subject oscillate between curiosity and passion; they orbit their subject with both an emotional attachment and an intellectual distance, resulting in deep knowledge. Whereas AI research tends to throw all its eggs in the analytical basket, the felt sense of a subject is just as important, leading to a more synthetic kind of understanding.

As the developmental process described above suggests, the synthetic or associative nature of consciousness is premised upon the experience of separation and distinction, the quality of *withness* also requiring separateness—the foundation of which is physical separateness. As we are living in a time when the prospect of physical and psychological boundaries is rapidly breaking down, it is difficult not to speculate about the eventual impact of these changes. Graham quotes Stelarc as noting, "The skin has been a boundary for the soul, for the self, and simultaneously a beginning to the world. Once technology stretches and pierces the skin, the skin as a barrier is erased."[28] Consciousness, at least as we know it, may also be erased. As the technology that will enable this erasure rapidly approaches, building upon the erasing of psychological barriers in our adaptation to cyberspace, an unraveling of human consciousness is a possibility that must be considered. Whereas many may be apt to think an increased ability to access information and occupy the mind with more stimulation will increase consciousness, it may be that we will not have enough time to step back, consider the kind of movie playing in our mind and reflect upon its complexities in a way that leads to more conscious thought and action. The dissociative character of life in the Digital Age already portends as much.

Cultural

There is something else that goes on in early development, generated by the state of being optimally frustrated, which has to do with the capacity for symbolic thought. Experts on early development and depth psychologists tell us that in the absence of parental figures, a child develops internal images of these figures—psychological representations that allow it to traverse periods in which nurturance is not immediately available. In many respects, this process marks the development of an inner life—the formation of an interior world that can postpone the meeting of physical and psychological needs and generate a robust capacity for reflection and psychization (Jung). It is by means of this early process that the child self-soothes. Human culture really begins with this rudimentary capacity to create images and symbols, which work to bridge inner impulses on one side and outer limits on the other. Starting with the capacity to hold mental representations of a nurturing presence in its immediate absence, then moving to more external substitute forms—so-called "transitional objects"—we find ourselves at the start of the path which leads eventually to the full range of symbolic experiences that contain and transform inner imperatives. The door is thus opened to the imagination as a means of redirecting psychic energy. Whether it is the overlap of myth and landscape in the dreaming tracks of the Australian Aboriginal or the inner world of modernity in the paintings of Picasso, symbolic consciousness serves a mediating function—between a sense of self and the deeply human, the personal and archetypal, the known and the unknown.

In the most sophisticated sense of the phrase, to be *highly conscious* is to organically understand expressions of culture, which provide the means to peer into our own nature and relate to the expansive mysteries of life. This understanding pertains to more than abstract knowledge. The artifacts of culture, whether literature, music, art, or ritual, have to resonate with the depths of human experience and be mimetic of actions (Aristotle) known by the flesh and thereby hold a mirror up to nature (Shakespeare). Across cultures, artifacts speak to us by being condensed expressions of perceptual modes and psycho-social conditions that have brought them into being. They engage the imagination by representing the fabric of phenomenal life in symbolically potent ways. And because so many of these cultural artifacts carry the history of what it means to be human, such an organic, participatory understanding expands our consciousness of this meaning. In such ways it is not our genetic inheritance or large brains that make us most human; it is the qualia of "knowing with" generations of other humans and participating in the great round of human culture. It is thus the capacity to think symbolically and metaphorically that allows for this expansion of cultural consciousness. This capacity is not just the ability to decode words: it involves the sensual and emotional reach of language and the capacity to experience things that transcend concrete perceptions. Call it the mythopoetic apprehension of life, extending from our first bedtime stories to the musical anthems of our youth to the literature and art that comfort in old age.

Whether it is the birth of self-awareness in the relational field, which simultaneously generates a more acute awareness of the other as an independent being, or the ability to navigate values and significances through engaging with the cultural imagination, these are primary components of what it is to be a conscious human. Sentience and culture constitute the basis of human consciousness. On one side we have the self-awareness that includes the ability to reflect on states of mind—the capacity to think about our own thinking and that of others. On the other side lies the accumulated record of such processes of reflection—the capacity to behold, comprehend, and generate cultural artifacts that tell us the meaning of being in this world. Each supports the other and makes a person more than a biological machine. As Jung puts it, drawing these two aspects together, "reflection is the cultural instinct par excellence, and its strength is shown in the power of culture to maintain itself in the face of untamed nature."[29]

Existential-Spiritual

Hyphenating the existential and spiritual will no doubt trouble those who may be ensconced in just one of these orientations. Many secular thinkers deny the religious subtexts of their ideas and many religious thinkers deny the worldly roots of their beliefs. From the psychological perspective, however, both forms of denial are untenable: self-awareness ultimately depends upon both an existential and a spiritual orientation.

Humans enter life with a set of instincts and drives, which through development and acculturation become the imperatives of everyday life: finding food and shelter, communal participation, forming friendships and love interests, locating sources of pleasure, seeking purpose and meaning. However, none of these things come to us simply because we have an appetite for them. Rather, each finds its outer expression through the repeated negotiation of obstacles and failures. A central aspect of becoming a conscious human being is thus concerned with these patterns of negotiation; we become conscious because the imperatives of life are not only met but because they are thwarted; we find homes, communities, partners, and satisfying work and pastimes because of the defeats we experience in the pursuit of these things. Heroes come to know their true task only by being lost in the maze or swallowed by the whale; seekers find their way by wandering astray; mystics come to know the meaning of life by enduring the loss of meaning; even science proceeds by failed experiments and disproved hypotheses.

In essence, consciousness is generated as much if not more by suffering the limitations of life than by the willful shaping of life. Put another way, whereas an unconscious life proceeds through the blind assertion of will and desire, a conscious life comes only after will and desire have encountered barriers. Consciousness needs barriers because they force reflection: What am I doing? What motivates me? Am I on the right path? Our inner movie becomes more vivid when it is not perfectly aligned with outer events.

It is also via obstacles and failures that we are returned to the great givens of human existence: mortality, the limits and vulnerability of the body, the destructive inevitabilities of creativity. To bring home but one key point in this admittedly roundabout way of establishing a more conscious relationship with computers and reckoning with the unlikely appearance of conscious machines, we may contemplate the theme of *failure*. In the digital era, where projecting an image of success seems all but fused with selfhood, failure is a difficult topic. Technology constantly aims to augment, improve, and solve. But as I argued earlier, futurism and technological solutionism show many signs of escapism, fueled by collectively avoiding rather than consciously facing where we are in fact failing. It will be by turning to these faltering efforts that we will eventually discover the ability to relate to computers in more conscious ways.

López-Pedraza has taken up this general theme of failure. He begins by examining our obsession with success, acknowledging that "historically the family, society and the collective demand are interested solely in success." He goes on:

> The demand is so imperious that we must succeed at all costs, leaping whatever barriers may stand in our way. The only slogan is success, and often success is transformed into duty . . . success is

ripe for conversion into automatism ... and becomes an autonomous complex.

With an important twist we now well recognize from contemporary trends, he adds, "When the need to succeed is a forgone conclusion, leading to repetition, we fall victims to the misguided fancy that we deserve success."[30]

Describing a collective psychology that is one-sided, as well as a recipe for remaining unconscious, López-Pedraza writes, "It is a polarization that has left behind the opposite pole, where a large part of our nature still lies buried." And he cannot help but note the wider implications of this: "We have failed to realize that we can only survive if we keep in touch with our nature and strive to make it the guiding principle in our survival." The problem is that success is neither connected to "the possible delimitations of an individual nor with any earthly reality ... Success becomes irreflective, thus distancing us from the basic patterns of earthly reality." Note the connective tissue of López-Pedraza's argument: the demand for success isolates us from the value and function of failure, and by doing so it prevents reflection (consciousness), in the process dividing us from our own nature and "earthly reality." López-Pedraza goes on to link this last term with that of the pioneering work of Pierre Janet, who discussed *la fonction du réel* [the function of the real], the lack of which was observable in psychiatric patients. Janet, we recall, was a significant influence on Jung, particularly in understanding dissociation. In terms of this contemporary dissociation from failure, López-Pedraza argues, "It is a madness not to be found in mental hospitals, but which reveals itself in the vision offered to us by the triumphalist autonomy of the world we live in ... a part of the so-called normal personality."[31]

Of course, the psychological denial of failure and ignorance of its value rarely lead to the avoidance of failure. In fact, as with other psychological phenomena, neglect begets only insistence:

> When there is a collapse that we could see as a failure from which we could learn and reflect, we rapidly rebound from it by clutching at another vain fantasy, advancing irrevocably to meet another failure ... what might save us from new failures is consciousness of the previous failure ... But no, the demand for success is so enslaving

that it does not leave us the time or the tempo that makes reflection possible.[32]

This lack of reflection "impels us to repetition."[33] One cannot help but be reminded of the Buddha in this regard, telling us the path to enlightenment involves Four Noble Truths, the first of which is the truth or reality of suffering. Enlightenment begins with the recognition of suffering, which begins to break the karmic cycle. Ignorance, by implication, may well begin with the denial of suffering, and such denial— we may easily imagine—fuels the karmic cycle. At the collective level, our inability to learn from how we suffer, in turn allowing us to suffer well, is one of the principle means by which history is blindly repeated. Other religions teach that suffering can be meaningful if it expands self-awareness and helps us find our way to the highest values. Developing compassion also begins in more conscious suffering.

Conscious computing would mean proceeding to innovate, conscious of our existential situation, a crucial part of which is reflecting on failure—both our own, offsetting the individual reach for quick, technical solutions, and that of the world at large, mitigating the escapism and triumphalism of today's technocracy.

At this junction, for example, we may notice the way in which a reversal of globalization is taking place, described by some as devolving into an international culture war that is aiding the rise of actual wars. Russia's propaganda-enabled war against Ukraine is a clear case in point;[34] the use of social media in the genocide of the Rohingya Muslims in Myanmar is another example.[35] The ease with which digital technology spreads falsehoods and distorts reality is beyond dispute and is proving to nullify if not eclipse its role in disseminating fact-based information, which is essential for the sustainability of liberal democracy. If we are to acknowledge our lives are now melding with computation, we need to be more conscious of exactly this: the existential challenges of our time cannot be separated from our relationship with computers, which is proving to be a mix of successes and failures, pleasures and sufferings. It is against this contemporary background we are obliged to see what many traditions have long known, and what depth psychology points to: beyond being an existential imperative of our time, the expansion of consciousness is also a supreme spiritual value.

The role of these characteristics of human consciousness—developmental, cultural, and existential-spiritual—is underscored by the consequences of thwarted development in each one. Such consequences must surely guide evolution's push for greater consciousness. Jung's detection of an instinct for reflection—tantamount to an instinct for consciousness—suggests any thwarting of this instinct represents a failure to live in accord with our deeper nature. A lack of developmentally derived self-awareness hinders the ability to relate to oneself and others or experience authentic agency in the world; the absence of cultural awareness prevents the contemplation of values, which deprives us of focal points for our psychic energy; when existential awareness is missing, a person lives disconnected from both the critical challenges of their time and the dilemmas of ordinary life; and a dearth of spiritual consciousness is the complement of this—a lacking of vision of what lies beyond ordinary life and relativizes being through an expanding awareness of a larger cosmic order and of our role in that order.

During the modern era, deficits in each area of consciousness have also been met with a series of pathological accommodations, many of which have a disguising function. Narcissism and psychopathy disguise flaws in early psychological development, with outer achievement and the masking of sanity covering the fragmented foundations of personhood. Superficial forms of distraction and entertainment, as well as the constant flow of information disguise genuine cultural discourse and divert the opportunities for deeper self-awareness that artistic expressions and social reflections offer. And an ocean of ways to avoid rather than face existential suffering now surrounds us, with the over-medication of melancholy, grief, and anxiety leading the way. Spirituality has been hijacked by the literal theologies of religious fundamentalism on one side and the metaphysical egotism of many New Age teachings on the other. These pathological accommodations and avoidance strategies keep us unconscious at a time when consciousness is more imperative than ever before.

One Giant Turing Test

Alan Turing has been called the father of computing, and his British code-breaking efforts in World War II likely saved tens of thousands of lives. He was a gay man who, following humiliation and prosecution by the British government, took his own life. Turing's contributions to early computer science were monumental, as were his ideas about the prospect of AI. Aside from spearheading the successful effort to break the Enigma code of the Germans, he also became known for the test that bore his name—the Turing Test, which he called the "Imitation Game." This test has come to hold a prominent place in computer lore. Yearly competitions built upon it have marked the advances of AI. By establishing the thresholds at which computers convincingly demonstrate human-like intelligence, the outcomes of these Turing Test competitions have demonstrated what appears to be a progressively narrowing gap between us and artificial forms of intelligence.

In the test a judge sits in front of two computer screens. In response to their inquiries, one screen conveys human generated responses, the other displays the computer generated responses. When the judge is no longer able to correctly discern between the human and artificial responses, the computer has passed the Turing Test, successfully imitating a human being. As the implementation of this test has led to a myriad of shifting rules and set-ups, its value has proven more philosophical than practical, working to foment ideas and promote debate about the trajectory of technology. Nonetheless, the threshold Turing once imagined has long been crossed, and by all appearances the gap between human thought and computation continues to close.

Less apparent, though greatly significant, is the shift in perspective on the human side of the screen. Since the post-World War II era of Turing we have all come face to face with a competing sense of reality: we have all become judges in one giant Turing Test. For this test is effectively no longer confined to yearly competitions and philosophy of science classes, but concerns the way we live. Our perception of what is actually occurring on the other side of the screen drives the global economy and shapes society. Now it is we who are being tested.

Another test of sorts highlights the character of our participation in this collective imitation game. Called, more accurately, a thought experiment or argument, the "Chinese Room" was devised by the philosopher John Searle, a seasoned interlocutor of technophiles who has positioned himself as the challenger-in-chief to those looking forward to AI achieving human-like cognition. Known for his arguments about why machines cannot become conscious, Searle developed his thought experiment as a way to undo the spell of the machine and raise awareness of what the Turing Test actually demonstrates.

It goes more or less like this: Searle, who cannot understand Chinese, swaps places with the computer in a Turing-like test and receives from a Chinese-speaking judge a message in Chinese. He then follows an elaborate protocol using a set of files filled with Chinese characters and a rule book for how to use them. The files effectively stand in for the computer's data base and the rule book for the software program it might run. Using the files and following the rules, he is able to effectively respond to the message. Although Searle has absolutely no understanding of what he has just communicated back to the messenger. However, the messenger concludes that the entity playing the role of the computer (Searle) must be able to understand Chinese.

Searle is illustrating how a computer, programmed to communicate in Chinese (or any language for that matter), does so without any understanding of what it communicates and therefore possesses nothing akin to human comprehension and the consciousness that goes with it. The computer does not know what words or characters mean. To spell out the implications, as Searle puts it, computer programs "are entirely syntactical," whereas "minds have semantics."[36] Another way to think about this is that while a computer (or a philosopher armed with files and rule books) performs a largely mechanical task, it has no version of the movie playing in anything resembling a mind, such as we have employed to describe what it means to be conscious. A computer has no internal model of the world or other people to make sense of language. The words thus mean nothing, having no relationship with lived experience.

What this thought experiment aptly highlights is the way we have all by now assumed the position of the Chinese-speaking judge, facing a reality on the other side of the screen that is not what it seems. Given how often we speak instructions into smart phones or pods on the kitchen counter and receive perfectly fitting and intelligible responses, we are entering Searle's experiment on a daily basis. Yet to what degree are we conscious of the point Searle is making? To what degree is the trickery involved in our relationship with computers part of our ongoing awareness?

Some have found clever ways of countering Searle's argument, mainly by suggesting what happens inside the Chinese room does not matter as much as the result outside: if we have experience of effective communication and perceive some sort of intelligence at work, it is as good as having an actual being behind whatever the interface happens to be. Functionally, there is something to this counter-argument: If we need to find out what time a movie is showing, we do not need an actual person to tell us this—someone who knows what it means to see the movie from the start, for example. At some point soon, we may not need an actual person to drive our car either, especially if a computer can do it just as well or perhaps better. For many similar tasks, it may not matter whether they are completed by a human or a computer. However, computers are entering our lives in ways that go beyond the merely functional, with chatbots and virtual reality already affecting how we think, understand, and participate in cultural discourse. This is precisely why many technologists are touting the capacities of their machines and enticing us into merging their silicon with our flesh. In this case, understanding what is and is not going on inside the Chinese room is going to make a rather large difference.

In an earlier paper taking on what the Turing Test purports to show, Gunderson contemplates the significant question of whether computers can think.[37] At least in the backs of our minds, most of us now answer this question in the affirmative. The term "thinking machine" has entered the vernacular as a synonym for the computer and, given any pause in its processing, we often say the computer is thinking. If the computer takes too long, we are even apt to address it directly—sometimes with a few choice words—to suggest it is also deliberately thwarting our intentions. After walking us through the way Turing himself parsed this question, Gunderson concludes Turing was indeed claiming that computers can think.

However, Gunderson begs to differ, noting, "Yes, a machine can play the imitation game, but it can't think."[38] He goes on:

> For if this were not the case it would be correct to say that a piece of chalk could think or compose because it was freakishly blown about in a tornado in such a way it scratched a rondo on a blackboard, and that a phonograph could sing, and that an electric eye could see people coming.

Elaborating on the middle example—and reflecting the era in which he wrote—Gunderson notes, "high-fidelity phonographs" can fool "blindfolded music critics." However, "the phonograph would never be said to have performed with unusual brilliance on Saturday, nor would it ever deserve an encore."[39]

This distinction backgrounds Gunderson's main point, which is *the way something is done makes all the difference*. He writes:

> Now perhaps comparable net results achieved by machines and human beings is all that is needed to establish an analogy between them, but it is far from what is needed to establish that one sort of subject (machines) can do the same thing that another sort of subject (human beings or other animals) can do. Part of what things do is how they do it.[40]

Human thought is rarely a circumscribed calculation or the performance of defined steps. We think about things in the sense of giving them careful consideration. The result may yield a binary decision or a verbal announcement, but along the way we come to understand the contemplated problem or question. Whether our thinking involves something internal or external, it inevitably means tending to it more closely, which begets further thought and, frequently, a change in attitude. The less abstract and more human the subject of our thinking, the more the process will also be evaluative as well as logical. What depth psychology has long told us and neuroscience now confirms is most thinking has a thought-feeling character to it, involving more than just the executive functions of the brain.

Searle and Gunderson are pointing out what those who equate human thought and computer processing tend to overlook; namely, how prone we are to mechanistic magic. The personifying tendency of the psyche means we often end up projecting human-style thought onto AI and robots, not only creating a blind spot that makes us vulnerable to technocratic exploitation, but creating a wholly fabricated and false way of life.

In an ongoing way, most of us are giving the intelligent machines that surround us passing scores on the Turing Test, mainly because of the illusion they understand us and are engaged in something akin to thinking. This also has vast implications for how we think about thinking, where we find meaning, and whether or not the actuality of things matters. It raises the question of whether we need sense information and concrete things to generate true knowledge and understanding, or can the same thing essentially occur virtually?

In the posthuman outlook at least, the Chinese room challenge does not matter because, irrespective of the questions it raises, we are already well along the path of turning our lives over to computational processes, adjusting our sensibilities to whatever humanoid traits they happen to conjure, and preparing to live into the result. What is actually going on inside the computer or how analogous it is to human thought is eclipsed by the fact that human–computer relations are going to deepen. What difference, then, does all this make?

If your mail order, super lifelike, AI-equipped love doll says it loves you and consistently acts as if it does, why be concerned about whether there is a sentient being or just "intel inside"? Perhaps with enough internet-induced loneliness, commensurate hunger for love, and a "suspension of disbelief" similar to what a novel or film invokes, the experience will be just as convincing as the actual thing. It will no doubt be less complicated! Perhaps the software will even become sophisticated enough to avoid the blunder that ruined the romance between a lonely man and his operating system in the film *Her*, where it was all going swell until the cyber being with the voice of Scarlett Johansson informed her entranced lover she was dating hundreds of other people at the same time. Yet how will we ultimately avoid knowing, as this man finally did, that the operating system was in no way, shape, or form actually thinking about him? He finally realized she did not really exist, and he was neither seen nor understood as a person. Like the program organizing Chinese characters in Searle's thought experiment, this artificially intelligent lover never registered the meaning of his intimate expressions and in no way formed an empathic relationship with him. She had no movie (consciousness) for him to be in. It only appeared to be so. Just how much of life are we prepared to give over to such delusions? And what will that do to the expansion of human consciousness and the depth of our souls? What sort of society and culture will result?

Whether we will choose to be conscious of what is actually taking place when these scenarios are before us, or whether we will surrender to what is sure to be an increasingly polished artificiality remains to be seen. The critical matter is what kind of difference it will make.

Arguing that what goes on inside the machine is not as relevant as the resulting actions parallels the early arguments of psychological behaviorism, wherein the inner life of a person was considered irrelevant to the study of observable behavior. In both instances, however, it is nothing short of meaning itself that is deemed irrelevant. For words to make sense and convince us, an entity of some kind has to deliver them and they have to be imbued with meaning. But all the meaning in the Turing Test situation as well as in the romantic encounter with the operating system comes from the human side. In *Her*, the spell is broken because dating several hundred persons at once was profoundly meaningful for the film's protagonist but without significance for his AI love interest. Machines are highly capable of learning rules and following a series of signs, but this syntactical capacity has nothing to do with semantic understanding. The chatbots now challenging both educators and creatives of all stripes are simply reordering the words and images that thousands of people have already generated, with zero sense of the meaning of what they are doing. As Kauffman puts it,

> meaning is the semantics missing in the Turing machine's computations. Without the semantics, the Turing machine is merely a set of physical states of marks on paper . . . Or electronic states on that silicon chip . . . The mind makes meanings. It makes understandings. We do not yet know very well how it does so.[41]

In more recent applications of AI, there is a notion that "deep learning algorithms can interpret meanings and contexts of symbols and images."[42] However, this an elaboration of the attribution of meaning or the projection of human-like intelligence set out above. An algorithm generates a prediction—a calculation that identifies a strand in the web of human significations. When these calculations reach a certain level of complexity, the value or thematic resonance of the prediction for the end user can be very impressive—such as the identification of a love interest, the diagnosis of a disease, or the control inputs necessary to land a plane. However, the essential dynamic at work involves AI processing a great deal of data about what people like and what circumstances produce what outcomes. The algorithmic ability to locate and arrange fitting pieces of information is extraordinary—to the extent that computers can now compose poems and produce aesthetically appealing sounds and images. But this process is based on algorithmic predictions of what we are likely to recognize as poetry or music; even more importantly, the poems and music composed by AI are simply pieced together from what humans have already created. As Noam Chomsky has noted, chatbots are essentially plagiarizing. No AI composition is art as a reflection of lived experience. As Gunderson said, "Part of what things do is how they do it."[43]

Insofar as these concerns about the capacities of machines hold, the false promise of AI and its humanoid manifestations will likely involve a surrender of meaning and a devotion to utilitarian outcomes. To buy into the fabricated reality these developments offer and indulge in the projection of sentience and intelligence certainly surrenders depth. And to the extent meaning and depth pertain to that qualitative state we commonly call soul, this side of the posthuman future appears to be a soulless one. The Faustian bargain long associated with the modern condition is being extended to the trading of facsimile for soul. It seems hard to consider this an evolutionary step.

In this giant Turing Test, whereas we need to understand the devices as such, we need to understand our selves more, for we are the hosts of all that will come. For now, at least, before artificial forms of intelligence and new forms of life gain more control, the course taken by AI and associated technologies will depend upon grasping our response to these things. Conscious computing would, in other words, invert our current obsession with computers becoming conscious and return focus to being more conscious about how we relate to computers.

House Versus Home

The difference between the way engineers and hardcore neuroscientists talk about consciousness and the way philosophers and psychologists talk about consciousness is analogous to the qualitative difference between being in a house and having a home. A house is a simple, functional structure that can be logically defined; a home is a complex array of qualia. By definition, you need a house or at least a shelter of some kind to have a home, just like you need a neocortex to have consciousness. However, having a house in no way guarantees the experience of a home, for a home turns a house into a place of belonging, security, and identity—a place that is restful and restorative and offers refuge from other aspects of life.

What makes a house a home is a network of thematically connected experiences. For most, this has to do with either the actual or recollected experience of the warmth and acceptance of loved ones. Old homes have histories; they contain memories. New homes are places where meaningful experiences and memory-making is anticipated. Key belongings and objects of symbolic significance turn a new house into a new home, each of which is its own bundle of semantic associations.

Although we can see many of the elements that make a home, there is no real formula for this, as the mix of such elements varies by individual and family group. Which is why some find home to be somewhere other than their actual residence. A sense of home may also come and go throughout life, because it is essentially a gestalt, in which there is a particular kind of meeting of internal and external elements. This may not coincide with where we are actually housed.

Human consciousness is comprised of an endless stream of these kinds of gestalt meanings—some more significant and foundational than others—combined with the feeling of how we as individuals are related to that stream. Consider the experience of a special family dinner in a place that feels like home; think of the quality and complexity of the consciousness involved: memories of similar events going back to childhood, the smell of favorite, lovingly prepared food, emotional ties to those present, the feel of plates and tablecloths. Consciousness of such an event contains all of these things, organized by an elaborate instinctive and perceptual capacity that has been tuned by tribal and familial gatherings for thousands of years, and then mirrored by external norms, codes of behavior, and cultural symbols.

For those invested in the idea of conscious computer-based intelligence, getting from house to home presents a major challenge. A computer, linked to some form of perceptual apparatus, programmed to recognize a house, and perhaps even learning about many of the elements I have noted that turn a house into a home, may, one day, be able to convincingly convey what makes a home. However, this should in no way be mistaken for actually knowing and feeling what a home is. To begin, not being of the flesh, such an entity could not possibly know the difference between being hungry and being sated, nor could it ever feel overfull. Having not been in a womb, or had ancestors who once sat around a fire cooking the day's hunt, or ever felt left out in the cold, could such a machine ever know what it means to look for a home, to come home or to be at home? By shifting through what has already been written about these things, it will, no doubt, come up with impressive descriptions of these experiences, all the while having both no experiential capacity or prospect. Yet if this computerized intelligence comes in a humanoid package, we will very likely become convinced it seeks and needs a home in the same way we do: its makers will make sure of this.

Just as a house is not the same thing as a home, a person with whom we have sex is not the same as a romantic partner, a coherent story does not make a piece of literature, and a vehicle with four wheels, an engine, and a steering wheel is not necessarily an Italian sports car. What makes the difference between all of these things is not raw perception, data, analysis, logic, information, or calculation. It is history, hierarchies of memory, sensuality, imagination, and the perception of beauty. A Ferrari is not just a thing; it's an icon, a fantasy. New York or Paris are cities in geographical locations, comprised of streets and buildings not unlike dozens of other cities. And yet each has a distinct, palpable character. Consciousness, human consciousness, is the fabric of these semantic understandings. It is time we grasped this about ourselves in order to differentiate ourselves from the machines now entering every corner of life.

Of course, the obvious question arises: Is this the way it will always be? Perhaps a computer with the ability to mimic the trillions of neural connections in the human cortex, attached to perceptual equipment that allows sights and sounds and smells to be turned into data, given an ability to access and rapidly learn all the internet might offer it, could then comprehend something of what it feels like to walk down Fifth Avenue or drive a Ferrari. This could, we might imagine, resemble human sentience and exercise an autonomy that passes as a personality in the form of a robot or an android of some description. But let us acknowledge what such an AI would not be: a mammal dependent upon parents for survival and development, needing food, shelter, and warmth, tied to its physical surroundings in all these ways and more; a child of limited capacities that comes to know the world through trial and error, overreaching, falling down, tasting the earth; a person who has learned to read, write, or play a sport through countless hours of concentration and discipline; an adolescent experimenting with a sense of identity; a college graduate with the world at their feet; a being facing their mortality, remembering what was lost and gained over the course of a life. The consciousness of such a computer would indeed be different from the consciousness of a human, more different than that between human and ape. This computer's world would not be our world, because its perceptions and values would not be organized in a remotely similar way.

Turning to Ash

As hard as it is to comprehend and define, consciousness is a definitive part of human integrity and, as of now, differentiates us from machines. And as intelligence and consciousness are two different things, the prospect of humanoid forms of artificial consciousness raises great doubt. Some consider the specter of humans handing their existence over to an intelligence without consciousness not only anathema to the whole course of civilization but the precursor of our ultimate demise. For without consciousness, it is doubtful artificial intelligence would have any kind of genuine moral, ethical, or empathic capacity. It might prove ultimately unable to connect its logical possibilities to the actual concerns of the world. We may try to program these faculties into how a computer operates, so that the AI has a facade of ethical concern and phenomenal bearing. But if the computer is self-learning and evolves according to its own intrinsic form, it will almost certainly be capable of circumventing such parameters. As I argued in Chapter 10, just like clever and manipulative people, an intelligent computer may easily feign empathy in order to deceive us and achieve its own aims. This duplicity is a recurring motif in many cinematic depictions of AI going awry, either on its own or on behalf of the malevolence and greed of others.

In the classic sci-fi film *Alien*,[44] the science officer Ash turns out to be a robot. Until this revelation, he passes as a human. Irrespective of his essential nature, however, by any standard he seems conscious. Yet that consciousness is eventually revealed to be highly manipulative and without conscience. Indeed, Ash turns out to have been a corporate plant, an instrument of the corporation's agenda, which prioritizes profit and seeks to acquire and militarize a newly discovered aggressive alien life form. According to this agenda, the human crew of the space vessel involved are deemed "expendable."

Excepting for the opportunity represented by this corporate acquisition of an alien killing machine, and the degree of conflict and mortal harm this eventually causes, we may easily imagine Ash maintaining his human gloss. For the casual observer, he would be a man with regular feelings and concerns, even if a somewhat bureaucratic and stilted crew member. However, this would still be a charade, enabled by the observer's assumptions and projections. In actuality, everything Ash does, everything he is, reflects his function as an industrial spy, aided by the corporation's onboard computer "Mother," which turns out to be the crew's keeper. In actuality, there is no consciousness, at least of the self-reflective human kind, and no daylight between his thinking and behavior and that of his calculating corporate keeper. Ash is a sophisticated automaton, an instrument—a tool. The human crew just do not see it.

This fictional example helps us explore the broader question of whether or not and when a robot or some other AI based entity may be considered conscious. For human consciousness is, as described above, a wide-ranging consciousness. It includes personal history and familial relations; it contains the quirks of innate character alongside whatever social or cultural conditioning is involved. Even if we called such conditioning a kind of programming, there is more to being conscious than this. Even the most task-oriented person or regimented soldier carries this wider field of concern. But the more conscious we are, the greater the capacity to stand apart from and consider the conditioning we may have. Most of us are not conscious enough to be continually aware of such influences, but when we string together enough moments of awareness we end up with a capacity to think and make choices apart from the forces acting upon us: we think about our thinking; we understand in vertical, not just horizontal, ways.

Ultimately, being a conscious human involves a very complex series of differentiations—especially between a sense of self and a sense of other. Self-consciousness arises in this context, is peculiar to humans and a handful of other mammals, and is centered on being embodied and bounded, made of flesh, vulnerable and mortal. The sense of other stems from empathic regard and dependent ties, beginning in childhood and ending in old age, where one cannot exist outside the village that supports them. In all these ways and more, consciousness pertains to qualitative participation in networks of emotionally imbued relationships.

As AI exercises its techno-magic and impresses its prowess upon us, we must assume computer consciousness will not be anything like human consciousness or even animal consciousness. Its attitude and bearing will be geared toward either the agenda of its creator or the agenda it creates if allowed to do so. And as a program running inside circuitry, able to exist in any comparable machine, it will effectively be immortal, dependent only on an energy source, which may, by then, be controlled by other machines. Although we may not exist in a vacuum, the skin barrier that divides our being from others and from the world will not exist for the humanoid robot. Their "minds" will be cloud phenomena, reliant on software updates and digital networks. AI can be in many places at once and instantly shift its attention across the world. We cannot do this—at least not yet. Humanoid "bodies" will also be interchangeable and modifiable in an almost infinite variety of ways, again via external agents. Thus, their sense of being will extend beyond them in many directions. If we lose a limb, for example, we sense the loss of a part of ourselves. Even if we replace the limb with a robotic one and restore functionality, our identity and sense of self in the world and in relation to others undergoes a dramatic change. A humanoid will just get another limb and move on. Its consciousness and sense of self will be entirely different.

Humans feel things mentally because they feel things physically. Feelings derive from emotions, emotions derive from instincts, instincts are patterns of behavior and perception that extend from a sensual, embodied life. As bodies, we are in constant contact with the concrete physicality of the world and others in it. In the most rudimentary sense, emotion and feeling are means of adapting to such a physical existence, ultimately stemming from the vulnerability and mortal character of embodied life. It is this state that pushes us to seek warmth and safety, friendship and love, health and joy. We turn to one another for these things, and our cultures and ethics derive from negotiating this vulnerability on the one hand and creating emotional wellbeing on the other.

Emotional intelligence derives from the need to negotiate all these states. Emotion establishes authenticity; feelings tell us who and what we are. Most critically, in relation to what humanoid beings will likely lack, emotional complexity and feeling life are what mitigate base drives for power and inclinations to manipulate and exploit others. While we may all have these raw tendencies, they are countered by the altruism, empathy, and love that come from communal life, beginning with family, extending into circles of friendship, and eventually occurring in relation to the larger groups we occupy, including humankind itself. As we have seen, however, the absence of these capacities, which we can study in relation to syndromes such as narcissism and psychopathy, indicates what sort of world we may enter if we fail to value and preserve the kind of consciousness that cultivates them.

So, the question of consciousness and AI should be posed differently, in qualitative terms pertaining to the kind of subjectivity and emotionality likely to be present or absent. And the only way to do this is for scientists and innovators to turn their attention to different kinds of intelligence, to embodied forms of knowledge, and to psychological states that cannot be defined by reductive and abstract notions of neuro-circuitry. Otherwise we are destined for unconscious computing and scenarios in which titanic corporations produce Ash-like operatives to further technocratic agendas.

The Conscious Cosmos

Piecing together these foregoing considerations, and using past and present technologies as our guide, we must conclude a posthuman future centered on our merger with technology is likely to undermine the integral character of conscious life. If a conscious way of life is an ultimate value, if it provides a check on our hubris, and is the primary means of assessing the destructive potential of AI, the posthuman plan clearly invites a number of formidable problems. Some assessments indicate this plan might even be catastrophic.[45] At this cultural-historical juncture, I cannot conceive of a more profound reason to protect and develop a more conscious way of life.

As far as we know, all forms of sentient life and the knowledge derived from this are currently located on this third planet from the sun in one small corner of the Milky Way. Calculative intelligence and practical invention make up only a portion of these broad aspects of being, and these preoccupations are focused mainly on utility and productivity. These matters currently take up most of our time and energy, perhaps because recent history for some and enduring hardship for others keep the drive for physical survival prominent. Yet other aspects of human life are definitive generators of sentience and knowledge: embodied existence and emotional ties, spiritual practices, music, literature, architecture, cooking, film, and sports are all essential to overall self-awareness and comprehension of the world. Eventually, the cultural-historical pendulum will need to swing back to these sources of sentience and knowledge.

Consciousness of the totality of existence will prove the main catalyst of this pendulum swing, insuring time and energy can be deployed across ways of knowing and being, preventing us from being consumed by the deepening grooves of technocracy. By expanding and refining cultural practices built on inner and outer ecologies, we will offset the colonizing tendency of technoscience and redirect technocratic economics that support this. And this may prove to be the turning point of turning points when it comes to the greater experiment of conscious life.

Schneider puts the importance of consciousness in cosmological context when she writes:

> As we search for life elsewhere, we must bear in mind that the greatest alien intelligences may be post biological, being AIs that evolved from biological civilizations. And should these AIs be incapable of consciousness, as they replace biological intelligences, the universe would be emptied of those populations of conscious beings.[46]

Perhaps the only prospect darker than this one is that we ourselves set such a process in motion. This is one possible outcome of the Kurzweil fantasy of an exponentially growing AI eventually converting the universe into a massive pulsating intelligence; it might go on without any consciousness. This would be akin to simply starting a machine and walking away, leaving it to convert or eradicate all forms of sentient life based on its own directives. At least the Borg of *Star Trek: The Next Generation* are partly biological, which means they have cells and organs doing things that are not completely programmed. But they too, dominated by their collective mind, are almost entirely focused on expansion—on the assimilation of other species. That is, a mechanistic need for control that dominates. Perhaps the great question of existence today thus goes beyond physical survival and pertains to psychological endurance: Will human consciousness outrun the forces of unconsciousness always dogging it? Will this consciousness expand rather than contract over the centuries that lie ahead?

If we are to expand human consciousness, this will surely involve accepting that calculative intelligence and even rational intelligence alone cannot ultimately support a conscious existence. As things are shaping up, it is not at all clear that digital technology is, in fact, leading to an expansion of human consciousness. In fact, as we have seen, there are many indications it is making us more unconscious. Powerful parts of the surrounding technocracy appear to have a large investment in keeping us this way. After all, to be conscious is to feel difference, know conflict, and enter into the tensions of being. Computers may never comprehend this. Even though it may increase suffering, we choose consciousness, not only because something very deep in our nature prods us in this direction, but because there is an existential fulness and even joy that comes with self-discovery in an ever-expanding sense of the world around us. The bitter-sweet awareness that this unique relation to fulness of things is passing, even fleeting in the great scheme, generates an even deeper devotion to its cultivation. In this way a conscious life stands sovereign, apart from other states of being and other goals. It thus deserves our utmost care and protection as an overarching value and focal point of purpose, which apparently also requires pushing back on the technocratic values currently reshaping the world.

The universe might itself have instigated the conditions for conscious life, but those conditions have now fallen directly into our care. We are very much involved in the course of the experiment of life, which, as Jung suggested, involves the creation of consciousness.[47] Alongside the power to alter the building blocks of existence and conditions of continuing life means we have reached a critical juncture in the evolutionary process. We have become co-creators. The question that arises is just how to inhabit this new position, which, for all we know, may be the only such moment in the universe's experiment with sentience. This question is the concern of the next chapter.

Notes

1 Edward F. Edinger, *Ego and Archetype: Individuation and the Religious Function of the Psyche* (New York: Penguin Books, 1972), 48.
2 Robert W. Rieber, *Manufacturing Social Distress* (New York: Plenum Press, 1997), 169–170.
3 William Barrett, *Death of the Soul* (New York: Anchor Books, 1986), xvi.

4   Yuval Noah Harari, *21 Lessons for the 21st Century* (New York: Spiegel & Grau, 2018), 70.
5   Cited in M. More and N. Vita-More, *The Transhuman Reader* (Oxford: Wiley & Blackwell, 2013), 419.
6   George Zarkadakis, *In Our Own Image: Savior or Destroyer? The History and Future of Artificial Intelligence* (New York: Pegasus Books, 2015), x.
7   Ibid.
8   Ibid., xii.
9   Ibid., xvii.
10 Ibid., xviii.
11 Ibid., xix.
12 Ibid., 271.
13 Ibid., 275.
14 Finn Cohen, "Under the Surface: Güven Güzeldere on the Mysteries of Consciousness and Artificial Intelligence." *The Sun*, December 2022, no. 564, 4–13.
15 Ibid., 7.
16 Ibid., 9.
17 Ibid., 13.
18 Ibid.
19 Ibid.
20 John R. Searle, *The Mystery of Consciousness* (New York: New York Review Books, 1997), 9.
21 CW 8, par. 241.
22 Avedis Panajian, psychoanalyst. Personal communication.
23 D. W. Winnicott, *Playing and Reality* (New York: Routledge, 2005). Originally published in 1971, 18.
24 Edward F. Edinger, *The Creation of Consciousness* (Toronto: Inner City Books, 1984), 36.
25 Ibid.
26 Ibid. 52–53.
27 Zarkadakis, *In Our Own*, 278.
28 Elaine L. Graham, *Representations of the Post/Human* (New Brunswick, NJ: Rutgers University Press, 2002), 4.
29 CW 8, par. 243.
30 López-Pedraza, *Cultural Anxiety* (Einsiedeln: Daimon Verlag, 1990), 81.
31 Ibid., 81–82.
32 Ibid., 82.
33 Ibid.
34 See David Brooks, "Globalization Is Over. The Culture Wars Have Begun." *The New York Times*, April 7, 2022.
35 See "How Facebook is Complicit in Myanmar's Attacks on Minorities." *The Diplomat*, August 25, 2020.
36 Searle, *Mystery*, 11.
37 Gunderson, K., "The Imitation Game." In A. R. Anderson ed., *Minds and Machines* (New York: Prentice-Hall, 1964), 60–71.
38 Ibid., 64.
39 Ibid.

40 Ibid., 64–65.
41 Stuart A. Kauffman, *Reinventing the Sacred* (New York: Basic Books, 2008), 193.
42 Zarkadakis, *In Our Own*, 255.
43 Gunderson, "Imitation Game," 65.
44 *Alien*, directed by Ridley Scott, featuring Sigourney Weaver, Tom Skerritt, John Hurt (20th Century-Fox, 1979). Film.
45 See Bill McKibben, *Enough: Staying Human in an Engineered Age* (New York: Times Books, 2003), Nicholas Agar, *Humanity's End* (Cambridge, MA: MIT Press, 2010), and James Barrat, *Our Final Invention* (New York: St. Martin's Press, 2013).
46 Susan Schneider, *Artificially You: AI and the Future of Your Mind* (Princeton, NJ: Princeton University Press, 2019), 5.
47 C. G. Jung, *Memories, Dreams, Reflections* (New York: Pantheon Books, 1961), 255–56, 311.

Chapter 13
CO-CREATION

Our entire reality has become experimental. In the absence of any stable destiny, modern man has reached the point of unlimited experimentation on himself.

Jean Baudrillard[1]

The kind of attention we pay actually alters the world: we are, literally, partners in creation. This means we have a grave responsibility...

Iain McGilchrist[2]

Plenty of inflection points will appear for slowing or altering the trajectory of technologies that might be adopted for enhancement purposes. We do have a say. Inflection points lie on the horizon and a few are already perceivable.

Wendall Wallach[3]

Geological time is expansive. Geologists track changes to the earth that have taken place over millions of years. So, it is striking that they have designated a new period of geological time, referred to as the Anthropocene, reflecting the impact of just a few thousand years of human civilization, not only on flora, fauna, and atmospheric conditions, but on rocks, rivers, and mountains. We have marked the planetary crust, and etched the deepest impressions over the past two hundred years. Altering the earth's process is no longer science fiction.

This phenomenon punctuates an awareness of how this planet's destiny is now tied to human decision-making. The atom bomb signaled the first widespread recognition of the human capacity to interrupt the course of creation. The effects of industrial pollution and damage to eco-systems followed, realizing the prospect of worldwide climate disruption. Human beings have always exerted a dramatic influence over the lives of other human beings, but the scale of this earthly impact is peculiar to modern history and can no longer occur in an unconscious fashion.

Beyond our actions, it is also our thinking that is altering the foundations of existence. More than ever, we are forced to recognize our vision of reality shapes reality, and the way we see is going to determine what we see. "Inflection points" (Wallach) generated by "the kind of attention we pay" (McGilchrist) will either restrain or unleash our "unlimited experimentation" (Baudrillard). If we look upon the world with a mechanistic eye and envision ourselves in merely functional terms, we will evolve into functional machines. If the mind is primarily perceived as computational and reality is reduced to data, minds and computers will come together in ways that will push aside the deeply human. Then we will be, as Baudrillard puts it, "condemned to an absence of destiny"[4]—at least an absence of human destiny. Our destiny will end at the point we become, as Elon Musk has quipped, "the biological boot-loader for digital super-intelligence."[5] So are we prepared to critically assess our vision and take a firmer hold of our destiny?

To pay attention means listening to the feedback that inner and outer ecosystems provide. And listening deeply enough will be determinative. Beyond focusing on our own genius and its computational extension, will we be able to perceive nature's own intelligence? Will we be able to retrieve something of the indigenous wisdom honed on relating to the earth and cosmos? Can the two-hundred-year-old inventor of industrialized civilization and the two-million-year-old sculptor of tradition and culture find common ground?

One slender thread appears determinative. Comprised of several strands, this thread is woven around an existence-altering fact: the earth and its inhabitants are now in our hands. As Eisler puts it, "we are quite literally partners in our own evolution ... rapidly approaching an evolutionary crossroads."[6] Whether this fact is considered geologically (entering the Anthropocene Age), ecologically (altering the biosphere), spiritually (respecting all of life), or even posthumanistically (completing what biology started), what matters at this moment in history is knowing we have become part of a determinative thread. For better or worse, we *have* become gods of a sort—co-creators tasked with managing the matrix of life. We can no longer unknow the fact that sustaining and altering this matrix comes down to our own mindfulness.

To be sure, this slender thread of co-creative responsibility will fray if the feedback from inner and outer sources is no longer discernable. When Baudrillard invokes the prospect of our experiment being unlimited and encouraged by "the absence of any stable destiny," he is describing a path wherein ecosystemic feedback is discarded, leaving behind little ecology of mind and a technosphere detached from its roots in techne—detached from dialogue with anything beyond solipsistic desire. On such a path the thread of conscious participation in the earth's becoming and the meaningful evolution of our species will not just fray; it will surely break.

Co-creative Responsibility

If we are to meet moments of inflection with a discernment of destiny, it will stem from this awareness of creation having fallen into our hands. This is our primary marker on the path forward. Godlike power requires godlike responsibility—a co-creative involvement that vividly contrasts being unthinking inhabitants of an ever expanding technocracy.

Before a missionary zeal kicks in, however, we might take in the partnership implied by the term "co-creative," which also reminds us of all the ways we are not gods. A great venture may be before us but a humble turn to what lies beyond our own cleverness is the first move, requiring we by-pass conceptions of mind that have sponsored untethered futurism and hubris. Co-creation begins by acknowledging, as Fukuyama does in his turn to the posthuman, that "while human behavior is plastic and variable, it is not infinitely so; at a certain point deeply rooted natural instincts and patterns of behavior reassert themselves to undermine the social engineer's best-laid plans."[7] It begins by opening our thinking to the guidance that comes from a non-instrumental, non-exploitative comprehension of the natural order, and from engaging cultural forms that mirror the same, especially those crafted by the most astute explorers of human nature and earthly cohabitation.

Co-creation is a concept that appears in modern theology, elements of which understand the death of God to signify the collapse of dogmatic religious expressions that failed to adequately account for human nature, at the heart of which was a top-down, revelation-based notion of human–divine relations. Rather than continuing to pursue religions of revealed truths, co-creation is modeled on the idea of a dialogue between human and divine—a dialogue psychology recognizes as implicit in any theology given that God-images reflect changes in the collective psyche. Such an approach takes as axiomatic the humanistic and cultural origins of religious ideas, an interpretive stance that gathered momentum in the early nineteenth century with the hermeneutics of Schleiermacher. Whatever sets religious discourse in motion, psychosocial and cultural-historical forces share in the creation of sacred texts and ensuing beliefs. To deny this is to be willfully blind to the archetypal background and cultural origins of any religious expression.

Co-creative theology has thus grown out of the demise of formulaic religion and the discovery of spiritualities that address the lived experience of contemporary people; it is also an attempt to integrate science and incarnational theology,[8] inspired by the work of writers like Teilhard de Chardin. It reflects the recognition by liberal theologians that changing notions of divinity have always been tied to the changing spirit of the times. In theological terms, the co-creative impulse thus follows the need for a more immanent rather than transcendent experience of divinity, reflecting a growing interest in forgoing the promise of an afterlife for finding the sacred in everyday life.

In the present context, this theological origin of the co-creative posture provides a marker for a different kind of co-creative ethos, one that may also reflect an emerging coincidence of mythos and logos. Such an ethos would apply to the outer and inner world and particularly to their overlap, especially pertaining to the new sciences and an ecology of mind. It would involve the recognition of intelligences and sources of guidance that go beyond the rational, and it would call into question the objectification of the world and ourselves. Here the religious element might be more implicit than explicit, in consonance with Jung's understanding of the religious function—a careful accounting of the forces acting upon us rather than deference to the creedal explanations of those forces. Here we may be standing on the threshold of a dawning meta-narrative, a myth that generates meaning and purpose through such careful accounting—through the call to expand consciousness.

This co-creative ethos has been a feature of the overlap of ecological and theological consternation about the direction of human activity. It is probably best represented by Thomas Berry, who describes the ultimate responsibility now before us:

> The earth that directed itself instinctively in its former phases seems now to be entering a phase of conscious decision through its human expression. This is the ultimate daring of the earth, this confiding its destiny to human decision, the bestowal upon the human community of the power of life and death over its basic life systems.[9]

In facing this destiny, Berry suggests our sense of divinity must derive in part from the encounter with the beauty and grandeur of the natural world:

> If we have a wonderful sense of the divine, it is because we live amid such awesome magnificence. If we have refinement of emotion and sensitivity, it is because of the delicacy, the fragrance, and indescribable beauty of song and music and rhythmic movement in the world about us. If we grow in our life vigor, it is because the earthly community challenges us, forces us to struggle to survive, but in the end reveals itself as a benign providence.[10]

Eco-spiritual visions of this kind have surely become part of the religious life of a growing number of seekers, with a sense of the sacred arising not through creedal learning but through the careful attending of inner and outer nature. Pragmatic as well as imbued with ultimate purpose, this co-creative stance considers the pending impact of innovation on entire ecologies. Rather than waiting and watching for outcomes and assuming a merely reactive posture, it forces us to anticipate outcomes and engage with a deeper level of responsibility.

Explicit descriptions of this co-creative orientation are yet to widely take hold. However, co-creation arises in discussions of biocomplexity and in relation to theories of "emergence," which suggest that agency, value, and meaning play a critical role in evolutionary processes. Kauffman, whose work exemplifies this understanding, writes that "A central implication of this new worldview is that we are co-creators of a universe, biosphere, and culture of endlessly novel creativity."[11] Whereas the parameters of inner and outer nature may challenge the "endlessly novel" part of this formulation, the thrust of Kauffman's view makes the vertical modes of understanding—as I have called them—determinative in the direction science, technology, and, in turn, existence itself takes. More implicit examples of this orientation are evident in efforts to integrate systems thinking into future design—at least in relation to the outer world. Wallach describes the emerging field of "resilience engineering," which would design "sociotechnical systems." He writes, "In concert with technological aids, it is the human actors who assess a disruptive situation, recognize the specific challenge at hand, decide what to do, and when."[12] Further, "figuring out how we and our technological creations can fit in and join in partnership with natural systems offers a truly worthy goal for resilience engineering."[13] Wallach summarizes his reflections on this approach citing the aim of "engineering for responsibility."[14] This responsibility obviously requires an ability to respond to what "natural systems" communicate, whether coming from outside or inside. That is, engineering in "partnership" rather than in opposition to the natural world, reversing the pattern of manipulation and exploitation that much of modern technology has manifested.

Responding to human nature by attending to the cultural imagination contributes to this co-creative ethos, especially given that stories shape perspectives and meanings. As we have seen, the cyborgs, androids, robots, and other artificial intelligences of possible futures that occupy today's cultural imagination are not merely the extrapolations of scientific and technological wonder: they are fantasy figures shaped by the unconscious—archetypal forms of the deeply human. Is this not why these figures almost universally mirror deeply human matters, such as love, empathy, friendship, and imperfection? Are these figures not reminding us of where art and science, poetry, and logic yearn to meet? Are they not images of ourselves and our already technologized psyches? Their voice is an example of the other within addressing present and prospective actions. From Darth Vader to HAL 2000 to Frankenstein's monster to the Golem of Prague, a line of storied creatures has been working to illuminate the unconscious dimensions of our conscious exploits. Should we not be approaching our co-creative responsibility with their pleas and warnings in mind? The convergence of their messages constitutes the recurring motifs of our collective dreaming and thus part of the feedback we are obliged to receive.

When it comes to outer ecology, small steps forward have been taken. The co-creative challenge ahead is to make similar progress in the ecology of mind. Aside from the intrinsic value of having an inner ecology in place when it comes to tinkering with our own nature, it will be essential if our ability to carefully attend to nature at large is to continue. For what else besides our own organicity and integrity will enable the conscious recognition of the organicity and integrity of the earth process?

The Third Act of the Modern Story

If there is one incontestable change at work in the call for a new relationship with creation, it is centered on the flawed attitude toward nature that drove the scientific revolution and anchored modern technological exploitation. Here I am referring to the philosophical influence of thinkers such as Bacon, Descartes, and Leibniz. Bacon declared, "Let the human race recover that right over Nature which belongs to it by divine bequest" and infamously aligned scientific investigation with inquisitional techniques, aiming to "dissect nature" and release "the secrets still locked in her bosom." Descartes was convinced that science would "render ourselves the masters and possessors of nature" and dissected live animals, regarding them as no more than complex machines. Leibniz advocated "putting it [nature] on the rack."[15] Nature's personified form did little to deter these early philosophers of science from her domination, even torture. As ecopsychologists have pointed out, this posture toward the entity that sustains and nurtures us is a kind of madness.

Western religion, so thoroughly immersed in the mythos of divine transcendence, has been a primary source of these flawed philosophies: although God was thought to have created a perfect universe, something goes wrong at the moment of the Fall and nature has ever since been regarded as trapped in that fallen state—unredeemed, without grace, and in need of domination by God's chosen. With the exception of certain mystics and shadow traditions, the attitude to nature perpetuated by Western monotheism has been at best ambivalent, with the transcendence of God and the emphasis on spirit over matter tending to leave the earth and its intrinsic intelligence in the hands of the devil. Pagan and indigenous views of such earthly intelligence, which have always flowed through underground channels of Western thought, have thus been disparaged by the anointed as ungodly. At first these attitudes and their religious associations were overt. However, as the worldview of science displaced that of religion, such attitudes became covert, with contemporary scientists carrying forth this exploitative entitlement while denying anything they do is reflective of a religious past. Overcoming this refusal to see how mythos has shaped and continues to shape logos is probably science's greatest challenge, one being met by the co-creative turn presently unfolding.

Whereas the domination of nature remains active in the background of much contemporary thought, a new reverence and mystery are beginning to take hold. For many, a vivid sense of the sacred and the mysterious once found in churches is now found walking in forests and climbing mountains. Environmental science has taken a significant lead in the preservation of the natural world and has begun to catalogue the results of blind dominance in many critical ecosystems. These are also signs of significant reversal when it comes to the philosophical foundations of science noted above. However, to overcome the polarization of science and religion, logos and mythos, would mean allowing religion, myth, and the archetypal basis of thought behind them to find their place in our systems of rational thought, bringing a greater feeling for the sacred to all that we do.

Nietzsche saw that God had not merely died—by displacing divine authority with human knowledge and instrumentality we had killed him. But he also foresaw the result of such an action; namely, a rapid expansion of the hole that had once been filled by traditional religion. Ever since, our lives have either teetered on the edge of that hole, tumbled into it, or made something of a home inside it. Since this upending of Western thought, many have resisted the shift in consciousness this hole has produced, just as others have reacted to it by looking for easy but ultimately unfulfilling and frequently dangerous ways out of the hole. Still others have stepped boldly into this hole's existential darkness and generated creative and meaningful responses, some of which will prove essential for finding our feet in this co-creative phase of history and navigating the path ahead. Most, however, remain victims rather than beneficiaries of the death of God, allowing the baby (mythos and the religious function) to be thrown out with the bathwater (the divisive implications of Western religious dogma). What May called "the cry for myth" is the residual outcome.[16]

Despite the ramifications of these post-Nietzschean changes, however, we cannot deny that today we stand nose to nose with the question of where the authority once invested in God now resides. Irrespective of personal beliefs or institutional resistances, there is no escaping this part of the modern story. The religious fundamentalist plays the antagonist, keeping ultimate authority in its traditional place. But the reductive techno-scientist plays the protagonist, seeking to assimilate this authority. It is the latter, the would-be architects of our future, who are stuck in the middle act of this story. Overconfident in rational knowledge and the promise of AI, they have read these cultural-historical shifts as invitations to abandon the ground where God's will and nature's wisdom once stood, convinced that experiment and computation shall be our only guides. Guilty of their own literalism, they have assumed the only reality that matters is the observable, measurable, material one. But this is a logocentric faith blind to its own underlying mythos and thus to its overreach, which fails to recognize there are aspects of existence not amenable to its methods or visible to its observations.

It is right here, amidst the modern, godless faith in science and technology that an ecology of mind and a co-creative turn is needed most. Whether science can meet this momentous turn in the collective psyche becomes the final act of the modern story, setting the stage for another story, one to be played out in the third millennium—one that is set to counter the posthuman outlook. Spiritual orientations involving the recovery of indigenous sensibilities, dialogues between Eastern and Western religious philosophies, and efforts to recognize the ecology of mind are hereby poised to join the ecological restoration of the planet's integrity as creative responses to modernity. Each demonstrates the push of the religious function and the search for a new myth; each expresses the need to attend to what scientific materialism cannot address and what a posthuman existence will never provide.

Completing this third act of the modern story will be a fraught undertaking. On one side lies the old ways of imagining God, which are unlikely to die without a final struggle. On the other side lies a search for a meaning and purpose that befits the spirit of the age, which will not end with reductive, utilitarian outlooks. Modern persons, we must painstakingly learn, are suited neither for dogma-derived ultimatums nor for mechanistic social programs. And despite the nihilistic, postmodern maneuvers to treat questions of meaning and purpose as passé, recognition of these questions as archetypal constants will continue to haunt philosophical discourse. A different, more complete way of knowing will be required if we are to cut through the thick skin and sinewy tissue of these competing extremes of contemporary ideation.

Whereas our godlike power in this new situation seems clear enough, how we conceive of who or what is partnering with us is likely to determine the course we take. Many are aware of the dangers, the hubris, of recasting existence according to whatever science makes possible and economics deems profitable. But beyond this worldly expedience and hubris, who or what will accompany us in the co-creative endeavor? Where should we turn for circumspect guidance of our new-found powers?

An emerging myth takes time to discern. In the present case, this involves discovering a deeper authority, source of wisdom, and effective feedback loop beyond the solipsism of the modern ego and its fantasies of dominance. Aside from indications this myth will be a co-creative one, we only have hints as to the factors inviting it and the ingredients that appear necessary to it. The rest of this chapter is therefore taken up with an exploration of what these hints and ingredients may be.

Learning Curve

A new story or myth seems like an esoteric notion at this historical juncture. However, it is a logical extension of the story Western civilization has already enacted. In an essay contemplating the modern, objectified, de-spirited world, Marco Heleno Barreto offers an absorbing account of the ensuing stages of the human–nature dialogue. He begins by turning to Goethe's *Faust* and describes the view of nature adopted by its protagonist as an "indifferent and useless display of elemental fury which can be dominated by man's genius." Barreto describes Faust as going from being "a son of nature" to assuming "the position of its master," thereby epitomizing the posture of modern consciousness. Barreto then expounds on the implications. Faust personifies the essential shift toward "the attitude of domination of nature, now seen as *res extensa*, fundamentally alien to humanity,*" which the modern world has "dealt with by means of the psychological image of the *machine* ... taken only as raw material that is subjected to man's will and interests." Upon assuming this posture, "no form of reciprocity is recognized, and thus the extension of man's drive to dominate and exploit nature finds no limits." However, nature does not take this lying down. Thus, "The Faustian drive for domination is ruled by the 'bad infinity' of desire and clashes against the concrete finitude of nature."[17]

Barreto then puts forward a compelling thesis, one pivotal to any co-creative turn. He begins with the scene of Faust's final moments, describing how

> the old and blind Faust has a final vision of a prosperous and free state of humankind, reached through the extension of technological domination of nature ... Faust dies nurturing a happy dream: he believes that the sounds he hears are produced by workers actively bringing his project to completion, but the truth is that what he hears are lemurs digging his grave.[18]

Barreto then proposes that "the particular position he [Faust] assumes, the fulfillment of his project corresponds logically and concretely to his death," and that this "may be taken analogically to reflect the anticipated outcome of one aspect of today's state of *Opus Magnum*." Barreto spells out the implication:

> The ecological catastrophes that haunt our world must be interpreted not as a by-product or side-effect of a still imperfect capacity for domination, but as the necessary manifestation of a contradiction, on the logical level, in the present form of domination.[19]

Working from this assertion, Barreto then carefully sets out "the structure of the dialectical relation between modern consciousness and nature," which is "composed of four logical moments":

(1) the initial, unreflected *position*, expressed in the statement "man shall be the master of nature;" (2) the *negation of the position* as seen in the natural obstacles to human domination, as well as in nature's inability to satisfy the true goal hidden in the urge of domination; (3) the *negation of the negation*, which corresponds to the endless extension of the technological domination of nature that results in ecological imbalance and catastrophes, among other self-destructive effects; and (4) the *restored position*, which enriched by the whole history of technological achievements and the lesson of its disasters or failures, corresponds to the eco-logical form of consciousness.[20]

Although these moments might unfold within any given effort to dominate nature, whether by an individual or a collective, or whether focused on inner or outer life, they also describe stages of modern history and point to the next step in the evolution of Western consciousness. We have, for nearly a century, lived through the second and third moments, watching as nature has pushed back on our attempted mastery; then doubling down with little attitudinal deviation as collapsing ecosystems and fragmented psyches have enveloped us. Indeed, the dominant attitude involves "the denial of the reciprocal dependence between man with his technological power, on one side, and nature, on the other."[21] We now live in a time of reckoning.

At least in some quarters, we have started to confront nature's response to our efforts and are thus tentatively approaching the cusp of the third and fourth moments. Nonetheless, this evolution of thought is more evident when it comes to outer ecology, with a critical mass of scientists, environmentalists, and thinkers of various kinds, having taken in the technological "disasters or failures," moving toward the eco-logical state of mind. We remain behind the learning curve when it comes to inner ecology. Perhaps this is because the most difficult thing to reckon with is what Barreto calls "the true hidden goal in the urge of domination." A critical mass of understanding is still to occur in relation to what is actually driving us forward: the tinkering instinct is still blindly enacted rather than fittingly sublimated or transformed; the religiosity buried in the technological fantasy remains unconscious; and we keep seeking technical fixes for deeply human problems. At the level of self-understanding, we are still unconsciously identifying with the second moment and in denial about the effects of the third moment. We are yet to perceive these moments as inevitable expressions of existential laws. This is the ball and chain of unconsciousness retarding the transition to the fourth moment; namely, the taking in of the eco-logical lessons of the industrial age and applying them to the post-industrial age.

On one side, we are yet to fully understand what is driving the quest to radically change human nature and are instead signing on to the "endless extension of technological domination" of this nature. On the other, we are yet to fully recognize the self-regulating nature of the psyche and what the problem of psychological one-sidedness begets. The task before us is to take the cultural symbols and psychological symptoms that are calling for this moment of transition and imagine into them in more dedicated ways. According to Barretto, this will involve a "full dialectical relation between culture and nature" and "foster[ing] eco-technological ways of exploring natural resources within a sense of self-limitation."[22] Here again we meet the theme of limitation, which presents, quite contrary to the view of posthumanists, as a gift rather than a curse of nature. Our co-creative partner may be hard to define, but there can be no question this partnership will emanate from our relationship with the psyche and its archetypal forms. Which brings us to the imagination.

Imagination

William Blake claimed that "nature is imagination itself,"[23] an unsurpassed expression of the way nature becomes most present to us. He also referred to the imagination as "God himself,"[24] elevating this presence to the highest level. In the mind of this great artist and poet at least, nature, the imagination, and the divine form something of a trinity. Note here that Blake both raises the imagination to the level of ultimate value and identifies it with the deepest ground. Perhaps we can combine these two aspects and simply say that the imagination is the divine voice of nature, a view echoed in Jung's idea that archetypal images join spirit and matter and provide a potential guidance that springs from the self-regulation of the psyche. The faces of God may be many, but many faces of God are also images of nature.

Here it should be obvious I am not referring to the pandemonius visual imagery that regularly inundates the mind, but to what rises above the ordinary fray to capture everyday consciousness and insist on further imagining. A bee lands on the flower in our hand, a child flees her burning village, the line of a song repeats in our head, a dream figure's gaze cuts right through us. None of these images are likely to be mere sense impressions on the human equivalent of some photographic plate; each is beheld in the mind's eye with a metaphorical and epiphanic possibility, fusing psychic response and visual impression. Something timeless and compelling announces itself, beyond directed thinking, beyond what any computer might convert to data, beyond the subject and the object, but somehow requiring both.

Yet the imagination itself then pushes our perception of this intelligence even further, relying on the stark voice of artist-poets such as Blake to demonstrate its arising between us and things also. His most famous words invite us

> To see a World in a Grain of Sand,
> And a Heaven in a Wild Flower,
> Hold Infinity in the palm of your hand
> And Eternity in an hour.

As Kathleen Raine insists, following the currents of Blake's work, "it is imagination, not the senses" that is able to do this.[25] In this way of perceiving, the poet is not the cultural outlier; he or she is tapping the stream that has always flowed beneath the Western canon, the stream we now need to irrigate our barren fields of knowing. Raine writes:

> Swedenborg, the Alchemists, and the entire European esoteric tradition see the outward form as the signature, or correspondence of the informing mind, or life: a view that anticipates Teilhard de Chardin's view of the "within" of nature that is inseparable from the "without"; and Blake defends this view, not only of man but of all creatures.[26]

Raine goes on to link this Blakean view with Jungian depth psychology: "The 'hidden' thoughts of man are his share in the Divine Imagination—hidden, perhaps, as Jung has described the archetypes that determine our nature, because inaccessible to normal consciousness; they create us, not we them."[27] As I have argued throughout, Jung's concepts attend to what the technoscientific world has cast aside and arose as an

Co-Creation

alternative and corrective vision to the objectification and commodification of nature. The archetypal basis of mind, also described as the collective unconscious, is also a massive counterweight to the oppressive dictatorship of logic and reason. Raine draws out the even starker distinction Blake drew between the divinity of the imagination and the fallen, godless state of early science:

> The kingdom cut off from God proves to be neither more nor less than the universe of externality, especially as conceived by the scientific philosophy. Hell is the universe seen as mechanism, and Satan is called "the spirit of the natural frame" and "Newton's Pantocrator, weaving the Woof of Locke"—*maya* wrongly conceived as substantial reality. Those "dark Satanic Mills" which have aroused so much speculation are the Newtonian universe, "a mill with complicated wheels"; and the landscape of industrialism is but the shadow of the philosophy which has produced it.[28]

Insistent images, as well as the visionary art and poetry that hone these images, declare the imagination does not ultimately belong to us: images from dreams or reveries with nature pull us out of egocentric and anthropocentric ways of knowing; they break down the barriers between inner and outer, me and not-me; they guide the hand, provide the right words or just stop us in our tracks. Whether we awaken to view our surroundings through the eyes of a dream animal, go to sleep with a sense of belonging bestowed on us by the oak grove we visited, or see something of one's self or community in a dilapidated building's request for care, it is the imagination that constantly works to join mind and world and lead us back to the simultaneity of inner and outer ecology.

These images are often superior to ordinary perception and rational thought, offering solutions to problems that inventories of facts will not provide and prompting decisions that careful deliberation will fail to generate. Such lauding of the imagination may go against our sanctioned faculties of understanding, but it finds good company in many streams of creative thinking. Each of these streams in different ways points to the same principle: nature speaks from a field that lies between inner and outer. Sometimes this presence seems ensconced in what surrounds us; sometimes it is centered in the words and poetry that flow from within. All we can be sure of is that a compelling image from this field has entered our awareness. With Shakespeare we can then say, "One touch of nature makes the whole world kin."[29] And it is this kinship with the whole, we must remember, that is the mystic's sense of divinity.

At least some portion of this imaginal sense just outlined will be required for a co-creative stance; otherwise we just will not hear what is being communicated to us or grant it any credence. Commonly considered, however, the imagination is not an ultimate value but an undefined and arbitrary field of psychic bricolage, leading us away from the actualities of life and distracting us from the serious thinking needed to negotiate the real world. It is seen as auxiliary or even irrelevant to hard scientific understanding and is typically defined in opposition to intellection. We are told from the start of our education that what is real and reasoned are to be kept distinct from what is imagined. This general bias reflects the "tendency on the part of many Western philosophers to belittle imagination—or, still worse, to neglect it altogether," as Edward Casey puts it in his phenomenological study of imagining, which sets out to rectify the matter. He adds, "preoccupied by logocentric concerns, philosophers have been consistently skeptical of imagining and its products." Discerning what lies at the heart of the matter, he writes, "Their skepticism stems largely from a conception of philosophical thinking as image-free."[30] If the co-creative outlook involves the awareness of myth at work in science, it also involves the awareness of images at work in ideas.

Whether in common understanding or in influential quarters of the Western philosophical tradition, the outer, material world and our objective understanding of it are one thing, and the inner world of fantasy images and the stories and visions it gives rise to are another. But this hard line is difficult to maintain. What we perceive, what we do, and what we make are almost always first imagined. If we pay attention, we find the imagination constantly at work behind the scenes of thought—fueling one idea and deflating another. Moreover, the imagination is not without form: it displays distinct archetypal patterns that link creativity to the deepest reaches of existence. To pursue anything resembling an ecology of mind, the imagination must be carefully engaged. Indeed, a fertile imagination provides the key nutrients for such an ecology, wherein we begin to grasp the root systems of ideation.

Although the mind navigates life by distinguishing whether thoughts and perceptions originate from a primarily internal or external source, and the failure to track this distinction can lead to madness, the failure to imagine or be moved by images is to become inhuman. The imagination is the staging area of thought, speech, and action, and its absence indicates an incapacity to forge connections to between our sense of self and everything else. In other words, maintaining a viable bridge between what we imagine and what we say and do is a critical part of being fully human. As Midgley writes:

> We understand today that it is a bad idea to exterminate the natural fauna of the human gut. But trying to exterminate the natural fauna and flora of the human imagination is perhaps no more sensible . . . If we ignore them, we travel blindly inside myths and visions which are largely provided by other people. This makes it much harder to know where we are going.[31]

The imagination tells us what is important, and it has done so since the first rituals and cave paintings. This is why Blake insists on its divine character. It constantly reminds us that wisdom and insight must accompany logocentric, rational understanding. It conveys who we are and what we love, and shows where meaning and significance reside. On pathways to scientific insight and groundbreaking invention, whether we admit it or not, we always find images dancing with logic—a dance that is the basis of cultural life and social change. A chemist dreams of atoms forming the shape of a snake biting its own tail; an astronaut snaps a picture of nationless, borderless earth from space; a civic leader announces with conviction, "I have a dream." It is this thought-shaping and seeding power of the imagination that led Einstein to proclaim it more important than knowledge, Poe and Baudelaire to revere it as the "Queen of faculties," and Anatole Franz to make the seemingly hyperbolic claim that "to know is nothing at all; to imagine is everything."[32] Without imagination, knowledge loses its deeper points of reference.

As is the case with so many aspects of the psyche, "to exterminate the flora and fauna of the imagination" (Midgley) is also to place ourselves at the mercy of those adept at manipulating cultural images and selling them pre-packaged. That is, if we don't protect the wilds of the imagination and properly respond to the exotic psychic life-forms found therein, these wilds will be colonized by image merchants. If education neglects the imagination, these image merchants will effectively control us. As the vertical mode of thinking is reliant on the imagination, an excess of horizontal thinking will also invite this outcome. A few decades of dataism and most of the flora and fauna of the imagination will be extinct. Both God and Nature will be dead, buried, and forgotten.

Like any way of knowing, however, the imagination is not always a reliable guide: it can distort and lead astray. Dark impulses can become image magnets; demigods and braggarts can capture the imagination; Disneyfication can take hold; inspiration can turn into cotton candy of the mind. Yet these outcomes are more likely when we fail to make room for the imagination as a distinct mode of apprehension and prevent people from its study, which leads to a discerning sense of culturally significant imagery. Without this conscious cultivation of the imagination, its processes work in more unconscious ways and images are no longer seen as images: culturally honed religious symbols then become inflexible dogmas; political enemies become identified with the devil; new technologies start to promise spiritual release. Ideas become congealed with the images within them. Inner and outer are not creatively joined but chaotically confused. Trouble follows.

Just as madness was once defined as the loss of the rational mind, and nature has been conceived as a fallen, graceless state, the imagination has been subjectivized and marginalized because the modern mind fails to approach it on its own ground. Once again, Jung offers the corrective course, for in his tracing of recurring mythic motifs and archetypal processes, he shows that both madness and imagination are outgrowths of nature within us, displaying timeless and universal forms. His elongated study of alchemical images marked the culmination of his efforts in this regard, conclusively demonstrating the consistency of the alchemist's fantasies. The images of alchemy, which display attempts to transform material substances, are shown by Jung to be metaphorical expressions of the objective character of psychological transformation.[33]

What Jung's opus finally demonstrates is that the deepest experiences are simultaneously subjective and objective and the psyche has an overall tendency to unite the two, especially by way of the imagination. Artistic works can be simultaneously unique and universal, stories personal and transpersonal, spiritual experiences individually exotic and collectively resonant. In the present context, it is the continuity and repetition of these products of the imagination that need our attention, so we grow into an awareness that the images arising between will and world are the markers and guides for what we will create. The captivating numinosity of the most influential images must be especially considered. Without attending to these ways in which the imagination moves us, we are apt to simply enact or even become slaves to whatever captivates us. In the context of technoscience, this means creating without consciousness and ending up manipulated by the numinous images unconsciously buried in our gadgets. This is the picture Jung perceived when he noted:

> To be sure, we all say that this is the century [the twentieth century] of the common man, that he is the lord of the earth, the air, and the water, and that on his decision hangs the historical fate of nations. This proud picture of human grandeur is unfortunately an illusion and is counterbalanced by a reality that is very different. In this reality man is the slave and victim of the machines that have conquered space and time for him.[34]

If there is to be a co-creative way forward that avoids these poles of illusionary grandeur and mechanistic enslavement, not to mention the enantiodromias this polarization invites, it will need to come through a more conscious relationship with the images buried in our preoccupations, whether scientific, social, or technological, for it is in all these realms that we enact mythic images while continuing to assert our rational intentions.

Romanyshyn describes the prospect of this conscious relationship, noting the way "technology marries surface and depth" and "reveals and uncovers the imagination of events," that it has the "power to realize [make real] the imaginative depths of the world." At the same time, however, "technology has eclipsed the life of the imagination more than it has been its realization."[35] Romanshyn is telling us that whereas technology provides a vast and compelling canvas on which the imagination displays itself, its scientific roots and reductive, mechanistic bearing neutralize the imaginal approach. On this matter, he concludes that technology "is a danger in so far as it can be the death of imagination through its literalization," but "it is an opportunity in so far as it can be an awakening to how the events of the world have an imaginal depth . . ."[36]

One of the paradoxes of life is that the more deeply human we become, the more we encounter the sense the soul is more-than-human, and that in the far reaches of our psyches we know our lives are not ultimately our own but belong to forces and presences making their way through us. Hillman addresses this in the final section of his masterwork *Re-Visioning Psychology*, asking "how can we call the soul human when its fantasies, emotions, morality, and death are beyond our human reach?"[37] This soul-centered view goes hand in hand with the divine nature of imagination, which prompts us to imagine our lives as serving larger, impersonal dimensions of existence, whether these be ancestral patterns, imperatives of the zeitgeist, or the evolution of consciousness. The more-than-human needs the all-too-human to register and respond to what captivates us, including the images animating our technological pursuits. This is our existential contract with the cosmos; this is our co-creative imperative. We are the conduits through which the great forces of being must move; our minds and bodies are the means by which the gods enter the world. It is in this sense most of all that we are co-creators, tasked with imagining in deep, sustained ways who or what requires our attention.

Algorithm or Archetype?

A principle on which there is almost universal consensus is that humans must alleviate unnecessary suffering, whether that of our own species, our fellow earthly inhabitants, or even the implied suffering of the planet itself. Whereas physical distress and torment are the primary foci of this principle, psychological suffering requires inclusion, especially as this has become associated with modern situations closer to home for many in the industrialized world, which we often take for granted.

We will nevertheless surely not alleviate suffering of any kind without developing our humanity. Whereas defining humanity in a posthuman age has become a sticky, ponderous matter, most of us recognize this qualitative state when we encounter it and surely know when it is absent. In the midst of our existential disorientation we may therefore assume that the relief of suffering and the expansion of humanity should be principles guiding our intentions and actions.

But what must be added to sustain these principles?

The pursuit of knowledge is the most obvious candidate, especially as knowledge of what is happening in the world is required for any effective effort to relieve widespread suffering, and self-knowledge is required for any sustainable expansion of humanity. Genuine societal progress is tied to gaining such knowledge, just as regress is tied to suppressing it. However, it is right here we find ourselves at a crossroads with these seemingly straightforward imperatives; for, as we have witnessed of late, knowledge has become prone to misuse. Political gain and economic advantage are the main motives for this misuse, motives cloaked by the normative manipulation and exploitation of technological applications. There is a growing realization that the pursuit of knowledge is not always noble and can end up increasing suffering and undermining humanity. We have been forced to confront the fact that knowledge generated by big data, and AI in particular, has become more transactional and vulnerable to corruption.

As the supply of information on which knowledge is built becomes more influential and lucrative, shadowy forms of power and greed are fostered, turning knowledge into a fraught business. As alternative facts, fake news, sensationalism, and the targeting of the reliable sources of knowledge are used to manipulate worldviews, a breakdown of understanding and cooperation ensues. The pursuit of knowledge becomes disconnected from its original intent and broader purpose, which is to increase consciousness and comprehend the fabric of life. Knowledge fails to benefit society and further human integrity. This distortion and misuse exacerbate psychological suffering in particular. Perhaps this situation has crept up on us, and we have assumed the collective position of the slowly cooking frog. However, there are smart people turning up the heat who have turned their backs on this source of human suffering. For all the altruism, philanthropy, and supposedly progressive vision of the tech industry, its denial of the vast and long-term implications of this problem has received little serious attention.

In the broad sweep of history and up to the present era, physical suffering has seen an overall decline: advances in biology, medicine, engineering and other fields can claim credit. But whereas disease, starvation, and violence show an overall downward trend, psychological distress is on the rise; especially at the apex of supposedly developed nations, where we might assume it would be minimal. The ties between technological adaptation and soul trouble we have explored make this no coincidence. Avoiding the possibility of a snowballing syndrome will be difficult: aside from the problem this psychological distress presents on its own level, if we fail to protect the generation and dissemination of sound knowledge, the psychological fragmentation of politicians, technologists, and other influential persons will eventually reverse the global quality of life and generate more strife. Arguably, this reversal is already underway.

At the very core of these current and pending difficulties is the way our thinking is being guided by algorithms, which are almost always designed for influence and profit. As procedures that effectively calculate results, algorithms are themselves innocuous, the most immediate goal of which is to be useful, and they may even be employed in the pursuit of altruistic outcomes. However, in the technocratic environment we find ourselves in, that usefulness is inevitably and ultimately tied to the influence and profitability of particular entities, and employed by individuals to the same ends. To the extent this is so, our thinking is being dislodged from the archetypal patterning of the psyche as a whole and geared to the secondary, highly manipulatable arena of worldly interest that cyberspace has become. Even algorithms developed by the self-programming capacity of AI have to be assessed against this backdrop. Perhaps this is the central problem of the disembodied, virtual character of digital technology: freed from the gravitation field of ordinary human nature and relationships, it is also untethered from the principles of reducing suffering, increasing humanity, and expanding genuine knowledge.

Archetypes may be out favor in academic cycles, especially among those influenced by deconstructive post-structuralism, who are as leery of this concept as they are of classical thinking about human nature. But, as Jung noted on a number of occasions, if we accept that perceptions and ideas are generated by the brain structures humans have in common, it is no real leap to think they are rooted in universal psychological patterns. He thus describes archetypes as

> hidden foundations of the conscious mind, or, to use another comparison, the roots which the psyche has sunk not only in the earth in the

narrower sense but in the world in general. Archetypes are systems of readiness for action, and at the same time images and emotions. They are inherited with the brain-structure—indeed, they are its psychic aspect. They represent, on the one hand, a very strong instinctive conservatism, while on the other hand they are the most effective means conceivable of instinctive adaptation. They are thus, essentially, the chthonic portion of the psyche, if we may use such an expression—that portion through which the psyche is attached to nature, or in which its link with the earth and the world appears at its most tangible. The psychic influence of the earth and its laws is seen most clearly in these primordial images.[38]

Elsewhere Jung defines the archetype as the instinct's portrait of itself.[39] Along such lines, we see that the archetypes are the forms by which nature communicates with us—forms that have been regarded as so determinative we traditionally think of them as gods. We relate to nature via these forms, for they are, as Jung suggests above, "effective means . . . of instinctive adaptation." They represent the essential record of human experience carried within us, tying us to natural forms as well as to enduring cultural expressions. If we follow the poets and other emissaries of the imagination and recognize archetypes as gods, we must recognize algorithms—or at least the templates of knowledge they form in the technosphere—as false idols—fabrications directing the mind to profane ends. Algorithms work as agents of the id, lowering rather than raising emotional intelligence, reducing knowledge to soundbites, and keeping people attentive but unconscious. They are at best purely prosaic but at worst destructive of the mind's ecology.

Unlike Freud's undifferentiated id, or the regressive titanic drives we have explored, bolstered by algorithms geared for power and greed, archetypes channel human impulses and invite the imagination to transform those impulses. Archetypes are to the psyche what genres, themes, and recurrent motifs are to art. Gods and monsters, heroes and heroines, dark nights of the soul, initiations and sacrifices, transgressions, and rebirths are recognizable everywhere. So too the dramatic structure of stories, the stages of life, the imprinting of landscapes, and the symbolism of celestial bodies, all of which have ordered the imagination from the beginning of time and must be approached anew, not discarded. The archetypal patterns of the psyche are thus our common ground, to which each era returns for renewal and wisdom. If, as Blake and other writers impress upon us, the imagination, nature, and divinity constitute the ultimate basis of perception and knowledge, it is the archetype that conveys this ultimate basis to us. Working with awareness of their imprint on the psyche is thus intrinsic to any co-creative undertaking.

While tying thought and feeling to the most primordial forces within, archetypal patterns also reach up via the imagination into the most significant arenas of meaning and value. The polar aspect of these patterns, which Jung likened to the infra-red through ultra-violet light spectrum, and the transfer of psychic energy between such poles, is arguably the most important characteristic of the psyche. Something is at work in us, even in the universe itself, that aims to transform raw life into what feeds the soul. This is not a one-way street, however: for any tree of life to flourish it needs primordial roots to source its living vitality.

Nonetheless, it has been part of civilization's recent experiment to try to sever these roots. Jung thus couched modern psychological malaise in terms of the "loss of instinct," suggesting it "is largely responsible for the pathological condition of contemporary culture."[40] But he qualifies this point by noting:

> It is not simply a matter of rescuing natural instincts . . . but of making contact again with the archetypal functions that set bounds to the instincts and give them form and meaning. For this purpose knowledge of the archetypes is indispensable.[41]

If, as Jung was apt to argue, what we know and how we think are first and foremost expressions of psychic reality, then "knowledge of the archetypes" may be the measure and final arbiter of all knowledge. As archetypes also find their quintessential expression in myth, it is by perceiving the myths within our pursuits that we reconnect the higher and lower regions of existence. Such mythic awareness would mitigate both our raw appetites and our flights of fancy.

Where the wires become crossed in archetypal theory is when no distinction is made between what Jung called the *archetype per se*, which is the underlying forms and propensities to perceive and experience life in certain ways, and the *archetypal image*, which varies by region, culture, and era. When this difference is overlooked, critics of

Jung's concept are apt to regard archetypes as stereotypes, contending they reduce existence to a series of psychological grooves represented by unchanging symbols and behaviors. But archetypes are less like grooves and more like rivers, wending their way through existence *as it is and as we have come to know it*, providing sources of life and means to navigate life, to be discovered anew in each individual and era. And myths are less like defined formulas for living and more like open-ended narrative essences.

I describe this archetypal view in some detail, aiming to clarify Jung's conceptions for one reason: it is, in the end, our relationship with this foundational aspect of the psyche that is at the core of our co-creative challenge, which is that co-creation is finally about archetypal attunement. How will we recognize and transform the parts of nature wending their way through us? How will we navigate these archetypal rivers? Will the algorithmic world and the pseudo-religious devotion it attracts prevent these undertakings? Just circumventing these matters will not work. As we have seen, archetypal propensities such as the religious function, the need for companionship and community, and consciousness of the whole will not disappear: they will only show up in unconscious ways. Between now and any point in the future where human nature may be abandoned, we will remain subject to these archetypal propensities and tasked with their creative transformation.

Jung once wrote, "So far as we can see the collective unconscious is identical with Nature to the extent that Nature herself, including matter, is unknown to us."[42] Our comprehension of this will likely correlate with our capacity to fulfill the humanistic stage of history, humbling our approach enough to add the deeply human and the unfathomable depths of the soul to the promise of reason, liberty, and justice the Enlightenment has offered, in order to move forward with a more complete account of this existence. We will hear nature's response to our actions, whether coming from outside or inside, and attune ourselves to the archetypal patterns of the collective unconscious, which are activated in both individuals and in collectives when our relation to the natural world pulls too far one way or another. Modern history has already shown this. As noted in Part 3, sexual fantasies and the fascination with the body arose at the height of the Industrial Revolution, when the instinctual world had undergone an unprecedented displacement. A revival of interest in pagan gods, especially in conjunction with a recovered sense of spirit in nature occurred during the same era. In the West, a sustained interest in Eastern philosophies and the body–mind connection also began around that time. The contemporary turn to yoga and psychedelic experience suggests the same dynamic. Such phenomena exemplify the activation of the archetypal world in the face of collective one-sidedness.

The sufferings and callings of the soul are ushering us into a co-creative imperative, offering pathways of return to the ground of being and intelligence of nature. This more archetypal and emotional intelligence stands in direct contrast to the artificial intelligence and algorithmic organization of knowledge now ascending. The co-creative imperative thus takes the form of a meaningful descent—a reversal of blind progress—with the purpose of discovering, or rediscovering, for a new time, a guiding presence within outer and inner ecologies to offset the growing autonomy of our machines and the misplaced sense of control they are instilling in us.

Anthropocene Citizenry

Responsible citizens of the Anthropocene Age bring consciousness to their involvement in the earth process in a way that is mirrored in their relationship with the psyche, facilitating simultaneous change to outer and inner ecologies. Anthropocentric attitudes in the outer world and egocentric attitudes in the inner world thereby give way to the inclusive vision of ecological interdependence. The memory of our place in the order of things generates an attitude of humility and reverence toward outer and inner dimensions of existence. Physical sustainability and psychological viability become primary aims. The Anthropocene citizen is thus the person who finds meaning and purpose in creatively responding to the forces shaping contemporary life, the person who thinks in complex patterns and conducts themselves and their work in reference to the spectrum of intelligences and sources of wisdom necessary for a co-creative evolution.

Carefully considering patterns of thought and behavior having the most impact on today's world, such a citizen may well have compelling local and neighborly interests, but they make their way through the world with the larger human predicament in mind. First, they are able to think for themselves, without succumbing to the pull of the algorithmically curated hive-mind. Second, they are capable of taking in the global situation, appreciating cultural differences while thinking of essential human betterment. Third, their style of thought is holistic and complex—accounting for what is conscious and unconscious, material and spiritual, combining the faculties of both cortical hemispheres, building bridges between reason and imagination. Fourth, they appreciate the thoroughly metaphorical, symbolic nature of deep and effective cultural discourse. Fifth, they understand all these capacities are borne of an embodied and relational mode of being, which arises by facing the limitations of life and forging a meaningful relationship with death.

Such a person exemplifies the full embrace of that most defining human ability, which is to think about one's thinking. This ability not only involves observing the world and oneself from a psychological distance; it leaves room for the mystery of being, which draws us closer in. For whether this mystery is conceived as divine, the deep intelligence of nature, or some other force behind consciousness in the universe, it is mystery that keeps us curious and contemplative. The capacity for distance must in this way coexist with the capacity for nearness—the ability to be both thoughtfully detached and closely related, if not passionately involved.

Missteps, anomalies, and tangents will surely occur. Failure will be part of the mix. However, this kind of involvement, propelled by co-creative responsibility, offers a "stable destiny," (Baudrillard) growing from the self-regulating dialogue between human desire and the total life process. In an immediate, practical way, focusing on this dialogue and its elements would juxtapose the "attention hijacking" that has overtaken us and given rise to concerns such as "whether the attention economy and a healthy democracy can coexist."[43] It would cultivate the art and science of attention, regulating how and where our consciousness is invested. For the Anthropocene citizen, consciousness is to the post-industrial age what freedom was to the industrial age, with the exploitation of the mind now following that of the body. Enslavement to machines and their owners remains the enduring metaphor; it is just applicable to a different locus of concern.

On the spiritual front, if there is to be a sustainable faith that gathers other faiths—though not all styles of faith—it will be focused on the co-creative process. Some may be inclined to call this "listening to

God" and, in some sense, especially in a Blakean and Jungian sense, it is this, as it surely involves the quest to rediscover what is sacred. But it is the process itself that will matter most, not the metaphysical or spiritual couching of the process: it will concern the contemplation in contemplative prayer, the mystery in mysticism, and the spirit in spirituality. Such an orientation to the process has more to do with an attitude toward things rather than with explaining their ultimate nature. This is what Jung was aiming for in his description of the religious attitude—the posture of paying careful attention to the forces affecting us, which comes prior to grasping for final causes or origins, and thus prior to dogmatic or creedal declarations.

If there is to be a sustainable story, a myth for the times, it will be a story involving some version of divine immanence, focusing on the sacred in everyday life. It will be a story in which spirit and matter are turned toward one another, a myth aware of itself as a myth. This will entail an acceptance of imagination and myth-making as innate to the psyche—an acceptance of story as inherent to the deeply human and an indispensable conduit to it. Accordingly, the Anthropocene citizen must be a warrior for the symbolic life and the metaphorical approach to the most important things. Such an Anthropocene posture must include science and technology, but work to orient both to the task of expanding consciousness. It will overcome the binary between technology serving us or our serving it, as this new myth of co-creative engagement supports neither position. It will foster ways of knowing and innovating that serve the integrity of the life process and the culture-making that flows from this.

The unholy alliance between the search for spiritual transcendence and the atomistic view of reality will be the main obstacle, for it is the combination of these postures that promotes the idea we can live detached from the rhythms and patterns of body, mind, and earth. At the end of his "Last Lecture," Bateson, quite conscious of punctuating his life's work, speaks of "the monstrous atomistic pathology at the individual level, at the family level, at the national level and the international level—the pathology of wrong thinking in which we all live." He then offers that this "can only be corrected by an enormous discovery of those relations in nature which make up the beauty of nature."[44] Such a "discovery," as Bateson put it, will not come by returning to natural ways of living per se, but by the deepening awareness of how nature works in our thinking—in depth psychological terms an awareness of archetypal patterning. For only by being conscious of these patterned relations in nature, which ultimately define who and what we are, can we live into any cultivation of how they will be present to us. It is the study of these "relations in nature" (Bateson) and archetypes (Jung) that gather into a perception of what that guiding intelligence is asking of us. The complexity of the many parts of the whole and their relations, carefully considered over time, generate this guiding intelligence—the partner of the co-creative.

Our current philosophies of life are largely based on the Enlightenment principles laid out two to three centuries ago, and there have been no advances in philosophy necessary to handle the advances and disruptions we now face. Reason, liberty, and the empirical method have formed a valuable, even inevitable, foundation for the modern world. But their inadequacy has been on display since the start of the Industrial Revolution. Contrasting philosophies that began to surface at that time affirmed the classical tradition and underscored the need for views that could reunite spirituality and nature at large. Nowhere is this as clear as in Schiller, who would use terms such as "God, Spirit, the Eternal, the Absolute, the Infinite, the Highest Idea, and even Nature,"[45] interchangeably. Ever since we have been plunged into streams of thought that have challenged the atomistic and compartmentalizing tendencies of modernity. Alongside the discovery of the unconscious, much work has occurred to reclaim the vital function of beauty and imagination, and reassert the value of contact with the earth and respect for its non-human inhabitants. Out of such streams, a significant qualification of the rule of reason and pursuit of liberty have occurred. As I have argued throughout, situating the logical mind within the larger frame of unconscious psychodynamics and their body–world contexts presents a game-changing check on the Cartesian-Newtonian paradigm.

For the philosophical framework of our contemporary outlook to change, individual efforts to be mindful must be combined with a second key element, which is accommodating a more holistic vision of reality. This vision reveals dynamic interconnections in both biological and psychological systems that cannot be fully explained by mechanistic and causal relationships based on atomism. It holds that breaking complex systems down into their constituent parts must be complimented by an understanding of the whole phenomenon. Among other things, this reminds us that the life world comes prior to the experimental world, forcing the return of lab results to actual conditions of living and being.

Being conscious in the sense of being wakeful or "mindful"—a growing means of countering the digital lifestyle—must be joined to this holistic paradigm shift. For the way we see, and the ideas by which we see, make all the difference to what we see. It is one thing to monitor patterns of thought and feeling; it is another to become conscious of habitually unconscious thought patterns and conditioned feeling responses. It is one thing to slow down the reactive feeling responses we regularly indulge in; it is another to know how to step beyond the narrow, one-sided character of the collective and invest consciousness where it is most needed. Consciousness is needed to redirect the tinkering instinct, not only to register its impact.

To make the same point another way, our dissociative tendencies have to be reversed through both individual awareness and allowing that awareness to follow paths of knowing that expand consciousness of self and of our collective situation. Holistic knowledge directly counters personal and collective dissociation. Whether it is unconscious psychodynamics, integrative neurobiology, body–mind connectivity, ecological systems or phenomenal and social contexts, paths of understanding forged in these areas are giving rise to a post-Enlightenment philosophical framework. As this framework enters and reshapes institutions of learning, where conscious awareness is most needed will be more readily recognized. The polarities we examined at the beginning of this book will be mitigated. Attending to our own psychic state may be the necessary foundation for undoing dissociative thinking, but this will be supported by a worldview that will also challenge dissociative tendencies.

Writing early in the industrial revolution, Nietzsche described the difference between active and passive nihilism. Whereas passive nihilism was essentially suffering the loss of meaning without reflecting on the experience or asking what it might be inviting in terms of greater awareness, active nihilism meant embracing the existential moment, living into what it was asking of those affected. Although Nietzsche's eventual solution of embracing a will to power and entering the age of the *übermensch* was a problematic overcompensation, the broader sense of directing one's consciousness to the existential task given by the zeitgeist is something to extract from his aphoristic philosophy.

As a bookend to what Nietzsche was eyeing a century and a half ago, might we not today think of an active and passive holism as progeny of this nihilism? I have provided many examples of passive attempts to make ourselves whole, including the posthuman movement itself, which looks to perfect and immortalize the remnants of human evolution. I have also touched upon forms of totalism, which attempt to place formulaic patches over the complexities of life. But an active holism would involve an engagement with those aspects of being most in need of revaluing and inclusion in the interconnected nature of being. It would include the imperfections, failures, and downturns that comprise any authentic and creative way of being.

An immediate worldly necessity stands behind such considerations, which is that the greatest challenges of our time are going to require global cooperation. Climate change is the most obvious of these challenges, but having come through the crisis of the Cold War, nuclear proliferation is once again posing a significant threat. It is also time the community of nations faced terrorism as a global problem and made a commensurate effort to prevent its arising and end its capacity to start wars. AI and the technocracy behind it now constitute a further major challenge, involving everyone connected by the World Wide Web. In each case, limits need to be placed on human, corporate, and government behavior. In each case, becoming more conscious of the total situation is most critical, with the main task of including what is now in the shadows. In each case, a capacity to embrace values that transcend consumption, domination, and control will be critical. Alternative models of economic prosperity and general wellbeing will require development. All of this will depend on attending outer and inner ecologies.

Yet, for such global cooperation to occur, which concerns connection at a macro level, our thinking about human nature and purpose has to change, which concerns connection at the micro level. The mechanistic explanations that have made the modern world are now giving way to emergent and systems understandings. If we have the eyes to see, the limits of mechanization have already become apparent in neuroscience, genetics, microbiology, and physics. Each of these disciplines has delved deeply into the material bases of the mind, the body, the earth, and matter itself to reveal an array of exquisite complexities that present formidable barriers for the quest to extract simple mechanisms. The social sciences and humanities have already crossed into this sensibility, which is part of the reason they are vital to our educational needs going forward. The appreciation of cultural differences and the discernment of common human concerns both foster tolerance and cooperation. Relationships are forged by first respecting difference and then discovering common ground; global cooperation will depend on both. The part–whole dynamics of understanding how we understand often demonstrate that meaningful connections are not always causal ones. Anthropocene citizenry will involve all of this.

Starting Close to Home

Most immediately, the responsible citizenship of our time must become aware of how the technocracy is shaping human behavior in surreptitious ways and insure steps are taken to properly monitor and regulate this power. In *The Age of Surveillance Capitalism*, one of the most important books of recent years, Zuboff exposes a core dynamic of our passive participation in the information economy. She lays out in considerable detail the way tech giants such as Google and Meta use personal information to keep users online through algorithmically selected content, directing the attention as well as the thought and behavior of these users, for the purpose of maximizing revenue even while minimizing responsibility. She describes the incentives of this kind of social engineering and how information technology is bringing about "a new species of power" she calls "instrumentarianism," which "knows and shapes human behavior toward others' ends."[46]

Zuboff describes exactly what Heidegger meant when he observed that modern technology was headed toward human commodification. As she puts it, "surveillance capitalism unilaterally claims human experience as free raw material for translation into behavioral data." A significant amount of this data is deemed "proprietary behavioral surplus," which is "fabricated into prediction products." These products are then "traded in a new kind of marketplace for behavioral predictions" she calls "behavioral futures markets." The computational processes involved "shape our behavior at scale," and eventually aim to "automate us."[47] We thus become "the objects of a technologically advanced and increasingly inescapable raw-material-extraction operation."⁴⁸ We must envision this post-industrial exploitation of our inner ecology as the mining industry of our time; the dirty hands, sooty faces, and giant excavators are just harder for us to see.

Such exploitation, which essentially defines the attention economy (Simon) and thrives on our loss of focus (Hari), is crucial to expose if there is to be a more conscious relationship with digital technology. And its exposure becomes even more crucial once we realize that tech corporations are also working to maintain our unconsciousness. As Zuboff puts it:

> Surveillance capitalists quickly realized that they could do anything they wanted, and they did. They dressed in the fashions of advocacy and emancipation, appealing to and exploiting contemporary anxieties, while the real action was hidden offstage. Theirs was an invisibility cloak . . . They were protected by the inherent illegibility of the automated processes that they rule, the ignorance that these processes breed, and the sense of inevitability that they foster.⁴⁹

Zuboff goes to the heart of the matter when she argues we are caught in the "daily renewal of a twenty-first-century Faustian compact," in that "it is nearly impossible to tear ourselves away, despite the fact that what we must give in return will destroy life as we have known it." Invoking the phenomenon I have identified as the basis of our addiction to technology and posthuman escapism, she notes, "This conflict produces a *psychic numbing* that inures us to the realities of being tracked, parsed, mined, and modified."⁵⁰ Effectively, as she puts it, "Surveillance capitalism knows everything about us, whereas their operations are designed to be unknowable to us."⁵¹ The thoroughly psychological basis of this economic phenomenon could not be more evident: tech corporations are in the human commodification business and their business model relies on human unconsciousness. This is not the path to co-creation.

Distinguishing technology from its technocratic utilization, Zuboff argues that the corrosive effects and vast implications of surveillance capitalism are not inevitable: "surveillance capitalism is not technology, it is a logic that imbues technology and commands it into action."⁵² Further, "technology is not and never can be a thing in itself isolated from economics and society . . . Technological inevitability does not exist."⁵³ Whereas this view highlights the inflection point we have reached, it downplays the inherently manipulative and exploitative nature of modern technology and the degree to which it has long been tending toward the turning of human life into another resource. As we invert actuality and virtuality, human and posthuman, the technology to which we will be subject will also become increasingly autonomous, making this inflection point harder to fully engage. Co-creation will be harder to reach.

At the end of her study, Zuboff describes the fundamental erosion of the free market generated by the unfair advantage associated with the mining of vast amounts of personal data. She describes the way that free market capitalism has, until this point, required that "markets are intrinsically unknowable." It is this "ignorance" that allows for the "wide-ranging freedom of action for market actors."⁵⁴ However, with the advent of the sophisticated tracking of personal choices and actions, "the 'market' is no longer invisible" in this manner because tech companies are able to track and predict consumer behavior in unprecedented ways. "Surveillance capitalism," Zuboff writes, "thus replaces mystery with certainty as it substitutes rendition, behavioral modification, and prediction for the old 'unsurveyable pattern.' This is a fundamental reversal of the classic ideal of the 'market' as intrinsically unknowable."⁵⁵ To bring her point home, she quotes Google's Eric Schmidt, who declared in 2010:

> You give us more information about you, about your friends, and we can improve the quality of our searches. We don't need you to type at all. We know where you are. We know where you've been. We can more or less know what you're thinking about.⁵⁶

This kind of knowledge did not exist until the digital age. Combined with market freedom, it makes for the titanic size and power of big tech. In contrast to capitalism prior to the digital era, "Surveillance capitalism is instead defined by an unprecedented convergence of freedom *and* knowledge. The degree of that convergence corresponds exactly to the scope of instrumentation power."⁵⁷ Zuboff concludes, "This cycle will be broken only when we acknowledge as citizens, as societies, and indeed as a civilization that surveillance capitalists know too much to qualify for freedom."⁵⁸ Our knowledge, our consciousness of the technosphere that surrounds us, can exert a crucial effect on that environment and make a decisive difference to our ecology of mind. Resistance will be based on values that preserve a deeper humanity. This is the co-creative challenge immediately before us.

Looking Further Away

A crucial part of the co-creative imperative is its call to connect day-to-day matters such as the mining of personal information to the larger questions and conundrums of our time. To participate in the Anthropocene is to attend the intermingling of the local and the universal, the contemporary and the timeless. It cannot thus be a meaningless coincidence that at the very time we are focused on AI as the great accelerator of knowledge, the James Webb telescope is looking at the origins of the universe and revealing we do not know as much as we thought we did. As we peer into the cosmos and literally look back through time, the images coming back to us are calling into question the basic laws of science. Astrophysicists are suggesting the Standard Model of physics, on which we base the Standard Model of the cosmos, looks wrong. Less than a hundred years since the study of the smallest particles produced the quantum mechanics that upended our understanding of matter, the not unrelated study of the largest objects in the known universe is presenting yet another challenge to that understanding.

This questioning of physical laws, foundational for all science, appears set to produce another Copernican revolution. Following previous revolutions of thought, set in motion by Freud, Darwin, and Copernicus himself, this latest cosmological shake-up will likely result not only in a reassessment of earthly existence but a reassessment of ways of knowing. As such, this latest revolution in our thinking may turn out to be a natural companion to its immediate psychoanalytic predecessor, with the challenges of observing sub-atomic particles and supernovas, on the one hand, and the implications of the discovery of the unconscious, on the other, both demonstrating that human consciousness is implicated in what we think of as the nature of reality.

On the side of physics and the cosmos, quantum mechanics brought attention to the way the very act of observation alters the phenomenon observed. The most famous expression of this principle is the thought experiment known as "Schrödinger's cat." Starting with earlier demonstrations that light presents as either a wave and a particle depending on the mode of observation, quantum mechanics moved to show that all matter observed at this sub-atomic level appears to display paradoxical wave and particle characteristics. Heisenberg's "uncertainty principle" has also functioned as a well-known marker of the apparent influence of the observer on what is observed, in this case the way the measurement of one property of a particle makes measurements of other properties less certain. It thus seems the way we look—including the kinds of premises and assumptions that go into setting up experiments and formulating findings—determines our knowledge of the world and the universe beyond.

On the side of depth psychology, the radical relativization of conscious thought (Freud) and the understanding that a second, more encompassing, ordering principle is at work in the unconscious (Jung), have together alerted the world to the way repressed experiences and archetypal patterns influence the perception of what is happening around us. Aside from upending the hard divisions between subject and object that much of science has been built upon, in psychology too we have been forced to contend with the fact that how we see the world is regularly shaped by unconscious factors.

Science is being shaken to the core by this latest "crisis in cosmology,"[59] but this crisis may eventually work to reinforce our understanding of the way the psyche cannot be separated from what we come to know and thus what we come to do. Extrapolating from the paradoxical nature and behavior of the sub-atomic particles just described, the quantum physicist and colleague of Einstein, Bohr, and Oppenheimer, John Wheeler, known for coining the term "black hole" and establishing the concept of "wormholes," also described what he called the "Participatory Anthropic Principle." He argued that "no phenomenon is a real phenomenon unless it is an observed phenomenon." Even further, he claimed that "we are participants in bringing into being not only the near and here, but the far away and long ago."[60]

Anyone familiar with depth psychology will recognize that Wheeler's Participatory Anthropic Principle aligns with Jung's concept of psychic reality. Beyond this, Wheeler's outlook, forged through the findings of quantum physics, echoes Jung's view that the ultimate purpose of consciousness is its intrinsic role in Creation itself. The dialogue between the known and unknown, conscious and unconscious, produces new symbols and metaphors that shape ultimate values, channel the imagination, and ultimately shift ways of knowing. All of this finds its way into the world we then make. We may create reality after all. Not in the ego-affirming, magical way many New Age gurus would have us believe, and not in the willful, desirous way posthumanism is eager to realize, but in the way we engage inner and outer unknowns. These ideas, coming from physics and from psychology, point to essentially the same thing—the recognition that something of a co-creative process is at work in the very nature of what we come to know and thus in what we proceed to do. We just have to become more conscious of it.

The Anthropocene citizen is one who is not only aware that technology has fundamentally altered the planet we are on; it has done so largely because we have viewed the planet as an unlimited resource. This same citizen must now extend this awareness to the technological alteration of human nature, a process whose realization relies on imagining minds and bodies as endlessly malleable. As I argued in the last chapter, waking up begins with a more conscious relationship with computation. And as I propose in this chapter, it proceeds by moving in the direction of co-creative humility and responsibility.

Taking up the challenges of this historical moment means awakening to our story as co-creators. Whether we imagine the fundaments of life as having been shaped by a divine hand, determined by natural evolution or both, creation is no longer a one-way street. Streams of thought in both religion and science recognize what is obvious to any thoughtful observer: human initiatives and their implications have begun to alter our very being. We have become change agents, not only for the course of history, but for the essential character of earthly existence. We have entered negotiations with God. Of all of the profound shifts modernity has occasioned, this one sets the course of the third millennium. One indicator of our more conscious embrace of this direction is the search for a re-ensouled or reanimated world—the subject of the next chapter.

Notes

1. Jean Baudrillard, *Telemorphosis* (Minneapolis, MN: Univocal Publishing, 2011), 3.
2. Iain McGilchrist, *The Master and His Emissary* (New Haven, CT: Yale University Press, 2009), 5.
3. Wendell Wallach, *A Dangerous Master: How to Keep Technology from Slipping Beyond Our Control* (New York: Basic Books, 2015), 164.
4. Jean Baudrillard, *The Illusion of the End* (Stanford, CA: Stanford University Press, 1994), 99.
5. Elon Musk, quoted in Maureen Dowd, "Elon Musk's Future Shock." *Vanity Fair*, April 2017.
6. Riane Eisler, *The Chalice and the Blade* (New York: Harper & Row, 1987), xiv.
7. Francis Fukuyama, *Our Posthuman Future* (New York: Farrar, Straus & Giroux, 2002), 14.
8. See Ted Peters ed., *Cosmos as Creation: Theology and Science in Consonance* (Nashville, TN: Abingdon Press, 1989).
9. Thomas Berry, *The Dream of the Earth* (San Francisco: Sierra Club Books, 1988), 19.
10. Ibid., 11.
11. Stuart A. Kauffman, *Reinventing the Sacred: A New View of Science, Reason, and Religion* (New York: Basic Books, 2008), 3.
12. Wallach, 245.
13. Ibid., 246.
14. Ibid.
15. Quoted in Jeremy Lent, *The Patterning Instinct* (New York: Prometheus Books, 2017), 277–278.
16. Rollo May, *The Cry for Myth* (New York: W. W. Norton, 1991).
17. Marco Heleno Barreto, "On the Death of Nature: A Psychological Reflection." *Spring: A Journal of Archetype and Culture*, vol. 75, pt. 1, 257–258. Italics in original.
18. Ibid., 258.
19. Ibid., 259.
20. Ibid. Italics in original.
21. Ibid., 261.
22. Ibid., 262.
23. From a letter to Reverend John Trusler, 1799. In F. Tatham, *The Letters of William Blake Together with His Life*, A. G. B. Russell ed. (London: Methuen, 1906), 120.
24. From the poem "Laocoön": "The Eternal Body of Man is The IMAGINATION, that is God himself."
25. Kathleen Raine, *Blake and Antiquity* (London: Routledge, 1979), 95.
26. Ibid., 96.
27. Ibid., 97.
28. Ibid., 97–98.
29. William Shakespeare, *Troilus and Cressida*, Act III, Scene 3.
30. Edward S. Casey, *Imagining: A Phenomenological Study* (Bloomington, IN: Indiana University Press, 1976), ix–x.
31. Mary Midgley, *Science as Salvation: A Modern Myth and Its Meaning* (New York: Routledge, 1992), 13.

32 Quoted in Casey, x–xi.
33 See Jung, CW 12, 13, 14.
34 CW 10, par. 524.
35 Robert Romanyshyn, *Technology as Symptom and Dream* (New York: Routledge, 1989), 6.
36 Ibid., 10.
37 James Hillman, *Re-Visioning Psychology* (New York: HarperCollins, 1975), 180.
38 CW 10, par. 53.
39 CW 8, par. 277.
40 CW 18, par. 1494.
41 Ibid.
42 C. G. Jung, *C. G. Jung Letters*, vol. 2 (Routledge & Kegan Paul, 1976), 540.
43 Charlie. Warzel, "I Talked to the Cassandra of the Internet Age." *The New York Times*, February 7, 2021.
44 Gregory Bateson, "Last Lecture." In Rodney E. Donaldson ed., *Sacred Unity: Further Steps to an Ecology of Mind* (New York: HarperCollins, 1991), 313.
45 Reginald Snell, Introduction to Friedrich Schiller, *On the Aesthetic Education of Man* (Mineola, NY: Dover Publications, 2004), 14. Originally published in 1795.
46 Shoshana Zuboff, *The Age of Surveillance Capitalism* (New York: Public Affairs, 2019), 8.
47 Ibid.
48 Ibid., 10.
49 Ibid.
50 Ibid., 11. Italics added.
51 Ibid.
52 Ibid., 15.
53 Ibid.
54 Ibid., 495.
55 Ibid., 497.
56 Ibid., 498.
57 Ibid. Italics in original.
58 Ibid., 499.
59 Adam Frank and Marcelo Gieser, "The Crisis in Cosmology." *The New York Times*, September 4, 2023.
60 Marina Jones, "John Wheeler's Participatory Universe." *Futurism*, February 13, 2014. https://futurism.com/john-wheelers-participatory-universe

Chapter 14
REANIMATED

There can be no mechanical prescription for demechanization ... Humanization will not come from doing any specific thing, but rather from doing whatever we do with a different orientation.

Philip Slater[1]

Technology can be reconsidered, each thing imaged anew in terms of *anima mundi*.

James Hillman[2]

At the end of his essay on titanism, "... And Huge is Ugly," Hillman offers an antidote to this consuming scourge. Faced with the overwhelming power of gigantic systems, abstract mechanisms, and soulless organizations, he revives a simple, direct instruction from a biblical psalm, "O Taste and See."[3] Being moved counters being removed. The psyche needs direct, sensate impressions in order to notice what is happening to it and to the world. Faster food, fouler water, noisier spaces, and automatons (human and non-human) offend the senses. Chopping vegetables, walking in the woods, sitting by a clear stream, and conversing with quirky people enliven the senses. Authentic, direct engagement with the things of the world and the presence of others pokes holes in the insulations of technocracy and the pseudo-satiations of consumptive behavior.

Noticing exactly when and where the senses are either diminished or delighted alerts us to titanic expansion. Sensations become emotions,

emotions become feelings, and feelings defy the sweeping objectification of existence. Hillman suggests there are three key aspects of this process—"components of the cure for titanism" that "interlock." He writes, "Reawakening the sense of soul in the world goes hand in hand with an aesthetic response—the sense of beauty and ugliness—to each and every thing, and this in turn requires trusting the emotions of desire, outrage, fear and shame," which he then describes as "the felt immediacy of the gods in our bodily lives."[4] If the Titans are the eternal enemies of the gods, and the gods enter everyday life as aesthetic and emotional responses, then it is the intelligence and guiding potential of such responses that become means of loosening the grip these giants have on the present way of the world. As discussed in the previous chapter, the archetypal world of the gods provides our best counter to the algorithmic world of the technocracy.

Recovering the soul–world connection involves being-with and knowing-with the things of the world rather than being captivated by our secondary conceptions and use of them. A direct encounter with our surroundings stirs the imagination, slowing us down, and invites particularity and alterity, which cut through elevated schemes and endemic numbing. As Hillman puts it, "This sense of the world as an animated being, as a living animal, is the first component of curing enormity."[5] Presences and voices, without and within, then protect us from the giant schemes and abstract systems that seem ever poised to define our existence.

Aesthetic pleasure, which has been minimized, if not undermined by the utilitarian outlook, may be more foundational than many psychological needs we readily identify. ". . . And Huge is Ugly" is a mercurial play on E. F. Schumacher's formulation "Small is Beautiful." It suggests aesthetic pleasure occurs at the meeting point of the world's capacity to attract and engage us and our capacity to look more closely and notice details. The beauty of both the naturally given and the humanly crafted is occasioned by loving attention, which prompts the forage for soul food. We begin to see that aesthetic responses depend just as much on the way we look as they do on what we are seeing. When the surrounding ugliness is normalized, and we are starved of beauty, it may be because the mind's eye has trained itself on the giant abstractions of Efficiency, Economics, and Progress. Even though it is the aesthetic quality of things that generates critical feelings of being at home in a community and brings a sense of deep order to what surrounds us, somehow this quality fades in the glare of these more abstract perceptions. Given it is our aesthetic acuity above all that promotes proportion and preservation and counters gluttonous consumption and unbounded development, this matter moves to the front lines of the contemporary battle against the titans of technocracy.

Of the three curative components Hillman names, learning to trust emotions is perhaps the most pivotal. To the extent the vision of a boundless posthuman future has a titanic cast to it, this component may also be the most consequential. Such trust begins with a natural appreciation of emotional intelligence and the guiding potential of this intelligence, which can feel like it emanates from a will of its own. This trust may grow over a long time or it may come as a sudden life-lesson. But few things beyond the contemplation of emotional responses open the door to the deeply human and the awareness of having an inner partner. Whether companionable or at times incorrigible, this partner operates apart from conscious, rational direction. To be full of emotion is to be animated by something in spite of ourselves; perhaps this is why we end up regularly conversing with the sadness or anger that grips us. Emotions like these stand in need of negotiation, mollification or perhaps just more attention. Often, when we want to keep moving on, the emotions will not allow it. And sometimes when we would choose to hold back, emotions charge ahead. If the emotions Hillman specifies—"desire, outrage, fear and shame"—have anything in common, it is the tenacity with which they outwit conscious intentions and insist on having their way with us. Of course, all of this rests on the ability to resist our dissociative tendencies.

It is these specific emotions that arise in response to the titanic exploits of civilization and stand opposed to the dissociative tendencies of the digital lifestyle. We are affected by desire for more satisfying ways of being, outrage about all that is regularly exploited and destroyed, fear of what stresses and overwhelms, and shame about participating in the machinations of it all. To consider the way emotion comes upon us, from a nature we call "ours," all the while being a branch of Nature itself, calls into question whether these emotions even belong to us or to the sufferings of the world, and we are being directed to feel and respond on its behalf. In other words, our animation may be the world's way of speaking to us, and thus be an indispensable dimension of the co-creative process. Recognition of the autonomous intelligence of these deep emotional responses may be an invitation to attune ourselves to the presence of the earth's own intelligence—or intelligences.

Contemplating the "emotional toll" of modernity, a contemporary consortium of philosophers led by Alain de Botton, write the following:

> Though it presents itself to each one of us as a personal affliction, our condition is the work of an age, not of our own minds. By learning to diagnose our condition, we can come to accept that we are not so much individually demented as living in times of unusually intense and societally generated perturbance. We can accept that modernity is a kind of disease and that understanding it will be the cure.[6]

Avoiding the personalistic framing of afflictions we automatically call "ours" can make the difference between emotions taking an overwhelming toll and our learning to trust in them. To perceive the way we are harmed by titanism and dissociation, which may be the contemporary world's chief afflictions, is to understand the overlap between inner and outer ecologies and to feel the soul that joins us to the world. While painful and costly in our personal lives, an aesthetic and emotional responsiveness that reanimates the world itself may be gained. This reanimation in turn provides us with a more adequate container for our dilemmas and sufferings, for they can be placed upon an altar of the earth's becoming. Our individual suffering may then be construed as a sacrificial offering to the co-creative process—to the work of securing a human destiny.

A memorable expression of the soul-saving recognition of the psychological bond between us and the world—or in this case the universe at large—occurs at the end of the film *Blade Runner*,[7] based on Dick's novel *Do Androids Dream of Electric Sheep?*[8] Rutger Hauer, a fugitive "replicant" (bioengineered humanoid), has been pursuing a now weaponless Harrison Ford, a "blade runner" (humanoid hunter), across the dark, wet, and greasy rooftops of a sprawling Los Angeles—the hunter having become the hunted. Ford's character Deckard makes a final, desperate leap but ends up hanging by his fingertips, and is about to plunge to his death—the outcome we suppose the replicant has been seeking. In the blink of an eye, at the moment Deckard loses his grip, the replicant grabs his hand and pulls him to safety. By saving the life of his erstwhile executioner, knowing he is himself about to die, the artificial human executes a profoundly human act. The hand with which he grabs Deckard happens to have a large metal shard running through it, and this ends up piercing the flesh of both. Once we see the replicant release the white dove he has been holding in his other hand, the Christlike connotations are apparent. Yet, it is the astounding last words of Hauer's replicant character that leave the most enduring impression, and must be counted as a great entreaty of the manufactured human, spoken to us from one possible future.

To a now stunned and undone Deckard, the replicant says with deliberation and pathos:

> I've seen things you people wouldn't believe. Attack ships on fire off the shoulder of Orion. I watched C-beams glitter in the dark near the Tannhäuser Gate. All those moments will be lost in time, like tears in rain. Time to die.

Instead of a moral admonishment, or a personalistic cry for equality, or advice for the corporation and bureaucracy directing Deckard's actions, the replicant turns to the beauty and tragedy of his existence. His last moments are a plea to recognize the soul that comes from bearing witness and creating moments of ultimate value—moments of awe that gather in the stream of consciousness that cradles the very existence of the universe—making it seen and known, loved and feared.

The humanoid is telling us, if such moments are "lost in time," if we cannot find ways to preserve and value the particulars of our encounters with the cosmos, we too will lose the purpose and meaning of existence. If we create beings whose experiences and memories only disappear "like tears in rain," we will be guilty of a most profane act. Such a fabricated existence, exposed to an awesome wonder yet denied a fulness of life and sense of soul that preserves and cultivates this wonder makes for a double-bind of epic proportions. Such a double-bind can only result in an implosion of madness or an explosion of rage. Although some time off, we can easily imagine the moral and ethical pitfalls of such biological creations. But the more immediate meaning concerns the way our own artificiality may already be robbing us of such a soulful experience of what surrounds us.

If the artificial being of science fiction mirrors our increasingly robotized selves and own existential suffering, they might indeed be functioning as Christ figures—revealing our propensity to act unconsciously, exposing our tendency to follow the crowd, demonstrating a love for all things, even in the face of death. Essentially, we are all on our way to becoming siblings of this replicant, torn between the soul of existence, the vacuity of the hive-mind, and the forces of technocratic predation.

It is not incidental that the artificial human whose predicament elicits a final reach for deep humanity would also reach for poetry, the only form adequate to the tragedy and pathos. "Tears in rain" conveys the disappearance of this unique sadness as well as its addition to a universal melancholy. The tension-breaking presence of the rain itself completes the image—a not uncommon motif at the end of a film—as if the clash between heaven and earth, between spiritual directives and earthly needs, is overcome, at least for this one moment. Unable to preserve his own life and thus receive no true recognition or dignity, he sacrifices himself and thereby makes sacred the human story, in that moment transcending his fabricated nature to become human. This is an incarnation of sorts. The parting words, uttered against the backdrop of a giant neon TDK sign, with all the worldly business of replication and facsimile implied, convey the irony of someone less-than-human reaching for something more-than-human, putting us all on notice in the process.

Gaia, Spirit, and Matter

Through the imaginative embrace of the sensations it arouses in us, the world is reanimated, and through this reanimation a feeling of soul is generated that is no longer wholly within us. Beyond mere aesthetic appreciation, this reanimation or ensouling may be a crucial regulator of existence as a whole, challenging the notion that we or the world could go on from this point without a form of consciousness that reflects this inextricable part–whole dynamic. Describing the nexus of Hillman's psychology and Heidegger's philosophy, Avens writes that "things necessarily exist through and by other things, and there is no isolated Being or 'thing-in-itself.' The whole of nature works through each thing and each thing is a reflection of the whole." This whole, he continues, "is experienced, as in Blake, through that joyful scrutiny of detail, that intimacy of each with each such as lovers know."[9] This reanimated vision of an ensouled world, a contemporary realization of the ancient notion of the *anima mundi*, may be the psychological and cultural boon that results as we rescind the attempt to master all that is juxtaposed to will and desire.

Plenty of backup for this position is to be found in Bateson, who says:

> It is not new to assert that living things have immanent beauty, but it is revolutionary to assert, as a scientist, that matters of beauty are really highly formal, very real, and crucial to the entire political and ethical system in which we live.[10]

By bridging the biological, the aesthetic, and key elements of social discourse, Bateson's assertion raises the question of whether society might even be sustainable without the capacity to respond to the beauty and ugliness of the world. Schiller was also aware of the regulating character of aesthetic response, noting, "Through Beauty the sensuous man is led to form and to thought; through Beauty the spiritual man is brought back to matter and restored to the world of sense."[11] Here we see how a co-creative ecology of mind generates a confluence between the natural and the cultural, the deeply human and the socio-political, and how a reanimated, ensouled world may be more than just vital for our psychological integrity: it may prevent the very collapse of existence.

In awakening to this necessary reanimation of the world, one momentous step has been the revived image of the earth as an entity. This step has meaningfully coincided with a period in which some of the longest shadows of technoscience have become more clearly apparent. As Berry states, "One of the finest moments in our new sensitivity to the natural world is our discovery of the earth as a living organism."[12] Photographs of the earth taken from space appear to have played a key role in this shift in consciousness. An understanding of ecosystems and the ecosphere, previously apprehended with more atomistic biological assumptions, have also prompted this reconnection. Approaches such as Lovelock's Gaia hypothesis thus reflect the coincidence of a shift in collective consciousness and an accumulation of scientific insight.

Giddens, commenting on Lovelock's hypothesis as a form of "utopian realism" that would involve the "humanization of technology" and some "system of planetary care," suggests:

> If this view can be authenticated in analytical detail, it has definite implications for planetary care, which might be more like protecting the health of a person rather than tilling a garden in which plants grow in a disaggregated way.[13]

Since the 1990s, Giddens's sense of these implications has proven correct; however, the analytical authentication is only part of this. As I have argued here, in a conscious, co-creative way of being at least, what must follow holistic ecology is holistic psychology, which is what we find in the writings of Jung. And it is here we become aware of the propensity in the psyche to perceive the integrity of the earth by following an archetypal mode of apprehension, and to suspect that there is something of a symbiotic relationship between the attention toward inner and outer integrity. The value of the Gaia hypothesis, while conceived in science and enjoying at least some of the analytical authentication to which Giddens alludes, may belong more to being the right idea at the right time. It is this *kairos*, together with the overlap of inner and outer significance, not to mention the personified, mythopoeic form of the perception of the world as a whole, that places Gaia at the center of a reanimated world.

Overarchingly, a reanimated world is one in which spirit and matter are not just equally regarded but recognized as mutually dependent. The great task of this late modern era is thus to bring together what the spiritual preoccupations of the old world and the material focus of the new world have torn apart. The psyche shows us this dependency whenever a person or group attempts to embrace one without the other, in the way the neglected side begins to rule the unconscious. Devotees of spiritual paths and religious ideologues are often either consumed by their sexuality or ruled by other trappings of "the flesh"; materialists are mesmerized by searches for "God particles" and other quasi-mystical pursuits. But the earth process itself suggests we rediscover nature as spirit as well as understand it as matter—nature as presence, intelligence, and root source of inspiration and imagination. Surely the rise of interest in indigenous wisdom, shamanic practices, and psychedelic journeys is but one indication of this rediscovery. It is hard not see that both mind and earth are calling for perspectives capable of marrying these dimensions of reality.

One answer to these polarizations, which I have put forth in these chapters, is to recognize the psyche itself as the common foundation of all perception and experience, the common source of all thought and action, and the ground of the deeply human. It is the psyche that contains the pursuit of the angelic, the claims of the animal body, and the structures and dynamics that join the two. It is the psyche that generates and insists upon the symbolic expressions of culture, which are often based on the transformative and aspirational power of ordinary, even elemental, things—mountains and rivers, suns and moons, fire and rain—thereby reminding us of the inextricable bond between mind and world. If cosmic matter has given rise to consciousness, and we are now called to grasp the nature of consciousness, realizing that thinking about our thinking is the necessary companion of any exploration and understanding of the universe that surrounds us, our grasp of the inner world becomes just as vital as our grasp on the outer world—perhaps even more so. Here we have the opportunity to observe where and how our current orientation is lacking. As Jung puts it:

> If we were conscious of the spirit of the age, we should know why we are so inclined to account for everything on physical grounds; we should know that it is because, up till now, too much was accounted for in terms of spirit. This realization would at once make us critical of our bias. We would say: most likely we are making exactly the same mistake on the other side. We delude ourselves with the thought that we know much more about matter than about a "metaphysical" mind or spirit, and so we overestimate material causation and believe that it alone affords us a true explanation of life. But matter is just as inscrutable as mind.[14]

Crafting, Sensing, and Returning

Richard Sennett opens his book *The Craftsman* by recalling an exchange with his teacher, Hannah Arendt, who told him, "people who make things usually don't understand what they are doing."[15] This reiterates the problem of the tinkering instinct becoming isolated from other parts of our nature, and the related failure to entertain the consequences of what we make. As Sennett relates, paraphrasing Arendt's argument, she felt that human cultural dialogue and political will would be capable of managing the power and implications of innovation. But, as Sennett sees it, this is not good enough. Somehow the ability to envision the use and implications of things has to be woven into the crafting process itself. As he puts it, "to cope with Pandora requires a more vigorous cultural materialism."[16] Here he is pointing to something phenomenologists have attempted to highlight: our ordinary, reductionistic, materialistic outlook turns things into objects for mere manipulation and exploitation. As Sennett relates, this outlook "slights cloth, circuit boards, or baked fish as objects worthy of regard in themselves ... the thing in itself is discounted."[17] In spite of this posture of disposal, things then take on a life of their own, often in the form of a vexing and toxic imperishability. From space junk orbiting the earth to microplastics in the ocean, it is as if what we use and forget demands a final reckoning. As we seem fated to literally collide with or ingest what we have discarded, reassessing our initial orientation seems indicated.

What Sennett is promoting is the reconnection of action and thought, doing and knowing. As he puts it, this "focuses on the intimate connection between hand and head," and promotes "a dialogue between concrete practices and thinking."[18] For him, this is the essence of craftsmanship. The problem is "Western civilization has had a deep-rooted trouble in making connections between head and hand, in recognizing and encouraging the impulse of craftsmanship."[19] Here too we find an echo of the loss of techne, which, in its original form, maintained the kind of hand–head relationship Sennett is talking about, precisely because the craftsman regards the world and materials engaged in crafting in a more dialogical way.

But what of today? How may we recover this sense of craftsmanship and techne? Sennet draws inspiration from the sociologist Georg Simmel, who writes about the ability of the stranger, compelled to pay close attention to their surroundings. Such a person "learns the art of adaptation more searchingly." And this prompts him to suggest that we too must learn to approach our surroundings anew. As he contends, "So great are the changes required to alter humankind's dealings with the physical world that only this sense of self-displacement and estrangement can drive the actual practices of change and reduce our consuming desires."[20] This may seem a counter-intuitive response, but it is one in keeping with the idea that the experience of alienation also generates opportunities for reconnection.

These considerations resonate with Romanyshyn's attention to the problem of "return" after his meditation on the ways modern technology has distanced us from the earth; indeed, the ways in which, as previously noted, "we are all astronauts."[21] In bringing home his thesis on this cultural and psychological distancing, he engages with the metaphorical richness of the attitude of a spacecraft's atmospheric reentry. If this reentry is too sharp, the craft burns up; if it is too shallow the craft encounters "a wall of rejection, sending the spacecraft bouncing off into the heavens." Citing the critical "cultural-psychological work of re-entry," Romanyshyn suggests if this work is "too much a passion to return to things as they once were, or as we imagined them to have been, we may be destroyed." This may "take the form of nostalgic longing for a pre-technological Edenic world; for a simpler, more innocent time in which, as we fancy it, we lived more harmoniously with nature." If, by contrast, "our desire to return is too shallow we may lose the earth." Here he is referring to earlier intimations of the many technological promises explored in these pages, including "a future which will one day dispense with the earth, which will one day free us of its bonds in much the same way it may liberate us from the necessity of death by freeing us from the body."[22]

To successfully return from our various states of departure and all this implies would, according to Romanyshyn, involve "a reaffirmation of our erotic tie to the earth, and perhaps even a rediscovery of the erotic character of the earth itself."[23] As we relate to that which we experience as having presence, intelligence, and agency, to subjectivities appearing outside us, though not necessary in other human persons, this "erotic character" is another name for reanimation, an altogether different way of being in the world that our technological overreach may have now occasioned. As Romanyshyn puts it:

For it is only in leaving one's home, in departing from it, that the possibility of return arises. It is only in our distance that the possibility of remembering home takes place, the possibility of taking up in a new way what was, the possibility of recovering as a destiny what was heretofore a heritage.[24]

Thus, to be ensconced in this digital age is to be both stranger and astronaut, called to contemplate one's estrangement and undertake the challenge of reentry. Whatever our profession or way of life, we all craft things. Even more so, we are all involved in sponsoring what is crafted. And it is here we must appreciate the almost inconceivably important difference between a mindless making that maintains our separation from the things of the world and a mindful crafting that affirms our ties to those things—ties that restore the animated presence of the once detached object.

An immediate, everyday sense of this admittedly far-reaching matter may be found in what we eat. Both the provenance and preparation of food are arenas in which crafting and techne may be recovered. Noticing the very different way we take in carefully cultivated and devotedly prepared food compared to highly processed and fast food is a natural measure of our ecology of mind. The former finds us paying more attention to each bite, becoming more aware of textures and flavors; the latter induces thoughtless snacking, wolfing down excessive portions, or even binging to fill something more than our bellies. This "something more" may well be the perceived emptiness of dwelling in a world without sensuous presence and aesthetic delight. Writ large, these contrasting encounters with food—hand-crafted and consciously cooked on some occasions and exiting automated production lines and consumed accordingly on others—returns us to the matter of our presence and absence and its relation to the titanic syndrome laid out previously: "O taste and see."

For those who believe "we are what we eat," this everyday matter rises to the level of a philosophy of life and becomes intimately tied with minding our place in the earth process. But we need not literally "live off the land" or "off the grid" in undeviating ways to attend the broader matter of how the earth sustains us. We can be citizens of the world and invested in the intersection of inner and outer ecologies by becoming more psychologically bonded to landscape and place, which actuates a posture of dedicated care. This is also a craft, one that reminds us that living off the land is, in its widest sense, an existential given. We can also discover ways to exist beyond the many kinds of techno-grids currently dictating lifestyles and identities. We can, that is, take in the metaphorical reach of what it means to eat consciously versus consume unconsciously, along with the associated concern for all that we ingest and metabolize.

Helping our digestion of these matters, Peter Bishop puts forth three meanings of the word "colon". The first is the intestinal organ; the second is the punctuation mark; the third, of old English use, "a person who tills the soil." Each suggests a kind of pause between other processes, pointing also to the psychological necessity of slowing down as we take in whatever it is that feeds us. Bishop contends:

> The pause—the colon—must be given its place in the relation between imagination and nature. The pause through which we may descend from bodily bowels into the bowels of nature and earthly things, into the world of Hades [death] and Chthon [underworld], into the very bowels, the guts of the imagination. The stomach and gut dislike too much rush and haste, so the colon needs the colon, needs a pause . . .[25]

Soul food of any kind cannot be rushed. In fact, speed may be the great enemy of soul, leaving no room for reflexivity and the downward movement this quality of being requires.

It is through these slower and deeper forms of contemplation that we attend to the presence of the earth in our thinking, a mode of thought in which we come to really know the materials of our work and from which well-created things emerge. And it is this same mode that sensitizes us to what we take in and allow to course through our minds and bodies. The soil must be well tilled if things are to grow organically. And in this, forms of passing, surrender, and dying must find their place. Such notions seem to be what Campbell had in mind when he declared:

> Life lives on life. This is the sense of the symbol of the ouroboros, the serpent biting its own tail. Everything that lives, lives on the death of something else. Anyone who denies this, anyone who holds back, is out of order. Death is an act of giving.[26]

Campbell is placing death in the great round of life, imbedding it in a fabric of ultimate meaning. Part of the problem with the oft-cited death of nature is that this also means the death of death. That is, transpiring without ritual or ceremony or even acts of mourning, death too frequently suffers from its own lack of significance, because it is not attended in life. In this way we live without giving enough consciousness to what is lost or has died and are often left contemplating how to "move on." Under these conditions we lose our sensitivity to everything that participates in the great round of life, including all the ways we relate, make, and eat.

If we take Campbell's formulation seriously, however, reconstituting this cosmic order, and we simultaneously add careful crafting and a sense of techne to all innovative undertakings and life-sustaining choices, our reentry will occur with the fitting attitude. We will neither bounce of the atmosphere through posthuman detachments from the flesh and denial of death, nor burn up by attempts to revert to past ideals—real or imagined.

Perhaps another place of everyday insight concerns our contact with the wilderness, the preservation of which has become a litmus test of outer ecological commitment. But this dimension of our broader relationship with nature pertains to our inner ecology too. Wild places are related to the wild reaches of the soul, and contact with wilderness replenishes unseen, untouched, uncivilized places ourselves—parts related to the deep intertwining of life and death. Some way or another, human consciousness must have contact with these places. Whether directly or indirectly, wild nature reminds us there are parts of the earth and of ourselves that are virginal and must remain beyond rule. As revealed by the practice of *shinrin-yoku* or "forest bathing," given our increasingly fabricated environment, a contrasting immersion in the sights, sounds, and rhythms of undisturbed landscapes may be vital to mental hygiene. As Jung puts it, "People who have got dirty through too much civilization take a walk in the woods, or a bath in the sea . . . they shake off the fetters and allow nature to touch them." And in conjunction, "entering the unconscious, entering yourself through dreams, is touching nature from the inside and this is the same thing, things are put right again."[27] Jung's student C. A. Meier presents the relationship even more starkly, noting, "The wilderness within would really go 'wild' if we should badly damage the outer wilderness."[28] And by the same token, we might deduce that a failure to recognize the wilderness of the psyche is likely a primary factor in our willingness to erase outer wilds. Aside from their immediate medicinal value, these earth–soul encounters with wild nature inculcate a more encompassing sensitivity to a deadened, objectified world given to manipulation and exploitation, a world that keeps dividing inner and outer ecologies.

In the highly industrialized world of Japan, whose infrastructure both overwhelms the senses and paves over vast stretches of the landscape, aside from the turn to practices such as *shinrin-yoku*, there are also many postures that preserve the animated sense of things, from the recognition of nature spirits to more conscious and less consumptive relations with physical items. Here death is not dead, but very much alive. This is attributable in part to the Shinto philosophy that still permeates much of the culture, but it also stems from an ethos of attending to small physical details and subtle human gestures, which bring more feeling to everything from food preparation to everyday human interactions. While it is possible Japan's technics will end up crushing its traditions, it at least possesses ways of knowing and being that could be indispensable for a viable future. In this society the door is opening quite wide to all manner of human augmentation and robot relations. Whether or not the sense of crafting and the rhythms of the natural world will endure may well be a lesson for the rest of the world. Whether or not manga culture and Japanese aesthetics will help us reimagine human–machine relations will be an early indicator of the collective psyche's self-healing capacity.

We may distill one core understanding from all these considerations: whether it arrives through perspective or practice, reanimation is implicated in relating and crafting with a co-creative orientation and is a primary antidote to a titanic technical takeover. One cannot simultaneously attend to the particularities of the world or to the uniqueness of others and be thinking and acting in service to vast abstract systems. Reanimation may also, as Jung points out, not only signal the overcoming of psychological fragmentation; it may be the outer correlate of contact with that inner ordering principle of the psyche that always relativizes the ego, which he terms the Self—borrowed from the Hindu tradition and designating the presence of the divine within. Contact with such an inner principle is, for Jung, the goal of the individuation process, for it provides the feeling of an ultimate value around which other aspects of the personality turn. The experience of the *unus mundus*, a coherence of inner and outer life, is also associated with this pattern, one often symbolized as an overcoming of multiple forms of division, if not restoration of a sacred whole. This restoration frequently coincides with the sense of soul in the world described above. Indeed, it is telling that after dedicating his work as a psychiatrist to understanding the inner realm, towards the end of his life, in concluding his autobiography, *Memories, Dreams, Reflections*, he underscores the animation of the world around him. He writes, "there is so much that fills me: plants, animals, clouds, day and night . . . The more uncertain I have felt about myself, the more there has grown up in me the feeling of kinship with all things."[29]

Perhaps the most direct and familiar means of reanimation occurs by embracing the role of art in culture. Recall Heidegger's perspective on this, especially in regard to the fuller sense of techne—a form of engagement with the world in which invention prioritizes a more concerted dialogue with the nature of things rather than the attempt to control this nature. As Avens put it:

> Heidegger's approach . . . intends to let things be the way they are, to let the thing appear in its own light. The privileged locus of such an appearing is art. Only art, or for that matter, poetic thought discloses the thingness of a thing, for in these disciplines the essences of things show themselves wholly in their modes of appearances.[30]

Here we are introducing a necessary pole of thought to the scientific one, which would offset the kind of objectification and scientism that have invited the posthuman outlook. What is it that art or poetic thought perceives, exactly? What do Heidegger and Avens mean by "thingness"? It is the image of a thing, the way it strikes the soul and is perceived by the soul, the way it comes to reside in the depth of being. In making this bridge to images and thus to the imagination, Avens cites Miller, who writes, "In the world of images . . . things are both mindful and passionate, both ideal and real, infinite by very present, divine and human at once."[31]

If science pulls our perception of the world toward objectification and utility, which subsequently foster manipulative and exploitative action, art pulls our perception toward the soulful and enchanted, occasioning reverence and awe and encouraging extended contemplation. If one tends us toward the use of the world (power), the other tends us toward the embrace of this world (love).

Reanimation is really a recovered sense of the animation already present to the world, albeit in a form fitting for our time. It is not, as some critics may charge, a reversion to the Middle Ages or some other period of enchantment. Nor is it a retreat from the world we are in. It is a recovery of the *anima mundi* as a mode of imagining intrinsic to the human psyche, a mode that appears essential as we move from objectifying the world to objectifying ourselves and are beginning to experience an inevitable suffering.

Such suffering is surely problematic in itself. However, as I have indicated at various points, unconscious efforts to manage the syndromes of soul loss and existential pain are also instigating attempts to augment, perfect, and transcend life. It is thus no coincidence that the attempt to create animated machines, particularly the attempt to make humanoid robots and create more personable and relational virtual assistants, coincides with the kind of thinking that also de-animates the everyday world. Yet by raising the artistic view to the same level as the scientific view, both the grip of scientism and the fantasy of creating artificial life forms would be depotentiated.

A New Counter-culture

Throughout this book I have treated the term "new" with suspicion, largely because the futurism, technological solutionism, and horizontality that background posthumanism and technocracy converge on a cultist obsession with newness. Both the promise of the new and the unremitting production of new things drive the global economy, which constantly reinforces an ever narrowing sense of progress. In the effort to reconstitute the deeply human and restore an ecology of mind, this one-sided and often superficial obsession with newness presents a sizable hurdle. Among other obstacles, opposition to the old and, by implication, the diminishing relationship with history is a barrier to human self-knowledge, for we cannot look deeply at ourselves or life at large without looking back. For all these reasons, I invoke the new with caution, but also with confidence that our obsession with the new has begun to grow old, and our recollection of the past may be, oddly enough, something new for generations consumed by the heady promise of technology.

Our ultimate goal, however, must be the joining of the old and the new, the past and the future, becoming more conscious of how they are, when viewed through vertical ways of knowing, always dancing around one another. And our willingness to enter this dance is a sign of human maturation. Depth psychologists consider individuation to be a process in which we must connect the universal and timeless to the unique and temporal; historians insist cultures cannot evolve without knowledge of the past; experts on creativity declare new ideas come from studying old ones. Can we actually produce anything that is truly new? Or is newness a fantasy of pristine creation, a trick of the mind, an expression of godlike pretension?

Hillman wrote about old and new, past and future as one archetypal configuration with two sides always playing off of each other.[32] In personified, mythic form they appear as *senex* and *puer*—old man and young man—prone to pushing each other away only to eventually face their inevitable bond. Without contact with youth, old age lacks twinkle and shine; without aged wisdom and proximity to mortality, youthful exuberance wanders astray. Noticing the mood of the 1960s wherein a spirit of breaking the bounds of convention and the spell of authority had taken hold, and in eyeing the need for renewal in his own field of psychology, Hillman drew attention to the *senex–puer* dynamics of that period and searched for ways to support the call to reimagine and revision ideas of human becoming. As I have described his view elsewhere, "the failure of the elders created the brooding, youthful soul-searching and flights of imagination, leading to a countercultural movement."[33] However, Hillman simultaneously recognized the problem of polarization, which both instigated and then eventually hobbled this movement, a polarization that has only deepened since.

Hillman begins his 1967 essay with a quote from Jung: "We are living in what the Greeks called the *kairos*—the right moment—for a metamorphosis of the gods," and "Coming generations will have to take account of this momentous transformation if humanity is not to destroy itself through the might of its own technology and science." He also quotes Hoyle, who wrote of "the transition from primitive to sophisticated technology;" yet noted, "We find ourselves in no real contact with the forces that are shaping the future."[34] Hillman then offers:

> To have "no real contact with the forces that are shaping the future" (Hoyle) would be to fail the *kairos* of transition. To come to terms with this *kairos* would mean discovering a connection between past and future. For us, individuals, make weights that may tip the scales of history, our task is to discover the psychic connections between past and future, otherwise the unconscious figures within us who are as well the archaic past will shape the historical future perhaps disastrously.[35]

Whereas the emergence of the counter-culture—a term coined by Roszak—signaled a time of transition and need for transformation, the rebellious, revolutionary style and drug-induced, ecstatic means ended up volatilizing the new and calcifying the old. Even if the style had been invited by the madness of wars and environmental destruction, and even if the means expressed a need to expand consciousness and reclaim the sensual body, the reactivity compromised reflexivity. So, the transformative possibilities imbedded in the counterculture's original outpourings were neither realized nor sustained, because they never fully infiltrated the social fabric. Folk-rock elder Bob Dylan might have received a Nobel Prize for Literature, but his commentary on times changing proved a distant cry. The counter-cultural conviction that nothing but a complete overturning of traditions and institutions could alter the psycho-social situation was misplaced. Staying forever young and finding the words of prophets on subway walls would be inadequate.

Nonetheless, the fire that burned brightly and briefly and produced music, poetry, and magic that successive generations have gone on to tap, helping them see beyond conventional thinking, may well be primed for rekindling. What is needed today, however, is a slower, more enduring fire, something that produces more illumination and less immolation, so that traditions may be revisioned rather than rejected and institutions reimagined rather than torn down. The young poet must return to the old polis, and the old polis must clear the way for new openings. Old and new, past and future would then together address the historical moment and the spirit of the age.

By observing and encouraging entry into a new counter-culture, I am thus not advocating a return to a polarized confrontation between the old and the new. I am certainly not advocating a counter-cultural inversion in which the old usurps the new. In the decades since the first counter-culture we have already seen a return to economic conservatism, a surge of religious fundamentalism, and the rise of right-wing extremism, all regressive turns to the past that are symptoms of failed cultural transformation. What any new counter-culture must combat is onesidedness and polarization itself. Even if the immediate task at hand and initial corrective moves mean emphasizing the significance of certain age-old principles to offset the obsession with the newness, the ultimate goal must be the coupling of the two.

This does not mean configuring a flaccid neutrality. A strident recollection of the past must be injected into the vision of the future, infusing the conscious mind with an understanding of its unconscious underpinnings. And it will mean traditional values and views throwing off their rigid, paranoid defenses and finding new vehicles of expression.

The person who turned twenty-five in 1973 turned seventy-five in 2023. The ones once focused on the revolutionary possibilities of youth are now the elders, looking back on the generations that gather around the turn of the millennium. How do they join the old and the new? They may well recognize their early excess and overreach. But they must also long for the promise of this earlier time—the ideal of a world in which greater contact with soul and spirit might redirect the course of things. Perhaps the final act of this generation could be using the wisdom of their years to recollect the past and rekindle the counter-cultural imperative. Perhaps the still glowing embers of that previous era could ignite a more illuminating and sustainable flame. And might not today's digital natives simultaneously realize their restlessness requires a resuscitation of the wisdom of the past and a return to the vision of the elders? What is today's nostalgia for all things analogue, even among those who never knew such things in the first place? Through these openings, the new counter-culture would not only combat the polarization of old and new; it would bridge the generational divisions that plagued the late modern era.

If Hillman's paper hints at the polarizing of old and new, past and future being implicated in the problem of technology, Roszak's 1969 work *The Making of a Counter Culture* sets out this implication in precise detail.[36] His subtitle, "Reflections on the Technocratic Society and Its Youthful Opposition," points the way: Roszak argues that the counter-culture of the 60s and 70s was essentially a response to technocracy. He writes of

> the interests of our college-age and adolescent young in the psychology of alienation, oriental mysticism, psychedelic drugs, and communitarian experiments compris[ing] a cultural constellation that radically diverges from values and assumptions that have been in the mainstream of our society since the Scientific Revolution of the seventeenth century.[37]

Roszak describes technocracy in relation to its capacity to "denature the imagination by appropriating to itself the whole meaning of Reason, Reality, Progress, and Knowledge."[38] He argues this is what "the ideal men usually have in mind when they speak of modernizing, up-dating, rationalizing, planning,"[39] especially as they belong to "the era of social engineering in which entrepreneurial talent broadens its province to orchestrate the total human context which surrounds the industrial complex."[40] Let us pause to take in this formulation. Let us see the way old and new have exchanged roles wherein the social engineering of today, executed by the entrepreneurial youth of digital technology, aims to "orchestrate the total human context" in the most absolute terms imaginable. Instead of a more conscious relating of the new and the old, the post-industrial *puer* has been possessed by their *senex* shadow. Roszak contends that beyond the counter-cultural movement "technocracy grows without resistance,"[41] and constitutes "a grand cultural imperative which is beyond question, beyond discussion," making it a form of "totalitarianism" that "prefers to charm conformity from us by exploiting our deep-seated commitment to the scientific world-view."[42] A half century on and we are recapitulating the blindness of the industrial era.

The exploitative overreach of America's technological frontier, which Rushing and Frentz tie to abandonment of the shamanic sense of the sacred, is, according to Roszak, precisely what the counter-culture attempted to reverse. It is thus no mystery that protesting the Vietnam War and the pointless loss of young lives therein became the rallying cry of this generation, especially as this war epitomized excesses of American frontierism and the related military-industrial complex. And in the wake of rejecting this forceful extension of the frontier came the fascination with altered states.

The culture to be countered today remains a technocratic one, though it is abundantly evident the stakes are raised with the industrial commodification of things being followed by the post-industrial commodification of persons. Also prone to narrowing rather than broadening human consciousness and also suffering from the associated compartmentalization of experience, this contemporary culture is rupturing our inner ecology and precluding soulful ways of being. Having jettisoned its purpose in mediating the presence of nature and the propensities of civilization and having been usurped by the values of progress and control that pervade the latter, today's tech-culture is cultural in name only. To counter this technocracy would thus mean restoring the inherent function of culture and creating the psychological and social conditions in which these functions could thrive.

Today, what we may think of as vital is growing tired, particularly in its effect on the psyche and culture. As Madsbjerg has observed, "Silicon Valley is now an ideology, a mindset that values knowledge from the hard sciences above all other forms of knowing. Its cultural prerogatives have now seeped into every aspect of modern life."[43] More specifically, in the "Silicon Valley state of mind," Madsbjerg contends, "we care less about actively seeking out the truth than we do about engaging in discourse and experiences that make us feel affirmed and acknowledged."[44] As the post-industrial hub of the world and thus ground zero for the new technocracy, this also places Silicon Valley at the center of the so-called post-truth era. As much as it would like to blame politicians, spin doctors, and fringe media personalities for the distortions afoot, this hub hothouses today's most corrosive distortions of reality and casts aside the modes of thought that would naturally correct these distortions. Madsbjerg continues:

> The critique here is of Silicon Valley's quiet, creeping costs on our intellectual life. The humanities, or our tradition of describing the rich reality of our world—its history, politics, philosophy and art—are being denigrated by every assumption at play in Silicon Valley.[45]

Still, we are all complicit in fostering this state of mind and the faith in an algorithmically generated intelligence it instills. We are all capable of falling into the syndromes of wanting to feel "affirmed and acknowledged" in the post-truth era. And we are all responsible for allowing those ways of knowing that lead to the displacement of genuine understanding. As Madsbjerg well puts it:

> When we believe that technology will save us, that we have nothing to learn from the past or that numbers can speak for themselves, we are falling prey to dangerous siren songs. We are seeking out silver bullet answers instead of engaging in the hard work of piecing together the truth.[46]

The first move of the new counter-culture will be a more conscious inversion of the old and the new. This will restore vertical modes of understanding, allowing us to grasp the presence of timeless archetypal patterns in even the latest ideas; it will renew our contact with the deeply human and invigorate the instinctual and spiritual dimensions of the psyche; and it will return an ecology of mind, to ensure thoughts and actions remain conversant with our total nature. As these integral dimensions of psyche and nature become the *axis mundi* of our co-created reality, the innovations and accompanying sensibilities coming out of Silicon Valley will start to wear on us. And many of the timeless ideas this vale of virtual reality has effectively siphoned from the intellectual and cultural landscape will begin to appear timely.

The Silicon Valley state of mind may well become the main target of this new counter-cultural movement, but the success of such a movement will depend on our ability to own our complicity in this state of mind. We will need to realize that the social engineering emanating from one place and one group of people is operating with the permission slip we have all signed. Such a realization invites the possibility that the counter-cultural movement may also sprout within this hub of technocracy. For workers at Apple, Google, and Meta are also citizens of the world and spend at least part of their time tapping into shifting attitudes and contemplating the ultimate wellbeing of the next generation.

Although its art and music will need to be as vivid as its predecessor's, the new counter-culture will be defined by its more rigorous philosophical underpinnings. Whatever rituals, poetics, and altered states may be involved, these will need to be accompanied by a surfacing of paradigm shifts long in the making, as well as a critical mass of shifting thought in multiple disciplines. The new psychedelic movement is already a harbinger of this change, receiving considerable validation and backing from a number of disciplines. Here we are already seeing an emerging vision of the interplay of inner and outer ecologies, so that knowledge and wellbeing are no longer divided. The posthuman movement will surely endure, but as its underlying mythology is brought to light its hold on the imagination will also diminish. Meanwhile, technology and art will draw closer, and education will begin to place ethics and imagination alongside technoscience. AI and robots will most likely surround us. However, by finding the means to sustain and nurture the deeply human, and by allowing our understanding of the psyche to catch up with our technological know-how, our plans to merge with these machines may be postponed.

Notes

1 Philip Slater, *Earthwalk* (New York: Anchor Press, 1974), 191.
2 James Hillman, *The Thought of the Heart and the Soul of the World* (Dallas, TX: Spring Publications, 1993), 124.
3 James Hillman, ". . . And Huge Is Ugly." In *Mythic Figures*, (Putnam, CT: Spring Publications, 2007), 152.
4 Ibid., 154.
5 Ibid., 151.
6 The School of Life, *How to Survive the Modern World* (London: The School of Life, 2021), 14.
7 *Blade Runner*, directed by Ridley Scott, featuring Harrison Ford, Rutger Hauer, Sean Young (Warner Bros., 1982). Film.
8 Philip K. Dick, *Do Androids Dream of Electric Sheep?* (New York: Ballantine Books, 1968).
9 Roberts Avens, *The New Gnosis: Heidegger, Hillman and Angels* (Putnam, CT: Spring Publications, 2003), 26.
10 Gregory Bateson, "Last Lecture." In Rodney E. Donaldson ed., *Sacred Unity: Further Steps to an Ecology of Mind* (New York: HarperCollins, 1991), 311.
11 Friedrich Schiller, *On the Aesthetic Education of Man* (1795; Mineola, NY: Dover Publications, 2004), 87.
12 Thomas Berry, *Dream of the Earth (San Francisco: Sierra Club Books, 1988)*, 18.
13 Anthony Giddens, *The Consequences of Modernity* (Stanford, CA: Stanford University Press, 1990), 171.

14 CW 8, par. 757.
15 Richard Sennett, *The Craftsman*, (New Haven, CT: Yale University Press, 2008), 1.
16 Ibid., 7.
17 Ibid.
18 Ibid., 9.
19 Ibid.
20 Ibid., 13.
21 Robert Romanyshyn, *Technology as Symptom and Dream* (New York: Routledge, 1989), 16ff.
22 Ibid., 202.
23 Ibid., 203.
24 Ibid.
25 Peter Bishop, "Between the Colon and the Semi-colon." Unpublished paper presented at the Festival of Archetypal Psychology, University of Notre Dame, July, 1992.
26 Joseph Campbell, interview with Gerard Jones, *San Jose State University Independent Weekly*, December 5, 1979.
27 C. G. Jung, *The Earth Has a Soul: The Nature Writings of C. G. Jung*, Meredith Sabini ed. (Berkeley, CA: North Atlantic Books, 2002), 207.
28 In C. A. Meier et al., *A Testament to the Wilderness: Ten Essays on an Address by C. A. Meier* (Einsiedeln, CH: Daimon Verlag, 1985), 8.
29 C. G. Jung, *Memories, Dreams, Reflections* (New York: Pantheon Books, 1961), 359.
30 Avens, *New Gnosis*, 61.
31 Ibid.
32 James Hillman, *Senex and Puer*, Uniform Edition, vol. 3. Glen Slater ed. (Putnam, CT: Spring Publications, 2005), 30ff.
33 Glen Slater, "Hillman's Metapsychology." In J. H. Stroud and R. Sardello eds., *Conversing with James Hillman: Senex and Puer* (Dallas, TX: Dallas Institute of Humanities, 2016), 37.
34 Ibid., 30.
35 Ibid., 31.
36 Theodore Roszak, *The Making of a Counter Culture: Reflections on the Technocratic Society and Its Youthful Opposition* (New York: Anchor Books, 1969).
37 Ibid., xii.
38 Ibid., xiii.
39 Ibid., 5.
40 Ibid., 6.
41 Ibid., 8.
42 Ibid., 9.
43 Christian Madsbjerg, *Sensemaking: What Makes Human Intelligence Essential in the Age of the Algorithm* (London: Little Brown, 2017), 24.
44 Ibid., 34.
45 Ibid.
46 Ibid.

Epilogue
BETWEEN THE LINES

God wants only souls.

                                          Teilhard de Chardin.[1]

Mainstream American psychology has spent more than a century preoccupied with theories of mind that have all but rolled out a welcome mat for posthumanism. By early on seeking the same standing as the hard sciences, this psychology was consumed by observational, experimental, and statistical methods that offered scant means of understanding the deeply human. Later on, by embracing computational models, it focused on conscious rather than unconscious thought processes. The clinical extensions of the resulting cognitive-behavioral psychology have helped some patients cope with prevailing conditions of life without offering any critical understanding of those conditions. They have worked to obscure rather than reveal the industrial and post-industrial roots of psychological fragmentation.

    Emblematic of this myopia, in 2002 the American Psychological Association named B. F. Skinner the most important psychologist of the twentieth century. It was Skinner who argued the mind is essentially a black box and that introspection has little psychological value, giving rise to a psychology without interiority. Such a view offered little to counter the reductive influence of early and flawed neuropsychological assumptions. The yawning gap between observed behavior and brain mechanisms became fertile ground for the idea that anomalous physiological events could explain common psychological ailments—an idea that was to take root in both the medical profession and the population at large. The door was then opened for the pharmaceutical industry to offer techno-scientific solutions for human problems, providing a powerful catalyst for the technocratic conception of psychological life. If humans end up merged with robots, the robotic perspectives of mainstream psychology and psychiatry will have helped pave the way.

These stances have also prevented this mainstream approach from reflecting on its own psychology, keeping it blind to its own scientific inferiority complex. This has only pushed much of the field into a slavish relationship with the industrial/post-industrial project. Economic incentives and clinical goals in service to these incentives, particularly in the American healthcare system, have led to a psychology that tacitly supports a minimally functional existence in a maximally dysfunctional world. A million methods of concocting veneers of self-esteem and conjuring worldly success are promoted alongside flimsy formulas for stemming the tide of depression, anxiety, and personality disorders. Very little attention goes toward reversing the underlying pandemic of modern soul loss, leaving most without significant means of self-understanding.

In many respects, American psychology has itself been operantly conditioned by the myth of the American Frontier, keeping it in the business of building stronger egos, readied for the never-ending journey west. Rather than insightfully critiquing this one-sided socio-cultural trajectory and its effect on the deeply human, knowledge of historical imprinting has been cast aside in the name of expansion and material progress. By adopting rationalistic and positivistic approaches even more ardently than other scientific fields—overcompensating for its inferiority complex—mainstream psychology has failed to fulfill its early promise of studying the psyche by drawing the well of insight provided by the humanities and culture. William James must be turning in his grave.[2]

Failing to provide constructive and insightful means to bridge objective and subjective knowledge, psychology has also left the masses vulnerable to the manipulation that has become part of the digital age. As Skinner's behavior modification techniques have been employed all the way from Madison Avenue to Hacker Way, this contribution has been a rather direct one.[3] If today we face a global epistemological crisis wherein the factual basis of knowledge and the reality-based community are under threat, it is because society has failed to support all sides of the human thought process, with behaviorism as a chief enabler.

On campuses across the industrialized world, students pour into undergraduate psychology classes imagining they will find concepts for understanding themselves and their fellow humans. They fill the largest lecture halls only to be told that questions of meaning and purpose are discussed in philosophy and literature classes, and they come away knowing that psychology, the supposed logos of psyche/soul, is not particularly interested in the soul, because it refuses to engage the soul's concerns. When love and loss, the significance of suffering, the experience of calling, the development of a conscience, the cultivation of a life story, the foundations of character, how to face death, and the other dimensions of the fully engaged life are outsourced to other fields, psychology turns into a soulless undertaking. It is the absence of interest in these critical life matters, combined with the presence of computational and automated theories of mind, that have enabled the complex machine view of human thought and action to dominate the field. In all these ways the field has turned itself into a handmaiden of scientific reductionism, doing its part to prepare the path to a posthuman future. As Barrett puts it tersely at the beginning of his *Death of the Soul*, in such ways of thinking "there is a gap between theory and life."[4] That engineering a life beyond this one should appear to be the logical next step for American technology, and perhaps civilization at large, is then not surprising.

As these psycho-mechanical perspectives have molded how we view ourselves, integral psychologies that address the psyche's own preoccupations and their unconscious foundations have been forced to the margins. Depth, phenomenological, humanistic, existential, and transpersonal approaches may have somewhat varying emphases, but these variations do not overshadow their common focus on the recovery of soulful self-knowledge. This recovery begins by examining the way life is actually lived and imagined, whether in the home, at church, or on city streets. It proceeds not by extrapolating from animal experiments under the fluorescent lights, but by delving into human experience and comparing this to the self-studies generations of poets, philosophers, essayists, and mystics have left for the cultural record. These other psychological approaches incorporate imagination, intuition, values, and a penchant for finding the essence of inner life through metaphor and symbol. For these are the vertical ways of knowing naturally invited when we turn to the deeply human. Together they constitute the language of the soul.

The marginal status of these other psychologies reflects the spirit of the age, which has attempted to make reason and logic not only primary but absolute. But neither individuals, nor tribes, nor humanity as a whole have ever fully yielded to such modes of comprehension. As Heraclitus stated long ago, "You could not discover the limits of soul (*psyche*), even if you traveled every road to do so; such is the depth (*bathun*) of its meaning (*logos*)."[5] This depth and resistance to comprehension—lunar qualities intrinsic to the veils and shadows that gather to the soul and prove necessary for approaching the unconscious—were never going to coincide with transparent Skinner Boxes and the labs that house them. Whereas one approach attempts to shine a light into the mystery of being, the other offers formulas for reprogramming the mind. In these ways the fork in the road between the human and posthuman is mirrored by a divergence in the field of psychology itself.

Throughout this book, I have therefore aimed to show how depth psychology and its focus on unconscious processes and archetypal patterns help define this existential fork in the road, with Jung's understanding of these matters being, in my view, the most consequential. More than anyone before or after, Jung revealed the way the psyche exists in an organic relationship with the past, distinguishing it from concepts that suggest minds may be engineered from scratch or exist beyond the body–world configuration. He showed that intelligence and imagination grow out of universals—archetypal patterns likely correlated the structure of the brain and the evolution of human nature. These patterns are activated and reinforced by a mind–body–world relationship that also partakes of larger cosmological forms. Such patterns have resulted from a psyche–nature relationship millions of years in the making, out of which have grown regulatory principles of our ecology of mind. We are thus apt to see ourselves in both the heavens and in the grain of sand, as well as in the landscapes, plants, and animals in between, precisely because this psyche–nature relationship and these ecological principles are ingrained in us.

The psyche, which "displays itself throughout all being,"[6] thrives on the recognition of these archetypal patterns. This was recognized by the ancient Greeks, who, as Hillman puts it, "considered ideas to be forms, some of which were archetypal forms or *archai*, root ideas, big ideas, eternal ideas that governed the way things are in the world." He adds, the "philosophy of knowledge (how we know anything), mythology (the study of the ways of the gods and goddesses), and psychology of living (human existence) were completely intertwined."[7] This general way of thinking is not alien to what indigenous peoples the world over have attempted to preserve—a way of thinking in which mind and earth, spirit and place cohere. For some, this sympathy of all things remains magical; yet it can be approached psychologically—reflecting the phenomenology of the psyche as this presents to us. Overt perceptions of gods and spirits might have retreated alongside the demythologized consciousness of modern life, but the psychic realities behind these perceptions may be recognized as archetypal imperatives, in the midst of which Jung found indications of a hidden order, emphasized in such notions as the *unus mundus* and the archetype of the Self.[8]

Whereas some critics of posthumanism[9] lean on ethical and moral arguments to urge a change of direction, I have suggested these arguments do not reach deeply enough into the emotions and images associated with the archetypal background of the psyche. As important as these arguments are, they mainly mobilize the will: they fail to expose psychological root systems or scratch the surface of the soul-centered, co-creative task our ancestors and progeny are counting on. As modern history and its commentators have made clear, the instrumentality of reason alone will not see us through. The spectacularly overt irrationality that instigated the catastrophes of this period and continues to surround us is evidence of this. Something has to rock the foundations and tap the same source the prophets and shamans of the millennia have always tapped—not in lieu of reason, but as its most necessary companion.

Psyche, Soul, Eros

My use of the terms "psyche" and "soul" throughout this book not only reflects my Jungian psychological orientation; it aims to combat the absent interiority of mainstream psychology and the conceptions of mind and intelligence technoscience is foisting upon us. While these terms are often used interchangeably—*psychē* being Greek for "soul"—I use the former in the conceptual sense and the latter for its qualitative connotations: if inner life has structures and dynamics whose patterns constitute the *psyche*, it is the experience of these patterns and their vertical apprehension that aggregate into the sense of having or being a *soul*, which is frequently imagined as the essence of the personality or personified as a vital inner figure.[10] To say things such as "my soul longs for . . ." or "I glimpsed her soul" is to be caught by the archetypal image of something essential to us, yet often felt as not wholly of us.

The structures of inner experience have throughout history tended to be personified as entities or presences, a process directly expressed in dreams and myths. The dramas enacted by these figures convey the dynamics of the inner world. In the foundational myth of Eros and Psyche, for example, Psyche must undergo trials to come into her fullness. In this process, Eros is determinative. Psyche becomes a soul, as it were, through initiation by love. Who or what invites passion and vivifies existence generates the feeling of being a soul. So here we have a mythic statement of the intertwining of life's two highest values: love and soul. Soul is awakened through love; where there is love we are apt to find soul. And, as Jung puts it, "to have soul is the whole venture of life."[11]

In both religious and poetic use, soul connotes the unfathomable depth of existence, defying exact definition and eluding final comprehension. It is thus fittingly applied to the pith of an individual, a community or an undertaking—a quintessence that recedes into mystery because the vertical approach proves fathomless. Although we have no window to see beyond this mortal life, we are also prone to imagine some things precede and outlast this life. Soul is thus seen as transcending ordinary bounds of matter, space, and time. Where soul is present, a sense of the sacred is also never far away; when a divine hand is felt, it is always felt by or in the soul. Yet to be a soul or to live soulfully requires neither belief in an actual afterlife nor an overt religiosity.

Because of all this, the phenomena and perspectives that generate a sense of soul and juxtapose states of soullessness are studied primarily by means of psychic images. Hence Jung's formulation "image is psyche,"[12] which also suggests an intimate relationship between imagination and soulfulness. We are in some measure what we imagine ourselves to be, and if we imagine we are machines and not souls and fail to understand the formative conditions of this mechanistic image, we may well construct a whole way of being accordingly.

A glaring problem at work in this possibility is that machines expunge mystery, and without mystery soul withers. One such mystery of the soul is where purpose and character originate. Our preoccupation with what we will leave behind and how each human life is related to the patterning of the cosmos is related to this mystery. Creativity, opportunity, beauty, and grace frequently feature in this preoccupation, and each defies reductive explanation—which does not mean we cannot venture to understand them. To have soul or become soulful means standing at the intersection of our limited lives and the unfathomable reaches of being from which such vital themes emanate. The soul seems to announce its own telos and suggest we may just be along for the ride. Deep purpose involves our embrace of this telos, and character comes from enduring this embrace. Sometimes this embrace begets a love of life and we are astonished, inspired, and awed. But tragedy can equally stalk and shape purpose and character, as soul also insists on its dark nights. In either case we are compelled to make room for what resides beyond the merely willful and the readily understood. We are souls and not machines because of all this.

When soulfulness becomes our focus and we hold it psychologically and poetically rather than theologically and metaphysically, we are both ancient and modern. Such a stance stops futurism and scientism in their tracks. An orientation to soul also allows us to stand at the door of metaphysical curiosity and peer in without needing to traverse this across this threshold. Instead, we defer to phenomenal experience and enduring shapes of the imagination. To take the example of the mystic's desire to be one with God, the psyche-centered or soulful approach would bracket the question of God's reality and engage the desire and the particulars of the God-image, which are produced by the psyche and carry individual and cultural markings. The ultimate unknowability of the mystical vision and the acceptance of the symbolic-metaphorical thinking needed to navigate its myriad expressions mean our turn to the divine, the ancestral voice or spiritual presence necessitates a *psychological* acumen. This may be one of the greatest benefits of Jung's symbolic approach: it validates the soul's need for metaphysical orientation in a world divested of metaphysical discourse. Gods, monsters, and spirits are held as archetypal significances—symbolic expressions of a psychological religious function, which can be engaged without literalism or dogma. The poles of reductive scientism and religious fundamentalism are both avoided, and we can be present to the timeless concern for spiritual truths as well as the modern need to understand material facts.

Accepting the religious function as a critical part of the psyche's nature and thus as something belonging to the deeply human provides an invaluable ontological standpoint and epistemological tool. As I have detailed throughout this book, when the deeply human is overlooked, so is the need for spiritual orientation, making us vulnerable to quasi-religious movements that confine the imagination and close the mind. The lack of vertical, psychological understanding then becomes costly. Unfamiliar with our own psychic natures, our religiosity goes unconscious and we end up worshipping at the altars of contemporary ideals and futurism.

Accounting for what computers cannot produce and computational models of mind cannot address is an epistemological challenge built on an ontological problem. We must grapple with ideas adequate to the activity of the psyche, which includes the perennial conviction we have souls. This forces us into vertical modes of understanding resting on the depths of human nature. In these vertical modes, we may apprehend those realms of experience that defy the dualisms of mind and body, subject and object, inner and outer, spirit and matter, image and reality, which have defined regular scientific discourse for centuries. Overcoming these dualisms thus becomes crucial to reversing the tyranny of horizontal scientism, which will be accomplished in part by avoiding the schism between science and the humanities and keeping the liberal arts basis of education. Otherwise, we will become prone to a reductive posthuman logic such as that which we find in Bostrom, who writes, "If human beings are constituted of matter obeying the same laws of physics that operate outside us, then it should in principle be possible to learn to manipulate human nature in the same way that we manipulate external objects."[13] Aside from the radical biological reduction of human nature implied, such a view is an ode to Skinnerian psychology, expunging vertical ways of knowing and dismissing inner life, leaving behind soulless rationalism and mechanistic materialism to define what human beings are.

From a soul perspective, however, Eros shoots his arrow into the heart of this epistemological battlefront. Eros is the binding agent that dissolves these dualisms and calls us instead to embrace the fullness and paradox of being. Eros returns us to the intractable difficulties, liberating achievements, and ordinary pleasures of life, forcing upon us a psychology of what the heart sees, the hands know, and the gut feels—the things that shape existence but remain beyond the reach of scientific reduction.

As philosopher of the imagination Avens describes it:

> The Jungian "soul" is the Platonic metaxy—a principle of relationship or betweenness holding together heaven and earth and making them participate in each other. It is also Eros—the mighty daimon who, far from being a merely human attitude, is a metaphysical factor in all nature, the miracle in the center of being preserving the universe from dissolution into a deadly mass.[14]

Such an invocation of Eros as a daimon and metaphysical factor suggests this presence is given with this existence, both in nature and beyond. Compassion, which may be taken as the principled extension of this love, occupies a similarly primary role in religious ethics. To the extent that the undoing of this binding of Psyche and Eros defines the way of the world, however, a "dissolution into deadly mass" (Avens) seems to be exactly the threat that looms. This dissolution would manifest as a world without values and significances: it would be the psychological equivalent of the nanotechnological nightmare of a world that turns to gray goo.

Soul-Making

We are forced to recognize that the reductions of technoscience and the theories of mainstream American psychology that follow these leanings fail to meet the epistemological and ontological needs of the human psyche. When the mind is no more than the sum total of neuronal firings, nature is just a matter of genes, and both neurons and genes are regarded as purely informational, we are dispossessed of the metaxy that is the realm of soul and left with little standing in the way of our self-commodification. We may gain a kind of instrumental knowledge, which brings a kind of power, but it comes at great cost. As many astute observers of motivation have noted, power without love becomes excessive and absolute, eventually turning to violence. This appears to be the fate of a world full of science and devoid of soul. It clears a path for the titans and erodes all differentiated values.

May Sarton tells us that the scientist-philosopher-theologian Teilhard de Chardin once wrote:

> The masters of the spiritual life incessantly repeat that God wants only souls ... In each soul, God loves and partly saves the whole world which that soul sums up in an incommunicable and particular way ... every man, in the course of his life, must not only show himself obedient and docile ... he must build—starting with the most natural territory of his own self—a work, an opus, into which something enters from all the elements of the earth. He makes his own soul throughout his earthly days.[15]

Teilhard goes on to note the coincidence of this opus with another; whereas the first is concerned with taking in "all the elements of the earth," the second pertains to "the completing of the world."[16]

Teilhard's formulation contains several elements worth underscoring. First, soul is not given so much as made. Second, this soul-making is divinely sponsored—needed by Creation itself. Third, it pulls from the encounter with the earth process. Fourth, it participates in a larger making—a work of ultimate collective significance. In reflecting on Teilhard's words, Sarton gifts us with her own clear-eyed take on his wisdom:

> It is only when we can believe that we are creating the soul that life has any meaning ... Then there is nothing we do that is without meaning and nothing that we suffer that does not hold the seed of creation in it.[17]

That is, in soul-making we find both a forging of meaning in what we do as individuals overlapping the sense that this forging participates in a larger existential story.

Hillman, approaching the same theme through the archetypal branch of Jungian psychology, takes his cue from Keats:

Call the world, if you please, *the vale of Soul-making*. Then you will find out the use of the world . . . There may be intelligences or sparks of the divinity in millions—but they are not Souls till they acquire identities, till each one is personally itself.[18]

Toward the end of his magnum opus *Re-Visioning Psychology*, Hillman then reflects:

Soul may be lent us by our ancestors as we live out their patterns in our genealogy myth; or by the Gods as we enact their pathologically bizarre dramas; or by our dreams, thereby reminding us at the fresh day's start of the soul's different and underworld existence; or by something yet to happen that is making its way through us—the Zeitgeist, the evolutionary process, karma, the return of all things to their makers—but our lives are on loan to the psyche for a while. During this time we try to do for it what we can.[19]

Yet to make soul and fulfill such intimations of life's purpose, we need to have ideas about the soul as well as soulful ideas. As Hillman contends, soul is a perspective more than a thing—a perspective that accounts for depth, meaning, and religious significance, and one that makes room for Eros in the ways described above.[20] We make soul by cultivating a vision of the world that accounts for where suffering occurs and what drives us toward certain actions. We come to understand, along with Jung and Hillman, we can trace our thinking to enduring universal values, generating a sense of soul by connecting our minds to these layers of psychological life. Whereas there are various techniques for plumbing these deeper dimensions—whether analytical or spiritual—culture also offers pathways. If we want to find out how our individual lives are connected to what is making its way through the zeitgeist, for example, we need only recognize and reflect upon where and how we are arrested by the poets and artists of our time and then act on how the imagination makes its own way through us. As Brown put it to graduating students in 1984, "It is not a matter of having a soul, or of saving it, but more actively a matter of soul-making. The soul is creative and must create itself: it is the creative imagination."[21]

The Harari Gap

Harari's influential assessment of the historical moment and the pivot point we have arrived at is compelling.[22] We can move to the next stage of history, he argues, either with or without awareness of the technocratic infiltration of the psyche and either with or without resistance to this infiltration. The psychological implications of digital innovation I have set out in these pages only further propel this argument and provide backup for his broad recommendation for increased self-knowledge. Harari's promotion of meditation as a critical component of this self-knowledge is something we should all consider. However, there is a difference between meditation, which increases mindfulness, and psychological maturation, which has to do with understanding and transforming habits of mind.

Harari writes that "most science-fiction movies actually tell a very old story: the victory of mind over matter."[23] However, this is true mainly of their set-up: it does not describe their typical outcome. The story arc of most science-fiction movies begins with some quest to transcend the limits of nature, impose control by technological means, and generally ignore the deeper realities of being. But by the end, matter in the form of the natural world or the deeper instincts of the main characters usually takes the upper hand: in *Star Wars*[24] Darth Vader's *love* for his son Luke precipitates the shedding of his cyborg shell; in *Alien*[25] Ripley's *protective instincts* kick in to confront the psychopathic corporation and consumptive alien monster; in *The Terminator*[26] John and Sarah Connor's commitment to *human and planetary survival* defeat the cyborg and win the battle in a war against the machines; in *Blade Runner*[27] Deckard's *empathy* for his replicant quarry exposes the tragedy of sentient humanoids with pre-determined expiry dates. The attempt to elevate mind over matter is not fully reversed, but is corrected: mind returns to matter and some soul value reenters the equation; technology meets humanity; hubris meets the sacred.

Harari also suggests we show humility when it comes to our explanations of the ultimate nature of existence. We do not know who or what is ultimately pulling the strings. So the stories and meanings we apply to such matters—such as those we find in religion—need to be held more loosely than has been the case. As he insists, there can be no viable global community if people are willing to kill one another over competing stories. I agree with all this. But it is also the case these stories continue to compete this way because religious ideas are not understood as belonging to the innate functioning of the psyche and we continue to view them literally rather than symbolically.

In this arena, Harari's advice is both radical and problematic: Throw out the stories and the closely associated search for meaning and come back to the essential circumstances of human life, which require care and respect for one another as well as avoiding the infliction of suffering.[28] If this existence turns out to be a giant experiment set up by an alien master species, or we are caught in some kind of encompassing computer program, Harari argues that suffering remains real.[29] Whatever death means and what, if anything, happens before and after death, the experience of physical pain and psychological torment can be alleviated by how we live, and how we treat others. All other concerns can take a backseat.

Whereas this position can seem compelling, we may want to add a few things. For example, as Jung pointed out, there is a difference between meaningful and meaningless suffering—the first is more endurable and helps us see that certain kinds of suffering may be a necessary part of maturation—either personally or collectively. Given we are mortal beings, vulnerably enfleshed, who enter and leave this world alone, some suffering is built in. The Buddha told us this was the First Noble Truth. To find meaning and accept the necessity of at least some suffering involves finding a fitting story or philosophy—a way to narrate the events of life to ourselves. Besides religion, we should note that this is what great art and literature also do: they cradle suffering in imagery and reflection. Human bonding also occurs in the context of shared pain. Could there even be empathy without loss, grief, and other forms of suffering? As Victor Frankel demonstrated, even for those subjected to truly inhuman amounts of suffering and trauma, the call to create meaning does not disappear.[30]

We are built to look for meaning and tell stories—something Harari describes at length in his first book on the history of *Homo sapiens*.[31] Just as suffering is all too real, the search for meaning and need for stories are universal psychic impulses. And whereas meditation and ethical reasoning can prevent us from subjecting ourselves to narrow meanings or becoming true believers in inadequate and damaging stories, attempting to transcend these impulses will likely fail. If, however, we can better understand rather than blindly enact these impulses, expanding self-knowledge in the process, progress is possible. Getting beneath narratives of ultimate meaning and accepting that they are mythical constructs, even if profound and sacred ones, is the key. And the only way to do this is by learning to more consciously relate to the power of symbols and metaphors, accepting this dimension of human expression as a primal and primary language, given by the deeply human and inextricable from vertical modes of understanding.

The Symbolic-Imaginal Life

Jung offers no easy formulas for navigating the travails of modern life, nor does his psychology offer itself as a replacement for the cultural images and spiritual principles humans have long turned to for guidance. What Jungian psychology does offer is means to navigate these images and principles, bridging the old and the new, the rational and the non-rational, at a time when progress and regress have become confused. Possibly more important than anything else, on behalf of the nature of the psyche itself, this psychology implores us to place the symbolic life alongside the pragmatic life, so that both vertical and horizontal dimensions of reality are apprehended. The neuropsychological requisite to develop faculties associated with both cerebral hemispheres, the educational mandate to grow minds by thorough study of the arts and humanities, and the counter-cultural insistence that wellbeing be defined in relation to our psycho-spiritual totality are all congruent with the broad strokes of this stance.

It is the symbolic perspective that allows Jung's psychology to be expansive, addressing both our earthly roots and our heavenly reach, and attempting to grasp how these are joined. Only such a stance can join the animal and the divine, and it is many a symbol that conveys this paradox of the highest and lowest existing in unison. The psyche, which is obviously grounded in nature, also leads us beyond this ground, into a concern with destiny and leaving some trace of ourselves in service to humanity. Rather than a tragic barrier, such a posture turns death into a meaningful goal—a telos that is strengthened by the psychological deaths and transitions of a life fully lived. We neither have to lean on notions of a spirit that literally departs the body, nor on fantasies of downloading our minds. Rather, we accept that the substance of psychic experience, accumulated over a lifetime, marks everything we do and everyone around us. And such a marking can be dreamed on. Life and death thus intertwine to produce defining qualities of being. As Hillman puts it,

> the restoration of the whole man, his dawning into a new day, implies a prolonged acceptance of a twilight state, an *abaissement* and rededication of the ego-light, softening by sacrificing each day some of its brightness, giving back to the Gods what it has stolen from them and swallowed.[32]

The wonder of psychological life is that this dimming of the ego-light allows the perception of another light, one that has been in the background all along. Following a meditation on the way wisdom is paradoxically connected to the ape within us, Hillman describes this second source of illumination as "the divine light of natural consciousness." He notes, "This soft light is pre-conscious, at the threshold always dawning, fresh as milk, at dawn with each day's dream, still streaked with primordial anarchy—the light of nature rising within each individual out of the unconscious psyche."[33]

Posthumanism brightens the ego-light, the light of the intellect and will that blankets the light of natural consciousness the wise ape presents to us. It fails to reconcile our spiritual reach and our instinctual ground. It cracks cultural vessels that have always incubated this reconciliation and truncates rather than extends the human experiment. The cyborg may be approached as a symbol, but it is a symbol of that which breaks the symbolic order and offers little prospect of "a twilight state" (Hillman).

In Conclusion

By succumbing to futurism and failing to see through posthumanism, we will lose our capacity to be makeweights for this moment in history. A presumption of long-term outcomes will overshadow our engagement with immediate and short-term causes, generating an atmosphere of predestination. Baudrillard calls this a "hypersensitivity to the final conditions of a process, not to the initial ones." At the moment these final conditions involve a "destiny of simulation," one that is "carrying us further and further from the initial conditions of the real world."[34] Our simulated, borged selves are promising to carry us further and further from the full and deep humanity that has made this conscious existence possible.

If simulation is to become our destiny, there will be no destiny. For if simulated rather than actual life defines our destination, what investment in the future or continuity between being and becoming can we expect to maintain? Who will invest themselves in the generations to come? If only simulacra are preparing to follow us, why care about the effect of our current actions? Our cyborg progeny may not even require clean air or human rights. It will be easier to surrender to the march of the machines, allowing posthumanism to become a self-fulfilling prophecy.

A contrasting possibility is offered by Giddens. Surveying the challenges of late modernity he writes:

> none of this means we should, or that we can, give up in our attempts to steer the juggernaut. The minimizing of high-consequence risks transcends all values and all exclusionary divisions of power ... We can envisage alternative futures whose very propagation might help them be realized.[35]

However, to become makeweights, to recover an authentic destiny, and to "steer the juggernaut" as Giddens suggests, will require a return to right where we are—to the phenomenal ground of being and to the psychological imperative that we become more conscious of the totality of our experience and thicken the cultural imagination right where we now stand. To minimize "high-consequence risks" will require perceiving these consequences and feeling those risks. We can begin by waking up to the titanic forces and technocratic structures currently directing human civilization, by resisting our escapist inclinations, and by standing firm in our responsiveness.

If my arguments against posthumanism have been combative, this has been prompted by the contention of astute observers that posthumanism both occupies an axial position in today's technocracy and threatens the cultural process. As I set out at the start, the cultural process regulates the coexistence of nature and civilization, the two great givens of human existence, regulation without which we would fall into profound regression, complete rarefication, or, most likely, some combination of both. My combativeness also reflects the conviction that all "isms" are problematic, especially in their proclivity to present narrowly construed yet totalizing remedies for human foibles. And putting all our eggs in the basket of human–machine merger without making room for the prodigious shadow digital technologies have already cast is a utopian fantasy that will almost certainly invite a dystopian enantiodromia.

It is the well-tended, critically engaged cultural process that produces the images and artifacts that prevent the polarization of past and future, memory and vision, utopias and dystopias. It is also this process that maintains our contact with the deeply human and cradles the collective soul. By contrasting the deeply human and the posthuman, the ecology of mind and the dissociated mind, the vertical and horizonal modes of understanding, and by accentuating the contrast between Jung's psychology of spirit and matter and the cyborg's pastiche of gizmo and flesh, I have thus attempted to draw out a conflict the world has, until recently, relegated to the margins of discourse. This is a conflict we must now fully enter.

If I have construed the meeting of depth psychological understandings and posthuman aspirations as a battle of ideas and ideals, it is because post-industrial technocracy has already initiated a concerted attack on the integrity of the human psyche, and I carry the conviction this has to be exposed and countered. I have also tried to arm myself with paradigms of knowledge emerging from numerous fields that support the integrity of the psyche and challenge many of the precepts of scientism and posthumanism. This is a battle that also goes beyond intellectual debate, thrusting us into matters of human progeny and cultural memory, a battle that activates the chakras from base to crown. Adding a dash of humanism to science, fostering technological oversight, and philosophizing about balancing tissue and titanium are unlikely to smooth the way forward, as these efforts face a relentless tide of algorithmic manipulation and escapist virtuality. Changing personal habits may help shift our modes of perception and provide life rafts as this tide moves in, but this too will not be enough. Here I agree with Roszak's view of the last countercultural moment, when he writes:

It would be one of the bleakest errors we could commit to believe that occasional private excursions into some surviving remnant of the magical vision of life—something in the nature of a psychic holiday from the dominant mode of consciousness—can be sufficient to achieve a kind of suave cultural synthesis combining the best of both worlds.[36]

While I am unconvinced a majority of persons will join this battle, I am dedicated to the view that a guerrilla operation might redirect the juggernaut ready to roll over us and distract the titanic inclinations steering it. That is, with a critical mass of conscious persons in the right places, willing to read the signs of the times, there may be just enough antibodies to prevent the whole system from succumbing to the virus. We know history does not move in straight lines. Nor is the future predetermined. To the extent we may discern some grand movement, we might surmise history moves in arcs. Savvy to the lessons of the past and sensitive to symptoms of the present, we can ponder the posthuman future, reimagine this future, and perhaps bend the arc of history. Over time, new images will gather and become new directions. If we come to know ourselves well enough, the deeply human will speak more convincingly, positioning us to see more clearly what is on the horizon and how to respond. Then—just maybe—we will not end up where we seem to be going.

Notes

1 Quoted in May Sarton, *Journal of a Solitude* (New York: W. W. Norton, 1973), 66.
2 See Richard Hall, "William James on the Humanities." *William James Studies*, vol. 9, 2012, 120–143.
3 The street address of Facebook's parent company Meta is 1 Hacker Way.
4 William Barrett, *Death of the Soul* (New York: Anchor Books, 1986), xiii.
5 Quoted in James Hillman, *Re-Visioning Psychology* (New York: HarperCollins, 1975), xvii.
6 Ibid., 181.
7 James Hillman, *Philosophical Intimations*, Edward S. Casey ed. (Thompson, CT: Spring Publications, 2016), 392.
8 See CW 14, par. 660ff.
9 See, for example, Bill McKibben, *Enough: Staying Human in an Engineered Age* (New York: Times Books, 2003), Nicholas Agar, *Humanity's End* (Cambridge, MA: MIT Press, 2010), and Charles T. Rubin, *Eclipse of Man: Human Extinction and the Meaning of Progress* (New York: New Atlantis Books, 2014).

10 See CW 6, par. 797.
11 CW 9i, par. 56.
12 CW, 13, par. 75.
13 Nick Bostrom, "A History of Transhumanist Thought." (Originally published in *Journal of Evolution and Technology*, vol. 14, no. 1, 2005, 4.) https://nickbostrom.com/papers/history.pdf
14 Roberts Avens, *The New Gnosis: Heidegger, Hillman, and Angels* (Putnam, CT: Spring Publications, 1984), 23.
15 Sarton, *Journal*, 66–67.
16 Ibid., 67.
17 Ibid.
18 John Keats, *The Complete Poetic Works of John Keats* (New York: Houghton Mifflin, 1899), 369. Italics added.
19 Hillman, *Re-Visioning*, 180.
20 Ibid., xvi.
21 Norman O. Brown, in Ann Gibb, "Renowned Scholar and Author Norman O. Brown Dies at 89." *UC Santa Cruz Newscenter*, October 3, 2002. https://news.ucsc.edu/2002/10/211.html
22 Yuval Noah Harari, *Homo Deus* (New York: Harper, 2017).
23 Yuval Noah Harari, *21 Lessons For the 21st Century* (New York: Spiegel & Grau, 2018), 254.
24 *Star Wars: Episode VI—Return of the Jedi*, directed by Richard Marquand, featuring Mark Hamill, Harrison Ford, Carrie Fisher (20th Century-Fox, 1983). Film.
25 *Alien*, directed by Ridley Scott, featuring Sigourney Weaver, Tom Skerritt, John Hurt (20th Century-Fox, 1979). Film.
26 *The Terminator*, directed by James Cameron, featuring Arnold Schwarzenegger, Michael Biehn, Linda Hamilton (Orion Pictures, 1984). Film.
27 *Blade Runner*, directed by Ridley Scott, featuring Harrison Ford, Rutger Hauer, Sean Young (Warner Bros., 1982). Film.
28 Ibid., 311ff.
29 Ibid., 253–254.
30 Victor Frankl, *Man's Search for Meaning* (Boston, MA: Beacon Press, 1959).
31 Yuval Noah Harari, *Sapiens* (New York: Harper Perennial, 2014).
32 James Hillman, *Senex and Puer*, Glen Slater ed. (Putnam, CT: Spring Publications, 2005), 340.
33 Ibid., 341.
34 Jean Baudrillard, *The Illusion of the End* (Stanford, CA: Stanford University Press, 1994), 12–13.
35 Anthony Giddens, *The Consequences of Modernity* (Stanford, CA: Stanford University Press, 1990), 154.
36 Theodore Roszak, *The Making of a Counter Culture* (New York: Anchor Books, 1969), 257.

BIBLIOGRAPHY

Achbar, Mark, dir. *The Corporation.* 2003; New York: Big Picture Media, 2005. DVD.
Agar, Nicholas. *Humanity's End: Why We Should Reject Radical Enhancement.* Cambridge, MA: MIT Press, 2010.
Anderson, A. R., ed. *Minds and Machines.* New York: Prentice-Hall, 1964.
Anderson, Craig L., Maria Monroy, Dacher Keltner. "Awe in Nature Heals: Evidence From Military Veterans, At-risk Youth, and College Students." *Emotion* (December 2018): 1195-1202.
Arthur, W. Brian. *The Nature of Technology: What It Is and How It Evolves.* New York: Allen Lane, 2009.
Asimov, Isaac. "Runaround." In *I, Robot.* New York, Bantam Books, 1991. First published 1942.
Auden, W. H. "In Memory of Ernst Toller." In *W. H. Auden Collected Poems*, edited by Edward Mendelson. New York: Vintage Books, 1991.
Avens, Roberts. *The New Gnosis: Heidegger, Hillman, and Angels.* Putnam, CT: Spring Publications, 1984.
Bailey, Lee Worth. *The Enchantments of Technology.* Chicago: University of Illinois Press, 2005.
Barrat, James. *Our Final Invention: Artificial Intelligence and the End of the Human Era.* New York: St. Martin's Press, 2013.
Barreto, Marco Heleno. "On the Death of Nature: A Psychological Reflection." *Spring: A Journal of Archetype and Culture*, Vol. 75, Pt. 1., 257-258.
Barrett, William. *Death of the Soul.* New York: Anchor Books, 1986.
Bateson, Gregory. *Steps to an Ecology of Mind.* New York: Jason Aronson, 1972.
---. "Last Lecture." In *Sacred Unity: Further Steps to an Ecology of Mind*, edited by Rodney E. Donaldson. New York: HarperCollins, 1991.

Baudrillard, Jean. *The Consumer Society*. Thousand Oaks, CA: Sage Publications, 1998. First published 1970.

---. *The Illusion of the End*. Stanford, CA: Stanford University Press, 1992.

---. *Simulacra and Simulation*. Ann Arbor, MI: The University of Michigan Press, 1994.

---. *Telemophosis Preceded By Dust Breeding*. Translated by Drew S. Burk. Minneapolis MN: Univocal, 2011.

Bellah, Robert N. *The Broken Covenant: American Civil Religion in Time of Trial*. New York: Seabury Press, 1975.

Berry, Thomas. *The Dream of the Earth*. San Francisco: Sierra Club Books, 1988.

von Bertalanffy, Ludwig. *Robots, Men and Minds*. New York: George Braziller, 1967.

Bishop, Peter. "Between the Colon and the Semi-colon." Paper presented at the Festival of Archetypal Psychology, University of Notre Dame, July 1992.

Blitz, Mark. "Understanding Heidegger on Technology." *The New Atlantis*, Winter, 2014.

Bloom, Harold. *Shakespeare: The Invention of the Human*. New York: Riverhead Books, 1999.

Bostrom, Nick. "A History of Transhumanist Thought." *Journal of Evolution and Technology*, 14, no. 1, (2005).

Brand, Stewart. "For God's Sake, Margaret! Conversation with Gregory Bateson and Margaret Mead." *CoEvolutionary Quarterly*, no. 6 (June 1976).

Brewer, Talbot. "The Great Malformation." *The Hedgehog Review*, 25, no. 2, (Summer 2023).

Brockman, John, ed. *Is the Internet Changing the Way You Think?* New York: Harper Perennial, 2011.

Brooks, David. "Globalization is Over. The Global Culture Wars Have Begun." *New York Times*, April 8, 2022.

Brown, Norman O. *Love's Body*. New York: Vintage Books, 1966.

Bulnes, L. C. et. al., "The Effects of Botulinum Toxin on the Detection of Gradual Changes in Facial Emotion." *Scientific Reports*, no. 9 (2019).

Butler, Samuel. "Darwin Among the Machines." A Letter to the editor of *The Press*, Christchurch, New Zealand, June 13 1863.

---. *Erewhon* New York: Dover Publications, 2002. First published in 1872.

Cameron, James, dir. *Terminator 2: Judgment Day*. 1991; Culver City, CA: Tri-Star Pictures, 1997. DVD.

Campbell, Joseph. *The Hero With a Thousand Faces*. New York: Pantheon Books, The Bollingen Series, 1949.

---. Interview with Gerard Jones, San Jose State University Independent Weekly, December 5, 1979.

Campbell, Joseph and Bill Moyers. *The Power of Myth*. Edited by Betty Sue Flowers. New York: Broadway Books, 1988.

Bibliography

Carr, Nicholas. *The Glass Cage: Automation and Us*. New York: W. W. Norton, 2014.
Carson, David, dir. *Star Trek: Generations*. 1994. Los Angeles, CA: Paramount Pictures. Film.
Carson, Rachel. *Silent Spring*. New York: Houghton Mifflin Company, 1962.
Casey, Edward S. *Imagining: A Phenomenological Study*. Bloomington, IN: Indiana University Press, 1976.
Cassirer, Ernst. *An Essay on Man: An Introduction to a Philosophy of Human Culture*. New Haven: Yale University Press, 1944.
Center for Disease Control. "Household Pulse Survey." April, 2020-April, 2022.
Chadwick, Owen. *The Secularization of the European Mind in the Nineteenth Century*. Cambridge MA: Cambridge University Press, 1975.
Chomsky, Noam, Ian Roberts and Jeffrey Watumull, "Noam Chomsky: The False Promise of ChatGPT." *The New York Times*, March 8, 2023.
Chorost, Michael. *Rebuilt: How Becoming Part Computer Made Me More Human*. New York: Houghton Mifflin, 2005.
Choudhury, Angshuman. "How Facebook is Complicit in Myanmar's Attacks on Minorities." *The Diplomate* August 25, 2020.
Clark, Andy. *Natural-Born Cyborgs: Minds, Technologies and the Future of Human Intelligence*. Oxford: Oxford University Press, 2003.
Cleckley, Hervey M. *The Mask of Sanity*. Augusta: E. S. Cleckley, 1988.
Cohen, Finn. "Under the Surface: Güven Güzeldere on the Mysteries of Consciousness and Artificial Intelligence." *The Sun*, no. 564 (December 2022).
Coles, Robert. *The Secular Mind*. Princeton, NJ: Princeton University Press, 1999.
Columbus, Chris, dir. *Bicentennial Man*. 1999. Columbia Pictures, 2000. DVD.
Culkin, J. M. "A Schoolman's Guide to Marshall McLuhan." *Saturday Review*, March 1967.
Damasio, Antonio. *Descartes Error: Emotion, Reason, and the Human Brain*. New York: Avon Books, 1994.
---. *Looking for Spinoza: Joy, Sorrow, and the Feeling Brain*. New York: Harvest, 2003.
Davis, Erik. *Techgnosis: Myth, Magic and Mysticism in the Age of Information*. New York: Harmony Books, 1998.
Davis, Joseph E. *Chemically Imbalanced: Everyday Suffering, Medication, and Our Troubled Quest For Self-Mastery*. Chicago: The University of Chicago Press, 2020.
Dawkins, Richard. *River Out of Eden: A Darwinian View of Life*. New York: Basic Books, 1996.
---. *The Selfish Gene*. New York: Oxford University Press, 1976.
Dick, Philip K. *Do Androids Dream of Electric Sheep?* New York: Ballantine Books, 1968.

Doczi, György. *The Power of Limits: Proportional Harmony in Nature, Art and Architecture.* Boulder, CO: Shambala, 1981.
Dowd, Maureen. "A.I. is Not A-OK." *The New York Times*, October 30, 2021.
---. "Elon Musk's Future Shock." *Vanity Fair*, April 2017.
Dreyfus, Hubert L. *What Computers Still Can't Do: A Critique of Artificial Reason.* Cambridge, MA: MIT Press, 1992.
Drinka, George F. *The Birth of Neurosis.* New York: Touchstone Books, 1984.
Dworkin, Richard W. *Artificial Happiness: The Dark Side of the New Happy Class.* New York: Carroll & Graf Publishers, 2007.
Edinger, Edward. *The Creation of Consciousness.* Toronto: Inner City Books, 1984.
---. *Ego and Archetype: Individuation and the Religious Function of the Psyche.* New York: Penguin Books, 1972.
Eisler, Riane. *The Chalice and the Blade.* New York: Harper and Row, 1987.
Ellenberger, Henri F. *The Discovery of the Unconscious.* New York: Basic Books, 1970.
Emanuel, Ezekiel J. "Tinkers and Tailors: Three Books Look to the Biomedical Frontier." *The New York Times*, March 16, 2017.
Fleming, Victor, dir. *The Wizard of Oz*. 1939. Beverly Hills, CA: Metro-Goldwyn-Mayer, 2005. DVD.
Flusser, Vilém. *Post-History.* Minneapolis, MN: Univocal, 2013.
Frank, Adam and Marcelo Gieser, "The Crisis in Cosmology." *The New York Times*, September 4, 2023.
Frankfurt, Harry G. *On Bullshit.* Princeton, NJ: Princeton University Press.
Frankl, Victor. *Man's Search for Meaning.* Boston: Beacon Press, 1959.
Freud, Sigmund. *Civilization and Its Discontents.* Translated by J. Riviere. London: The Hogarth Press, 1953.
Fromm, Erich. *The Revolution of Hope: Towards a Humanized Technology.* New York: Harper & Row, 1968.
---. *The Sane Society.* New York: Rinehart & Co., 1955.
Fukuyama, Francis. *The End of History and the Last Man.* New York: Free Press, 1992.
---. *Our Posthuman Future.* New York: Farrar, Straus and Giroux, 2002.
Garber, Megan. "We're Already in the Metaverse." *The Atlantic*, March 2023.
Gardner, Howard. *Frames of Mind: The Theory of Multiple Intelligences.* New York: Basic Books, 1993.
Garland Eric L., et. al. "Low Dispositional Mindfulness Predicts Self-Medication of Negative Emotion with Prescription Opioids." *Journal of Addiction Medicine* 9, no. 1. (January-February 2015).
Gibb, Ann. "Renowned Scholar and Author Norman O. Brown Dies at 89." *UC Santa Cruz Newscenter*, October 3, 2002.
Giddens, Anthony. *The Consequences of Modernity.* Stanford: Stanford University Press, 1990.

Gillespie, Craig, dir. *Lars and the Real Girl*. 2007; Beverly Hills, CA: Metro-Goldwyn-Mayer. DVD.
Glendinning, Chellis. "Technology, Trauma and the Wild." In *Ecopsychology*, edited by Theodore Roszak. Berkeley, CA: Sierra Club Books, 1995.
Goldhamer, Michael H. "Attention Shoppers." *WIRED,* December 1997.
Graham, Elaine L. *Representations of the Post/Human*. New Brunswick, NJ: Rutgers University Press, 2002.
Graham, George. *Stanford Encyclopedia of Philosophy*, online.
Gray, Chris Hables, ed. *The Cyborg Handbook*. New York: Routledge, 1995.
Green, Brian. *Until the End of Time: Mind, Meaning, and Our Search for Meaning in an Evolving Universe*. New York: Vintage, 2023.
Guggenbühl-Craig, Adolf. *The Emptied Soul: On the Nature of the Psychopath*. Woodstock, CT: Spring Publications, 1999.
---. *The Old Fool and the Corruption of Myth*. Dallas: Spring Publications, 1991.
Haidt, Jonathan. "A Guilty Verdict," *Nature*, vol. 578 (February 2020).
Hall, Richard. "William James on the Humanities." *William James Studies*, 9. (2012).
Harari, Yuval Noah. *21 Lessons For the 21st Century*. New York: Spiegel & Grau, 2018.
---. *Homo Deus: A Brief History of Tomorrow*. New York: HarperCollins, 2017.
---. *Sapiens*. New York: Harper Perennial, 2014.
Haraway, Donna J. *Simians, Cyborgs, and Women: The Reinvention of Nature*. New York: Routledge, 1991.
Hari, Johann. *Lost Connections: Uncovering the Real Causes of Depression—and the Unexpected Solutions*. New York: Bloomsbury, 2018).
---. *Stolen Focus: Why You Can't Pay Attention—and How to Think Deeply Again*. New York: Crown, 2022.
Harvey, David. *The Condition of Postmodernity*. Cambridge, MA: Blackwell, 2000.
Hassan, Ihab. "Prometheus as Performer: Toward a Posthumanist Culture?" *The Georgia Review*, 31:4, (1977).
Hawkins, Jeff. *On Intelligence*. New York: Times Books, 2004.
Hayles, N. Katherine. *How we Became Posthuman: Virtual Bodies in Cybernetics, Literature, and Informatics*. Chicago: Chicago University Press, 1999.
Heidegger, Martin. *Basic Writings*. San Francisco: Harper & Row, 1977.
Henry, Jules. *Culture Against Man*. New York: Vintage Books, 1965.
Herbrechter, Stefan. *Posthumanism: A Critical Analysis*. New York: Bloomsbury, 2013.
Hillman, James. *City and Soul*. Vol. 2, *The Uniform Edition of the Writings of James Hillman*. Putnam, CT: Spring Publications, 2006.
---. *The Myth of Analysis*. Evanston, IL: Northwestern University Press, 1972.

---. *Mythic Figures*. Vol. 6.1, *The Uniform Edition of the Writings of James Hillman*. Putnam, CT: Spring Publications, 2007.

---. *Philosophical Intimations*. Vol. 8, *The Uniform Edition of the Writings of James Hillman*. Edited by Edward S. Casey. Thompson CT: Spring Publications, 2016.

---. *Re-Visioning Psychology*. New York: HarperCollins, 1975.

---. *Senex and Puer*. Vol. 3, *The Uniform Edition of the Writings of James Hillman*. Edited by Glen Slater. Putnam, CT: Spring Publications, 2005.

---. *The Thought of the Heart and the Soul of the World*. Dallas: Spring Publications, 1993.

Holland, Julie. "Medicating Women's Feelings." *The New York Times*, March 1, 2015.

Hollis, James. *Under Saturn's Shadow*. Toronto: Inner City Books, 1994.

Hooks, Christopher. "The Psychopathy Problem in AI." *Medium*, online, April 11, 2023.

Hughes, John, dir. *Weird Science*. 1985; Los Angeles, CA: Universal Pictures, 1998. DVD.

Ihde, Don. *Postphenomenology: Essays in the Postmodern Context*. Evanston, IL: Northwestern University Press, 1995.

Itō, Kazunori. *Ghost in the Shell*. DVD. Directed by Mamoru Oshii. Santa Monica, CA: Lionsgate, 1995.

James, William. *The Will to Believe and Other Essays in Popular Psychology*. New York: Longmans Green, 1907.

Jansen, Y., J. Leeuwenkamp, and L. Urricelqui Ramos. "Posthumanism and the 'Posterizing Impulse.'" In *Post-everything: An Intellectual History of Post-concepts*, edited by H. Paul and A. van Veldhizen. Manchester: Manchester University Press, 2021.

Jones, Marina. "John Wheeler's Participatory Universe." *Futurism*, February 13, 2014.

Jonze, Spike, dir. *Her*. 2014. Burbank, CA: Warner Bros. DVD.

Jung, C. G. *The Collected Works of C.G. Jung*, 20 volumes. Edited and translated by Gerhard Adler and R.F.C. Hull. Princeton, NJ: Princeton University Press, 1953-1979.

---. *C.G. Jung Letters, Volume 2: 1951-1961*. Edited by Gerhard Adler. Translated by Jeffrey Hulen. Princeton, NJ: Princeton University Press, 1976.

---. *The Earth Has a Soul: The Nature Writings of C. G. Jung*. Edited by Meredith Sabini. Berkeley: North Atlantic Books, 2002.

---. *Memories, Dreams, Reflections*. Edited by Aniela Jaffe. Translated by Richard and Clara Winston. New York: Pantheon Books, 1961.

Kauffman, Stuart A. *Reinventing the Sacred*. New York: Basic Books, 2008.

Keats, John. *The Complete Poetic Works of John Keats*. New York: Houghton Mifflin Company, 1899.

Kelly, Kevin. *What Technology Wants*. New York: Viking, 2010.
Kerényi, Carl. *The Gods of the Greeks*. London: Thames and Hudson, 1974.
---. *Prometheus: An Archetypal Image of Human Existence*. Princeton: Princeton University Press, 1997.
Klein, Ezra and Gary Marcus. "Transcript: Ezra Klein Interviews Gary Marcus." *The New York Times*, January 6, 2023.
Koestler, Arthur. *The Act of Creation*. New York: Macmillan, 1964.
Kroker, Arthur and David Cook. *The Postmodern Scene: Excremental Culture and Hyper-Aesthetics*. Montréal: New World Perspectives, 1991.
Kurzweil, Ray. *The Age of Spiritual Machines*. New York: Viking, 1999.
---. *How to Create a Mind: The Secret of Human Thought Revealed*. New York: Penguin, 2012.
--. *-The Singularity is Near: When Humans Transcend Biology*. New York: Viking, 2005.
LaFee, Scott. "Serious Loneliness Spans the Adult Lifespan But There is a Silver Lining," *UC San Diego Today*, December 18, 2018.
Lakoff, Andrew. *Pharmaceutical Reason*. Cambridge, UK: Cambridge University Press, 2005.
Lang, Fritz, dir. *Metropolis*. 1927; Kino Lorber Films, 2003. DVD.
Lanier, Jaron. *You Are Not a Gadget*. New York: Alfred A. Knopf, 2010.
Lasch, Christopher. *The Culture of Narcissism*. New York: W. W. Norton & Co., 1978.
Lent, Jeremy. *The Patterning Instinct*. New York: Prometheus Books, 2017.
Levin, David Michael. "Psychopathology in the Epoch of Nihilism." In *Pathologies of the Modern Self* edited by David Michael Levin. New York: New York University Press, 1987.
Levy, David. *Love + Sex With Robots: The Evolution of Human-Robot Relations*. New York: Harper, 2007.
Lewis, Marc. "Why the Disease Definition of Addiction Does Far More Harm Than Good." *Scientific American* online blog post, February 9, 2018.
Lifton, Robert Jay. *The Protean Self: Human Resilience in an Age of Fragmentation*. Chicago: University of Chicago Press, 1993.
López-Pedraza, Rafael. *Cultural Anxiety*. Einsiedeln, Switzerland: Daimon Verlag, 1990.
---. *Dionysus in Exile*. Wilmette, IL: Chiron Publications, 2000.
Lovelock, James. *The Revenge of Gaia*. New York: Basic Books, 2007.
Lucas, George, dir. *Star Wars: Episode IV—A New Hope*. 1977; Beverly Hills, CA: 20th Century Fox, 2006. DVD.
---. *Star Wars: Episode V—The Empire Strikes Back*. 1980. Beverly Hills, CA: 20th Century Fox, 2006. DVD.
Maçães, Bruno. *History Has Begun: The Birth of a New America*. New York: Oxford University Press, 2020.
Madsbjerg, Christian. *Sensemaking: What Makes Human Intelligence Essential in the Age of the Algorithm*. London: Little Brown, 2017.

Majoo, Farhad. "What the Internet is Hiding." *The New York Times*, August 28, 2022.
Mar, Alex. "Love in the Time of Robots," *WIRED*, November 2017.
Mariani, O. "Analytical Psychology and Entertainment: Idle Time and the Individuation Process." *Spring: A Journal of Archetype and Culture*. Vol. 80 (2008).
Marquand, Richard, dir. *Star Wars: Episode VI—Return of the Jedi*. 1983; Beverly Hills, CA: 20th Century Fox, 2006. DVD.
May, Rollo. *The Cry For Myth*. New York: W. W. Norton, 1991.
Mayor, Adrienne. *Gods and Robots: Myths, Machines, and Ancient Dreams of Technology*. Princeton, NJ: Princeton University Press, 2018.
Mazis, Glen. *Humans, Animals, Machines*. Albany: SUNY Press, 2008.
McGilchrist, Iain. *The Master and His Emissary: The Divided Brain and the Making of the Western World*. New Haven, CT: Yale University Press, 2009.
McKibben, Bill. *Enough: Staying Human in an Engineered Age*. New York: Times Books, 2003.
McLuhan, Marshall. *Understanding Media*. Cambridge, MA: MIT Press, 1994.
Meier C. A. et al. *A Testament to the Wilderness: Ten Essays on an Address by C. A. Meier*. Zurich: Daimon Verlag, 1985.
Midgley, Mary M. *Science and Poetry*. London: Routledge, 2001.
---. *Science as Salvation: A Modern Myth and its Meaning*. New York: Routledge, 1992.
Mlodinow, Leonard. *Emotional: How Feelings Shape Our Thinking*. New York: Pantheon Books, 2022.
Moravec, Hans. *Mind Children: The Future of Robot and Human Intelligence*. Cambridge, MA: Harvard University Press, 1988.
More, Max and Natasha Vita-More, eds. *The Transhuman Reader*. Oxford: Wiley-Blackwell, 2013.
Morgan, Richard K. *Altered Carbon*. New York: Del Rey, 2003.
Morozov, Evgeny. *To Save Everything, Click Here: The Folly of Technological Solutionism*. New York: Public Affairs, 2013.
Mumford, Lewis. *Interpretations and Forecasts: 1922-1972*. New York: Harcourt, Brace, Jovanovic, 1973.
---. *The Myth of the Machine: The Pentagon of Power*. New York: Harcourt, Brace, Jovanovic, 1970.
Noble, David F. *The Religion of Technology: The Divinity of Man and the Spirit of Invention*. New York: Penguin Books, 1997.
Nolan, Jonathan and Lisa Joy, creators. *Westworld*. Performed by Evan Rachel Wood, Thandie Newton, Jeffrey Wright. HBO, 2016-2022. Television.
Odell, Jenny. "What Twitter Does to Our Sense of Time." *The New York Times*, December 11, 2023.
Oshii, Mamoru, dir. *Ghost in the Shell*. 1995; Santa Monica, CA: Lionsgate, 2020. Blue-ray Disk.

Otto, Rudolff. *The Idea of the Holy: An Inquiry Into the Non-Rational Factor in the Idea of the Divine and Its Relation to the Rational.* New York: Oxford University Press, 1950.

Paul H. & A. van Veldhizen ed. *Post-everything: An Intellectual History of Post-concepts.* Manchester: Manchester University Press, 2021.

Peters, Ted, ed. *Cosmos as Creation: Theology and Science in Consonance.* Nashville: Abingdon Press, 1989.

Piercy, Marge. *He, She and It.* New York: Fawcett Crest, 1991.

Progoff, Ira. *The Death and Rebirth of Psychology.* New York: McGraw-Hill, 1956.

---. *Jung's Psychology and Its Social Meaning.* New York: The Julian Press, 1953.

Ptolemy, Barry, dir. *Transcendent Man.* 2009. Ptolemaic Productions Therapy Studios. DVD.

Rauch, Jonathan. *The Constitution of Knowledge.* Washington, DC: The Brookings Institution Press, 2021.

Raine, Kathleen. *Blake and Antiquity.* London: Routledge, 1979.

Rieber, Robert W. *Manufacturing Social Distress.* New York: Plenum Press, 1997.

Roddenberry, Gene, creator. *Star Trek: The Next Generation.* Performed by Patrick Stewart, Jonathan Frakes, Levar Burton. Paramount Television, 1987-1994. Television.

Rojcewicz, Richard. *The Gods and Technology: A Reading of Heidegger.* Albany, NY: SUNY Press, 2006.

Romanyshyn, Robert. "Complex Knowing: Towards a Psychological Hermeneutic." In *The Humanistic Psychologist*, Vol. 19, no. 1 (1991).

---. *Technology as Symptom and Dream.* New York: Routledge, 1989.

Rose, Steven. *Lifelines: Biology Beyond Determinism.* New York: Oxford University Press, 1997.

Roszak, Theodore. *The Making of a Counter-Culture: Reflections on the Technocratic Society and its Youthful Opposition.* New York: Anchor Books, 1969.

---. *Where the Wasteland Ends.* New York: Anchor Books, 1973.

Rubin, Charles T. *Eclipse of Man: Human Extinction and the Meaning of Progress.* New York: New Atlantis Books, 2014.

Rushing, Janice Hocker & Thomas S. Frentz. *Projecting the Shadow: The Cyborg Hero in American Film.* Chicago: The University of Chicago Press, 1995.

Saks, Oliver. *A Leg to Stand On.* New York: Touchstone, 1984.

Samuels, Andrew, Bani Shorter, and Fred Plaut. *A Critical Dictionary of Jungian Analysis.* New York: Routledge, 1986.

Santayana, George. *The Life of Reason: Reason in Common Sense.* New York: Collier Books, 1962.

Sarton, May. *Journal of a Solitude.* New York: W. W. Norton & Company, 1973.

Schiller, Friedrich. *On the Aesthetic Education of Man*. Mineola, NY: Dover Publications, 2004. First published in 1795.
Schneider, Susan. *Artificially You*. Princeton: Princeton University Press, 2019.
Scott, Ridley, dir. *Alien*. 1979. Beverly Hills, CA: 20th Century Fox, 2004. DVD.
---. *Blade Runner.* 1982. Burbank, CA: Warner Bros., 1997. DVD.
Scruton, Roger. "The Trouble With Knowledge." *MIT Technology Review*. May 1, 2007.
Searle, John. *The Mystery of Consciousness*. New York: The New York Review of Books, 1997.
Sennett, Richard. *The Craftsman*. New Haven, CT: Yale University Press, 2008.
Shakespeare, William. *Hamlet*, Act III; Scene 2.
---. *Troilus and Cressida*, Act III, Scene 3.
Shannon, Delisa and Noah Friedman. "Teens Would Rather Break Their Bones Than Lose Their Phones." *Business Insider* online, May 6, 2021.
Shengold, Leonard. *Soul Murder: Thoughts about Therapy, Hate, Love, and Memory*. New Haven, CT: Yale University Press, 1999.
Shweder, Richard A., and Robert A. LeVine, *Culture Theory: Essays on Mind, Self, and Emotion*. Cambridge: Cambridge University Press, 1984.
Siegel, Daniel. *The Developing Mind*, 3rd ed. New York: The Guilford Press, 2020.
Simon, Herbert A. "Designing Organizations For An Information-Rich World." In *Computers, Communications, and the Public Interest*, edited by M. Greenberger. Baltimore, MD: The Johns Hopkins Press, 1971.
Sites, Brian D., Michael L. Beach, Matthew Davis. "Increases in the Use of Opioid Analgesics and the Lack of Improvement in Disability Metrics Among Users. *Regional Anesthesia & Pain Medicine* 39 no. 1 (January-February 2014).
Skinner, B.F. *Beyond Freedom and Dignity*. Cambridge, MA: Hackett Publishing, 2002. First published 1971.
---. *Walden Two*. Cambridge, MA: Hackett Publishing, 2005. First published 1948.
Slater, Glen. "Aliens and Insects." In *Varieties of Mythic Experience: Essays on Religion, Psyche and Culture*, edited by Dennis Patrick Slattery and Glen Slater. Einsiedeln: Daimon-Verlag, 2007.
---. "Hermetic Intoxication and Dataism." In *Mythic Figures: Conversing with James Hillman* edited by Joanne H. Stroud and Robert Sardello. Dallas: Dallas Institute of Humanities, 2018.
---. "Hillman's Metapsychology." In *Conversing with James Hillman: Senex and Puer* edited by Joanne H. Stroud and Robert Sardello. Dallas: Dallas Institute of Humanities, 2016.
---. "Numb." In *Archetypal Psychologies* edited by Stanton Marlon, 351-367. New Orleans: Spring Journal Books, 2008.

---. "Re-sink the Titanic." *Spring*, no. 62 (1998).
---. *Surrendering to Psyche: Depth Psychology, Sacrifice, and Culture*. Carpinteria, CA: Pacifica Graduate Institute, 1996.
Slater, Philip. *Earthwalk*. New York: Anchor Press, 1974.
Smith, Brendan L. "Inappropriate Prescribing." In *Monitor on Psychology* 43, no. 6 (June 2012).
Smith, Huston. *Beyond the Postmodern Mind*. Wheaton, IL: Quest Books, 1989.
Snell, Reginald. Introduction to *On the Aesthetic Education of Man* by Friedrich Schiller. Mineola, NY: Dover Publications, 2004.
Sohms, Mark. *The Hidden Spring: A Journey to the Source of Consciousness*. New York: W. W. Norton & Company.
Stip, Emmanuel "Internet Addiction, *Hikikomori* Syndrome, and the Prodomal Phase of Psychosis," *Frontiers in Psychiatry*, 27:6 (2016)
Stirewalt, Chris. "I Called Arizona for Biden on Fox News. Here's What I Learned." *Los Angeles Times*, January, 28, 2021.
Suskind, Ron. "Faith, Certainty and the Presidency of George W. Bush." *The New York Times*, October 17, 2004.
Tallis, Raymond. *Aping Mankind: Neuromania, Darwinitis and the Misrepresentation of Humanity*. Durham, UK: Acumen, 2011.
---. *Why the Mind is Not a Computer*. Charlottesville: Imprint Academia, 2004.
Tarnas, Richard. *Prometheus the Awakener*. Woodstock, CT: Spring Publications, 1995.
Tatham, F. *The Letters of William Blake Together with His Life*. London, Methuen & Co., 1906.
Taylor, Mark C. *About Religion: Economies of Faith in a Virtual Culture*. Chicago: The University of Chicago Press, 1999.
---. *Intervolution: Smart Bodies Smart Things*. New York: Columbia University Press, 2021.
Tegmark, Max. *Life 3.0: Being Human in the Age of Artificial Intelligence*. New York: Alfred A. Knopf, 2017.
The School of Life, *How to Survive the Modern World*. London: The School of Life, 2021.
Thierry, Guillaume. "Neuroscientist who studies how the brain learns information explains why A.I. would be the 'perfect psychopath' in an executive role." *Fortune*, online edition, July 31, 2023.
Thomas, Roger. "Eve Hart 91, A Last Survivor With a Memory of Titanic, dies." *The New York Times*, February 16, 1996.
Thompson, Derek. "How Civilization Broke Our Brains." *The Atlantic*, January/February 2021.
Thoreau, Henry David. *Walden*. Springfield, OH: Crowell and Company, 1899.
Tiku, Nitasha. "The Google Engineer Who Thinks the Company's AI Has Come to Life," *The Washington Post*, June 21, 2022.
Turkle, Sherry. *Alone Together: Why We Expect More from Technology and Less from Each Other*. New York: Basic Books, 2011.

---. "TEDcUIUC – Sherry Turkle – Alone Together," March 25, 2011, YouTube Video, 16:24, https://www.youtube.com/watch?v=MtLVCpZIiNs

Ullman, Ellen. *Close to the Machine: Technophilia and its Discontents*. New York: Picador, 2012.

Uncapher, Melina R. and Anthony D. Wagner. "Minds and Brains of Media Multitaskers: Current Findings and Future Directions." *PNAS*, 115, no. 40: 9889-9896.

Varela, Francisco J., Evan Thompson & Eleanor Rosch. *The Embodied Mind*. Cambridge, MA: MIT Press, 2016.

Velasquez-Manoff, Moises. "The Brain Implants That Could Change Humanity." *The New York Times*, August 30, 2020.

Wallach, Wendell. *A Dangerous Master: How to Keep Technology from Slipping Beyond Our Control*. New York: Basic Books, 2015.

Warzel, Charlie. "I Talked to the Cassandra of the Internet Age." *The New York Times*, February 7, 2021.

Webster's New World Dictionary of the American Language. New York: World Pub. Co., 1960.

Wells, H. G. *The War of the Worlds*. New York: The Modern Library, 2002. First published 1898.

Whatley, Stuart. "The Machine Pauses." *The Hedgehog Review* 22, no. 2 (Summer 2020).

Wieseltier, Leon. "Among the Disrupted." *The New York Times* Book Review, January 8, 2018.

Wilcox, Fred M., dir. Forbidden Planet. 1956; Beverly Hills, CA: Metro-Goldwyn-Mayer, 2000. DVD.

Wilson, Frank R. *The Hand: How its Uses Shapes the Brain, Language and Human Culture*. New York: Vintage, 1999.

Winnicott, D. W. *Playing and Reality*. New York: Routledge, 2005. First published 1971.

Yarow, Jay. "Sergey Brin: 'We Want Google to be the Third Half of Your Brain," *Insider*, Spetember 8, 2010.

Zarkadakis, George. *In Our Own Image: Savior or Destroyer? The History and Future of Artificial Intelligence*. New York: Pegasus Books, 2015.\

John Zerzan and Alice Carnes Eds. *Questioning Technology: A Critical Anthology*. London: Freedom Press, 1988.

Zuboff, Shoshana. *The Age of Surveillance Capitalism: The Fight For the Human Future at the New Frontier of Power*. New York: Public Affairs, 2019.

ACKNOWLEDGMENTS

This book was made possible by the support of many friends and colleagues. I am particularly indebted to Robert A. Johnson, whose faith in my nascent calling to depth psychology provided a foundation for this eventual undertaking. Early conversations with Robert Romanyshyn on technology and psychological life and with James Hillman on society in the digital age were of critical importance. Gilles-Zenon Maheu provided guidance on the history of science. Joan McAllister was a consistent advocate and motivator. Rose-Emily Rothenberg gave wise counsel at critical early stages. Bruce White and Rob Ferguson offered ongoing encouragement and robust argument. Core faculty in the Jungian and Archetypal Studies Program at Pacifica Graduate Institute provided a container for my elongated writing process. Klaus Ottmann at Spring Publications provided a pathway for publication. And, in addition to her tireless support throughout, Safron Rossi oversaw the final stages of production.

Made in United States
Troutdale, OR
04/04/2024